The Story of Prosecco Superiore

The Story of Prosecco Superiore

Susan H. Gordon

ROWMAN & LITTLEFIELD
Lanham • Boulder • New York • London

Published by Rowman & Littlefield
An imprint of The Rowman & Littlefield Publishing Group, Inc.
4501 Forbes Boulevard, Suite 200, Lanham, Maryland 20706
www.rowman.com

86-90 Paul Street, London EC2A 4NE

Grateful acknowledgment is made to the following for permission to reprint previously published material:

The One-Straw Revolution by Masanobu Fukuoka, Larry Korn, ed. New York: New York Review Books, 2009. Used by permission of the Estate of Masanobu Fukuoka.

From *An Unfinished Woman: A Memoir* by Lillian Hellman, introduction by Wendy Wasserstein copyright © 1999. Reprinted by permission of Back Bay Books, an imprint of Hachette Book Group, Inc.

Material and Mind by Christopher Bardt. Copyright © 2019 MIT Press. Reprinted with permission of MIT Press.

Epistenology: Wine as Experience by Nicola Perullo. Copyright © 2020 Columbia University Press. Reprinted with permission of Columbia University Press.

From *Six Memos For The Next Millennium* by Italo Calvino. A new translation from the Italian by Geoffrey Brock. Copyright © 1988 by the Estate of Italo Calvino. English translation copyright © 2016 by the Estate of Italo Calvino. Used by permission of HarperCollins Publishers.

British Library Cataloguing in Publication Information Available

Library of Congress Cataloging-in-Publication Data

Names: Gordon, Susan H., 1968- author.
Title: The story of Prosecco Superiore / Susan H. Gordon.
Description: Lanham : Rowman & Littlefield, [2024] | "Most DOCG labels . . . hold the word 'Superiore' to indicate Prosecco that is fully sparkling, defined as at least 3.5 times the average air pressure at sea level at 59 degrees Fahrenheit or 15 degrees Celsius, and in practice closer to 5 times"—Introduction. | Includes bibliographical references and index. | Summary: "Prosecco is often thought of as a cheaper alternative to Champagne. This story instead shows it as a uniquely Italian expression by tracing 150 years in two small hilly zones in the Veneto's Marca Trevigiana where the top Prosecco wines have long been grown"— Provided by publisher.
Identifiers: LCCN 2023048080 (print) | LCCN 2023048081 (ebook) | ISBN 9781538191262 (cloth) | ISBN 9781538191286 (ebook)
Subjects: LCSH: Prosecco (Wine)—Italy—History—19th century. | Prosecco (Wine)—Italy—History—20th century. | Wine districts—Italy—History—19th century. | Wine districts—Italy—History—20th century. | Wine and winemaking—Italy—History—19th century. | Wine and winemaking—Italy—History—20th century. | Veneto (Italy)
Classification: LCC TP548.6.S65 G67 2024 (print) | LCC TP548.6.S65 (ebook) | DDC 663/.20094509034—dc23/eng/20231205
LC record available at https://lccn.loc.gov/2023048080
LC ebook record available at https://lccn.loc.gov/2023048081

To Ben, who gets the impossible done.

L'errore se fatto senza cattivi fini, costumo perdonarlo, perché anch'io ho errato, erro, e sbaglierò finché non cesserò di lavorare.

—Antonio Carpenè, *L'Enotecnico.*

(The error if not made toward malicious ends, it is my habit to forgive, because I too have erred, I err, and I will make mistakes until I am no longer working.—Antonio Carpenè, *L'Enotecnico.*)

Contents

Introduction

> Before researchers become researchers they should become philosophers. They should consider what the human goal is, what it is that humanity should create.
>
> —Masanobu Fukuoka, *The One-Straw Revolution*[1]

The movement of this book about Prosecco Superiore is toward this: by late spring 2022 in the third year of the pandemic, the number of bottles made of this sparkling wine from northeast Italy remains on track to match the 121 million bottles produced in 2021, more than the almost 107 million produced the first year of the pandemic which was just about 1 percent fewer than those made in 2019, the unsuspecting and record-setting year before.[2]

In the United States, *Prosecco* is now a household word. *Prosecco Superiore* less so, and a term with its own specific story. Fame for all of these wines is a recent phenomenon, the so-called boom with first if widely imperceptible stirrings in the later twentieth century then swinging into full force more or less during the opening decade of the twenty-first. With this recognition comes misunderstanding that is both categorical and local. Like most Italian[3] wines with a history, Prosecco overall is a place where the wines that bear the word are made. Unlike many such wines, the name is not one, say, village, one homogenous-enough place to which you can address a postcard. It is instead a legal indication of areas whose history and terrains have together defined Prosecco as both a wine and, until 2009, a grape variety, those two categories themselves, we will come to see, a complex Prosecco juggling of things more plural. There are now more than 650 million bottles of Prosecco sold across the world each year—which is as of 2021 a record number. If you were to read the label on each bottle, you would see that more than half a billion of them include the acronym *DOC* (as of 2019, about 50 million of which, unlike any of the others, are pink wines), communicating to you that these wines are grown and made within a large area, mainly flatlands, in Veneto and Friuli-Venezia-Giulia, as of 1947 two regions in Italy's northeast and until its fall in the late spring of 1797 inland Terraferma provinces of the maritime Republic of Venice. One-sixth of the labels announce instead the acronym *DOCG* and at times *Valdobbiadene*, *Conegliano*, or *Asolo*, to let you know that these wines alone were grown in the three hilly areas of Veneto's province of Treviso, long noted as the places where

Prosecco's top-quality wines are made. While a small number of these are still[4] or lightly fizzy wines, most DOCG labels also, and uniquely among the 650 million, hold the word *Superiore* to indicate Prosecco that is fully—defined as at least 3.5 times the average air pressure at sea level at 59°F or 15°C, and in practice closer to 5 times—sparkling.

Each acronym—the first for *Denominazione di origine controllata* (denomination of controlled origin), the second adds *garantita*, for *guaranteed*; both issued by Italy's Ministero delle politiche agricole alimentari, forestali e del turismo[5]—carries a set of specifications for making the wine inside that bottle, of grapes grown in defined territorial boundaries, with specified methods and yields, and based on the wine's historical presence in its territory, including that its name be one long and widely used there. A DOC or DOCG, colloquially a *denomination* and Italy's system comparable to France's AOC (Appellation d'Origine Contrôlée) and Austria's DAC (Districtus Austriae Controllatus), is a legal thing, a juridical drawing of a wine and its place come into being on a specific date to serve and illustrate specific needs and therefore also shorthand and hints for other things. Each is equivalent to a map, one that if read well is a history and story of the wine it delineates. Each is also a legal description of quality, a notion that serves as basis for a DOC or DOCG existing at all and included therein as official term: this document about which grapes may be used, in which percentages, corresponding traditional, historical, topographical information, and the very specific boundaries within which production, is an intention to protect and promote the wine in question precisely because it is a *quality* wine. On April 2, 1969, the zone, grapes, and history running from Treviso's Conegliano hills west to the range of Valdobbiadene's were made a DOC by decree of the president of the Italian Republic. And by decree, on July 17, 2009, the DOC was reshaped to indicate instead the larger area where this wine had more or less long been made while, new, two DOCGs were enacted for the historically celebrated hills. The Conegliano-Valdobbiadene Prosecco DOCG and, out of 1977's broad Montello and Colli Asolani DOC, the Asolo Prosecco DOCG newly broken out from its previous quieter role within the historical territory are each "a complex space where several forces need to coexist; the productive drive of growing requests (global and local) of Prosecco, as well as rural representation based on local habits and a concrete hilly landscape."[6] The wine-defining boundary of each territory is an outline not drawn but written, complete with starting point and directions until arrival back, a walkably delineated zone. After that 2009 update, the Prosecco trail skips. Increasing popularity had already brought a need to decrease the number of steadily encroaching imitators: to prevent those growing these grapes outside the area from misleadingly labeling their wines with the word *Prosecco*, the denominations changed the grapevine variety from *Prosecco* to *Glera*,[7] the grape's local name (and so a *synonym* in grape identification research) around the Frulian[8] town named Prosecco (to which from "beyond Sessana the train descends," notes an 1877 travelers' handbook[9]), the former name turned into legal word for only the wines made of it in their territories. Slowly, particulars come in. *Valdobbiadene* can be on a Prosecco DOCG label, and with a 2019 update, *Rive*, for the old most steep places. The backing is historical, Prosecco a patrimony of here, pours into unspoken tales that are components of such labeling words, including that steepnesses are understood definitions within the DOCGs. All of which is to say that this is a book about the two, chosen as basis precisely because of that *DOCG*. This is a book about what each of their bottles is bringing to you, which is a wine that traces from the first years of Italy, a wine traditional not only to hills and grapes but also to the rising role, with particularly potent contributions from the Venetian republic,[10] of chemistry and technology in a new nation, taking its place within what was by then Europe.

Add together DOC and DOCG to see that Prosecco bottles (a DOP[11] wine must be sold bottled, and never sfuso[12]) are now sold at twice the rate of Champagne's—245 million bottles sold in 2020, down from 297.5 million in 2019[13]—and hold stronger, even (especially?) during a pandemic. Although the comparison with the performance and flavors of Champagne, the great sparkling wine of northern France that is often erroneously (see those denominations and appellations) used as synonym of *sparkling wine* as indeed *Prosecco* is today throughout Italy, is a common one, it is not immediately obvious why it should be. I began to think about that day in 1797 when Napoleon and his army arrived in Mestre, the Venetian lagoon's abutting and soon industrializing mainland steps outside the city of Venice that was head of the thousand-year-old Serenissima republic, on his sweep through the nation-states of the Italian peninsula and islands. On May 12 as Napoleon waited, with just past half of its one thousand members the Serenissima's ruling Signoria gathered for the last time and voted to dissolve the besieged republic. "We will therefore close, as we should," the heartbroken 120th and last Doge Ludovico Manin announced to the Venetians from the balcony of the Palazzo Ducale, "by reminding you to always turn to God the Lord and to his most holy Mother."[14] And that he read while the crowd yelled back against him, *viva San Marco*! Simply, France entered the city two days later, the first foreign army to do so, then gave Venice to the Austrian empire, who saw these new territories as wide and exploitable border, land to be used for military strategies, complete with livestock to be counted and inhabitants to be controlled but never governed. Treviso was to remain stubborn and proud subject of the Austrian occupiers until 1866,[15] when former Terraferma provinces voted to join the nation of Italy, formed five years earlier after the Risorgimento (period of planning, writing, eruptions, secret societies partitioned roughly across the three wars of independence, 1848–1849, 1859, 1866) finally (to set off with nationmaking stories: under the military and political maneuvering of Giuseppe Garibaldi and the thinking, writing, plotting, fighting of many, many others during the preceding decades of the nineteenth century) worked.

"Italy for centuries past has been a field open to the struggle between French and Austrian influence," the March 15, 1860 *New York Times* quotes then French Minister Thouvenel as saying the year before the nation minus old Venice (and Rome, to be added in 1870) formed under a piemontese rule. "It must be so no longer."[16] "Had it been possible for the arms of France to carry the Italian tricolor in triumph to the Adriatic, without involving all Europe in a mad, chaotic war," adds the reporter, "Italy would hardly have been to-day so near the morning of her lasting independence as she now is."[17] Among the many, many others was a pharmaceutical student born four decades after the fall of Venice and who from the city of Conegliano at the Trevisian countryside's jaggedly steep hills fought first for an Italy in the piemontese army then joined instead Garibaldi's forces: soon after his work as soldier, Antonio Carpenè will become a chemist, too, and then as such a winemaker, writer, maker of wine machines, organizer, and he will use each of those tools to promote, experiment with, and believe in a Prosecco spumante, sparkling wine for which neither public nor necessary winemaking tools yet existed. He is a basis, too, a point of convergence for unification of Italy and the notion of Prosecco spumante, a driving figure, stubborn visionary with unyielding curiosity and love of his land both near and nation. Time-wise, it is more or less at this point in the Trevisian hills of Conegliano, Valdobbiadene, and Asolo that this book that is one wine's mythology further complicated and populated begins. For fixing the Veneto, as its borders are fluid over this time period and I am heading into now from there, I mean mostly the region as it stands now, with the understanding that as we move through time and context, those borders shift into a portion of Venice that includes the city itself out to what is immediately further protected by the Prealps. The province of Treviso stands much as it is now from the outside, with renewed understandings of what it meant in relation to Venice both empire and city.

Today, the equivalents of about three billion euros are handed over from around the world in return for Prosecco wines. The grapes themselves look laden: Glera's berries are the definition in sight and touch of roundness, soft thin skins the warm and complicated colors of their landscape, swelled seductively in varying diameters, ashine with yellow's green lights and warm with orange moments. They are famously giving. Delicious even when, unlike most other grape varieties whose tastiness depends on the plant being manipulated into producing fewer grapes, plenty of them are allowed to grow on each vine, so long as that happens upon hills and under further conditions based on weather and human hands. Just as generous in expression, Glera offers a range of aromas that are trustworthy clues to which parts of the DOCG hillsides it grew in the year it did so. In variations by the weather, wines from the broadly twisted Coneglianese hills are likely to taste bigger-bodied and fruitier than the sharper floral-scented wines from the more plumb hills of Valdobbiadene on the west side of the Soligo river where crooked squiggled vineyard rows do their best to smooth the kinks of these steepest of the three areas, hills stacked a corde along some hewn x-axis, and where many grapevines grow on plots so steep they've come from the nineteenth century to the twenty-first unscathed by the tractors and mechanical harvesters that arrived to more accessible sites long after they were planted. The grapes from Asolo, if you can say *brawny* of these delicate wines and fruitier still, come from hills softened by the southward move across the Piave river from the two above and wilder with forest and Treviso's hazy yellow lights. In the sixteenth century, painter Tiziano Vecellio from Veneto's province of Belluno where northern Treviso ends harvested the light another way and filled his paintings with it, set them beating with the hills' fitful shine. I do not know what it is like to work on these hillsides in either grapes or paint, but I have walked them over six visits from my first one in mid-September 2015 to this book's last one in July 2021 and know that to do so is to be sharply aware of balance, to be in constant negotiation with geology as both shifting formative event and millennia-long adjustments and responses to this northern wrinkle swept by winds arriving from the crumpled geological afterthought that is the Prealps, northeastern releases of the massive push that majestically and catastrophically formed the Alps which hold all of Italy in varying degrees of Mediterranean climate, even in its north. On my first visit, in the earliest days of that year's harvest, these sparkling wines seductive in their delicacy and subtlety were an obsession I had not expected.

Before that first visit, hilly vineyards, city where Carpenè had begun to imagine his spumante, I had at least that I remembered never tasted top-quality Prosecco. From the moment in which I did and in the same instance spoke with their producers, I was drawn in in a way that has been slowly telling me a kind of story, but that was for a while an impulse I could not make sense of. Later, in following it came also one of my reasons for writing about this. As a descendant of families of Italy's pre-nation Kingdom of two Sicilies and Papal State, I am of the first part of the peninsula to become Italy, however reluctantly (historian John Duggan[18] notes that for the first years of the nation, two-thirds of its army was stationed in the south to destroy anti-Italy rebellions and the towns they came from). I am not of what is now the Veneto. The disconnect I feel there is greater than the tie, but the latter is oddly and undeniably there, and I am pushed beyond the beauty of the hills and the powerfully dainty lacy perfection of their wines to wonder, or better to follow a curiosity into what that connection might be made of. Of what it makes of *Italy.* Of my families in what are now the regions of Abruzzo and Campania turned co-citizens of families of the Venetian territories, three of many loyalties and languages that would now be one something, all citizens of Italy ruled by the French-speaking king of Piemonte's Savoia family, under an Italian language that was Toscana's, of a nation until 1871 governed from the capital city of Firenze, what does unification mean 150 years and five generations later? Three generations after my ancestors left, or rather sort of, since I myself was born and raised through middle school in

Rome? My great-great-grandparents and Carpenè were contemporaries in time and fragmented peninsula and so suddenly conationals. Perhaps, what are the ways to ask into the lives of my distant relatives? Does it help to imagine myself through my differences with them? Unrelated, I am nevertheless connected to the hills of Prosecco, to these ancestral wines made with willpower and steel and inquiry into their makeup because of the work and thoughts and writings of the risorgimentali. And of their wines.

By 1870, the Risorgimento was officially over, successful in that administratively united for the first time since Rome, Italy had been transformed into nation, if one filled with mixed expectations and very immediate needs. Agriculture had for centuries supplied most of the land's work, of inhabitants' living if a meager one, and their history, ancient shapes of what was now not just Italian but *national*. Poverty, a mix of suppressive foreign rule and then destruction of local systems under Piedmont's new Italian government was, to vastly understate[19] both conditions and causes, a very big problem that would take a long time to solve. "Starting in 1969, when [the area] became DOC, everything changed," Trevisian food and drink historian Danilo Gasparini told me of the hills' land and livelihoods. But "even in the 1970s, emigration still happened here." And for decades cases of pellagra, diet-induced low-niacin disease commonly said to be the fault of a diet that consisted pretty much of only corn, had been settling over the province. To address these problems, viticulture and viniculture, Carpenè had insisted one hundred years earlier, were crucial, holders of identity and pride, healthful nourishment, bringers of financial means. To make a prosperous, proud nation of its many parts, its ancient and many wines must be counted on, improved, presented to the world, he said again and again. Treviso, the hills of Conegliano and Valdobbiadene and Asolo were now Italy, and their greatest wine could now, with improved growing and better winemaking be one of the new nation's greatest, too: Prosecco, ancient but by this time grown only by a few believers, is, potential understood, a wine of Italy.

"Why is Prosecco so popular?" is the refrain today, repeated with rising intonation. There was no advertising campaign, not even a general surge forward by the world's sommeliers, currently one of the most potent tastemakers and movers of wine markets. Indeed many of that camp often misunderstand the difference between the historically exalted DOCG bottlings and the less-responsive Prosecco bottles that can be found among the DOC flatlands and, cheaper, are far too often bought instead. How we got here is, to somewhat quote creativity practitioner Questlove,[20] a question that is itself in question. What then are some paths to Prosecco? Follow the industrialists back, see them dissolve and then Carpenè dreaming in materials like newly unbreakable glass and coming stainless steel, a vision in analytic razionalismo of clear wines barely translated from grapes, delicious epitome of mind over matter, with a declaration of intent against wine-growing *empiricism*, a word he used often in reference to most of the contadini (see Glossary), who were most of the wine-growers. Stainless steel is possibility and arrives here with its own story. In this book neither it nor academic knowledge are yet the power, land, and knowledge domination we assume of both now, and still in the mid-twentieth century when literary television personality Mario Soldati "ignorant of enological science"[21] and on his "viaggi d'assaggio"[22] through Italy to learn its "genuine"[23] wines meets academic, writer, and ampelographer Italo Cosmo,[24] he immortalizes the latter's love of viticulture in his resulting anti-industrial pantheon of nationally forgotten wines, *Vino al Vino.* Step alongside the contadini and their descendants, in bloodline or in spirit, whose non-tuscan languages included these hills, observation and a listening intuitive of what had come before, know-how and history passed down in talking and listening. Stare at some producer's latest advertising campaign. The vineyard size, bottle numbers, money made in annual reports. Mix. Or, think of Prosecco as an answer, then: what are all the ways to get to it? If Prosecco is pleasure, then marketing, tradition, winemaking, viticulture, dozens of stumbling hills crushed together into something stitched of steepness in fossil-white calcareous marl, tiny never-ending vineyards grabbing hold of it, sites that look as witch-spelled as one of Italy-smitten Hans Christian Andersen's harsher visions. If Prosecco is contadini and a modern wine whose roots double through the family stories that kept land and grapes protected and known about before the two DOCGS' official Consorzi created in 1962 and 1985 to do so, did, then count Verdiso, Bianchetta Trevigiana, Perera, the different Prosecchi varieties here because they always had been here which is a place whose earth of ancient-volcanic weathered red and of white left by the sea over what would be the nation's northeast must be counted, too. If marketing, then in that also the wine media that is today's inheritor of Soldati and wine philosopher Luigi Veronelli's missions which is wine knowledge in places like New York City where I live, in, too, informed and demanding imaginations with wine shop shelves to match. Unification is a backdrop and explainer in that it is a tying of disparate but linked regions into one, coinciding in time with old and new empiricisms and technologies, with academies, oral history, and the press each a powerful mode of knowledge dissemination. To the soon-to-come decades of industrialization add the problem of the fascist Ventennio. All where the lines of Prosecco DOCG meet as abruptly as grapevine nodes within which the buds that each year give new shoots for the growing season's fruit and the following year's wood and are stories no less vitally connected.

For a lot of its history, *Prosecco* had been a word of several wines and many grapes, a term most accurately reflective of its areas' winemaking places and traditions. The Prosecco that Carpenè envisions adds something else, an alignment of chemistry, quality, and nation defined mostly by one grape variety although his wines included other local cultivars. Making the wine known today will take some changes, across many fields. The term *spumante*—that measurably fully sparkling through carbon dioxide gas and liquid water tightly bound together—is about pressure and will take almost a century to resolve, and to make official. "Prosecco and Prosecco Spumante," wrote Treviso's next great spumante scientist Tullio De Rosa in 1970 just as some of these wines were fitting this land's vision, "is to be drunk a breve scadenza[25] after buying it." His technological ancestor would have agreed with part of that. Lastingness of both nation and wine was a crucial feat, a fundamental and definitive trait, and Prosecco was as nation both ancient and just setting out. Having founded his winery as a reconceived continuation of the multimember Trevisian Società Enologica he had cofounded in 1868 and of which he was equally technical director and devotee, Carpenè planned to continue the first's commercial production of cognac and spumanti, write historians De Leo and Favero, while also expanding production of "'luxury' sparkling wines, which he planned to make by the champenois[26] method of fermenting in bottle starting with the Pinot [Bianco][27] of Austrian Trentino, which gave a better result than the Prosecco he had used first." But with the new enterprise he could also experiment. He did so with "spumantizzazione in artificial ways by adding carbonic acid to Prosecco cooled to low temperatures, using a system he patented in 1882," awarded by the Istituto veneto di scienze lettere and presented with acclaim at the Esposizione italiana di belle arti in London in 1888. Fitting and achievement, the word *spumante* brings glee on its own. At the Conegliano Valdobbiadene consorzio, Giulia Pussini whose degree is in marketing sparkling wines told me of her early days' choice of studies: "I didn't have a lot of knowledge about the grapes and the different production methods. It was just a type of wine that was magical, celebratory, fascinating." Here, spellbinding spumante was to be technology tied to what had long been here, and that was Prosecco. Wine and tale of chemists, of the details of lands through Italy once again granting some kind of Venetian, of building Italy's first wine-making school, of Carpenè who wanted to crown it all with that greatest achievement, sparkling wine from quality base, spumante to rival France's new sparkling sensation, to show off Italy's land and prowess in chemistry and character. His winery would be Prosecco spumante's first and foremost house and though he did not live long enough to reach the technologies of autoclave and stainless steel that have now been allowing Prosecco spumante its most fitting form for decades, his was the prototype that these needed to catch up to and that if not for it most readers and wine drinkers today would never know of any other versions. Prosecco Superiore is a voice from a technological age that only most closely resembles ours, a hill of future and rationality that they were walking up and whose downward slope we are still sliding on now that we have also weathered the tragedies that would so closely follow and that they did not know they would need to protect against. His writings on wine growing and making and selling, plural in an outlook that is Treviso-wide and can include the nation, are the papers of a patriot, love of his land with open-eyed criticism, a walking the Treviso hills in space and time and thought that is an overlooked strain against the nationalism and worse soon coming up.

"In the aftermath of the unification of Italy, the numerous[28] inchieste [*investigations*, with a *counting* and *official* twist[29]] offer us the chance to trace a complete panorama,"[30] writes Gasparini. The panorama matters. During my first few months of composing this book, in the summer of 2019, after a springtime that brought Prosecco producers more than twice as much rain than the past twenty-five years' while leaving other months at driest levels,[31] the DOCG areas were officially recognized as cultural patrimony by the United Nations Educational, Scientific and Cultural Organization (UNESCO). Soon after, in February 2020, what would be the pandemic came to the Veneto and then just as quickly to New York City to which I had just returned from the hills. March is traditionally a busy wine time in New York, with an early springtime flurry of wine fairs at which are offered tastes of the latest releases of some of Italy's most anticipated producers, those included in the annual guides by Gambero Rosso and Slow Wine (organizations founded against the rise of fast food in Italy in the 1980s and with a twentieth-century reiteration that gastronomy and good wine be available to working classes as well[32]). Among the last visiting Italians I saw in New York before both Italy and we shut down were Primo and Annalisa Franco, second generation now retired and having handed it over to the next generation, of one of the historic Prosecco-making houses—"grape-growing was a second job," Primo told me during lunch, of the winery's earliest days when the family's first business was silkworms and silk. Days later came the orders to stay home both here and, more strictly, there. Flights that were not canceled mostly just stopped carrying people. For months, the borders between Italy and the United States were almost uncrossable, even should one choose, as I did not, to come within six feet of most people; and most movement and outside activities banned throughout Italy. Five hundred years earlier, painter Giovanni Battista Cima was called da Conegliano for the city in which he was born and which he missed as he gazed northward and imagined into the hills he thought he could see from Venice where he had gone to earn a living, his fromness fuel for a kind of longing. Stranded, my project become one of longing, too, for the hills, for the cities of Rome and Conegliano, into understandings hazily defined during my first encounter with these wines. And then the protests came and lasted for months, first in New York and then in Italy and for a long time I hoped that everything here would at last change, as I continued to research and to write and it was impossible to not tie hunger and death and protest here to the nineteenth-century hills, the terrifying police response and constant helicopters to a shaking nation of the 1920s and '30s and '40s. To do something with what was becoming an overwhelming separation, I moved to something methodical: I could reach there by drinking these wines at home, by remembering my walks in the hills I had taken just weeks and months before and the ones that I had planned, by imagining that Carpenè's longsightedness was gained on such walks, too, by learning what chemistry he had, by following my longing.

That chemistry is a part of wine is of such common place as to escape unexamined. The compounds offhandedly called terpenes that we process as citrus, herbal, and floral aromas; bubble-granting carbon dioxide; color-full anthocyanins are common fact today, chemical presences that are never called so, *chemical* often used instead as synonym for *harmful*, and *not natural*. Even less considered, not to mention analyzed, is the world chemistry imagines into being and the ways in which it decomposes what it observes into maneuverable parts, each to then be studied and defined and its reactions known on its own terms, or a substance defined as its components. In place of wishful origin story,[33] Prosecco Superiore, the second term added to the disciplinare in 2009 and before that used mainly for only one top area in the hills, is a tale of a late-nineteenth-century formation of the chemistry knowledge twenty-first-century readers take for granted, if lacking mechanical and historical context, from elementary school. Technical wines, the top Prosecco Superiore DOCG bottles are also points where a particular place and a time in chemistry and its applications meet. "Antonio was the Steve Jobs of Treviso," Carpenè Malvolti's general manager Domenico Scimone said to me at the end of 2019 when I visited the winery now in its fifth generation. In the early 1870s, a thirty-five-year-old Carpenè had a succinct story on the state his region was in: "You must not therefore denounce the populations if they have not yet reached that progress that ensures the wellbeing of the Provincia,"[34] he wrote in a book covering Treviso's wine-grape growing and winemaking. Italy's popolo (see Glossary) was, he saw, also the way to its dreams and to its future.

The so-called technical sheets that we in the business assume will accompany each sample bottle we taste are stories told by way of soils, wine composition, winemaking, and grapevine training, a small sign of victory of the wine chemists like Carpenè. But were he here today he would note what is missing even in the most beautiful[35] of today's sheets. "The carts that bring the grapes are received under a canopy," he wrote in technical description, warning away the sun. "Dedicated containers are filled with grapes, and by means of a simple mechanism are brought to the height of the first floor of the building, where is there is a door through which two workers receive them, they place them on a scale and take note of the weight. The weighed containers are emptied of grapes, which are poured into dedicated storage bins, placed near the press. Once the grapes have been received, we proceed with pressing."[36] And, opinion and specificity on the most technical: "Potassium acid tartrate is the most important salt that wine contains," he shared in his textbook that is still read by winemakers today and just as enjoyable to a reader who is not one. "It affects not only the sapidity and finesse of the wines, but also, greatly, their hygienic properties."[37]

He is a chemist, technology moved, proponent of a rationalism that was still new enough to be both explained and mysterious, but he is also, in passion and in deep and modest and powerful search for understanding, a precursor to grapevine researcher Cosmo, to literary Soldati, to collector and guide Veronelli, all applying their particulars to researching and promoting what they identified as Italy's true wines. Carpenè had writing, personal and of chemistry and wine, but he took as just as much his subject and his medium the building of institutions for people, was a member of more than twenty organizations and universities over his lifetime—the title page of the fifth edition of his textbook announces him as Doctor of Chemistry, Member of the permanent council of the Società Generale degli Agricoltori Italiani, honorary partner of the Società Promotrice della Popolare Istruzione in Govone in Piemonte, honorary partner of the Comizio Agrario di Treviso, honorary member of the Società di Scienze Mediche in Conegliano, member of the Commissione Ampelografica Governativa, socio effettivo of the Società Veneto-Trentino di Scienze Naturali, and correspondent of the Reale Società Economica del Principato Olteriore in Salerno, in the heavily winemaking south. An inquiry into Prosecco spumante must begin with and follow him, and so is both a study of locality and a way to listen to the new nation of Italy itself, purposely, consciously, and determinedly intertwined. "We offer to winemakers this small guide to the chemical analysis of musts and wines, flattering ourselves that it will become suitable for them," he began the second volume of that textbook, "far however from the intention of addressing a lack, which we see in Italy, of a treatise on enochemistry, and which, in addition to the notions of chemistry applied to viticulture and wine technology, contains descriptions of the most rigorous and recommended processes for the complete analysis of what is directly related to viticulture and to the art of making and aging wine."[38] In 1923, twenty-one years after he died, the Champagne region moved to legally limit its name from being used on any wines that were not of the Champenois grapes and land, ending the practice in both Italy and the United States of labeling their sparkling wines *Champagne* of the place they were grown and made.[39] In 1924, his son Etile—*Ethyl*, a subtraction of one hydrogen atom from an ethane molecule and now three times a Carpenè family name—designed the bottle label that changed everything, turned his home's spumante wine inside into the word that these bottles would wear still more proudly on the outside, with a label of *Prosecco* and *Conegliano* make these hills' wines a new worldwide name.

Though it is dotted with pronunciation and word differences, in addition to the remaining home dialects most but not all Italians now speak a shared language—after centuries of debate, tuscan, already the language in academic and most written forms, was set as the common tongue of everyday Italy. Many Trevisians still, also again as younger generations find new pride in it, speak dialects Venetian or related and which I of course do not speak, though I've gotten better at listening my way into them, both spoken and written. Carpenè must have spoken at least enough of what the hills' wine-growing contadini spoke—as they thereby in practice shaped mostly an oral language and absolutely a local one—as near 70 percent[40] of Veneto's population does today. But he wrote, as most businessmen and official writers did, in Italian, creating a philosophy of wine through chemistry and nation, and I found myself just as drawn to his words, ones that instruct in the most laboratory of senses and yet include the people who will use those tools and those techniques. Perhaps nowhere else in Italy is the meeting of contadini and industry so tightly bound across history, objectives, methods, individuals, and grapes as it is in the hills and their leading city Conegliano, wherein those laboratories of the nation's first winemaking school. "Without the notion to make the winemaker an enochemist," Carpenè wrote cheerfully, encouragingly, "we recognize the enormous utility that he can obtain from the knowledge of the chemical processes used for the determination of quantity of the principal components of musts and wines, while he must devote himself, in addition to this, to the study and solution of more difficult and delicate problems, research which requires much more knowledge, scientific furnishings, and time."[41]

I was also worried.

The social and economic destruction brought by Venice's end, the lives and buildings (including much of Conegliano, the winery, Carpenè's more personal items,[42] only the first two rebuildable) crushed by the two world wars and flustered by emigration[43] make these hills a place to imagine back to, and into a difficulty. The fascism that the new nation would make "did not fail to leave some enduring traces on many of its subjects, as anti-fascists knew quite well."[44] This story is weighted toward the bright dream of unification and industry and wine and Italy, against what will begin in the decade after Carpenè's death: World War I (and therein another comparison to Champagne where the blood of soldiers in the world wars is respectfully ignored component of the vineyards; twenty years later as a front line in World War I, these hills and rivers must be made also of the same), then a terrible transition ("A few months later, in Mestre," writes Trevisian oral historian Alessandro Casellato, "workers demonstrated, yelling: 'bread, work, death to the contadini, what is being done by the government of Mussolini.'"[45]), then World War II.

But, analyze, it is not of it.

There were checks or at least the possibility of recovery built in. Carpenè wrote and in his just shy of one hundred articles and books and machines and instruments there were foolproof instructions, patient and diligent instructions, unshakable belief, grand outpourings of words on what was so clear to him, testament and requirements to this new nation that had been promised for so long, through so much suffering. Italy would overcome the tragedies yet to happen, too. The new Italians might compare themselves unfavorably to others, an "inferiority complex due to foreign propaganda,"[46] notes historian Silvana Patriarca; but this was habit not defect, Carpenè must have thought of it, looking inward to future glory, to wine not war, unpopular for the decades to come, but not for long. Change yes, but not to become something else, instead to express and show themselves. "The fascist experiment"[47] that was to rise, an Italian invention to be fair, soon created as well an enduring and visceral anti-fascist character. I needed to remember this when I inevitably ran into fascist lines. For the industrialization that was to follow, the attempted "mastery or lordship over raw materials" that was to become of agriculture, "an imposed system which . . . denatures it and falsifies it,"[48] would not have been recognizable to these chemists, whose work was firmly trained on what natural (even the word meant something else, closer and more at hand) processes did and how they worked.

An introduction is a chance for an author to reflect, a going back to let the reader in on some things I didn't know when I started to write this book. It is less a chance to roam than the book itself is, so let me explain something of what you are about to read. This is a story that is not to *capture*, *discover*, *explore*, words that have little to do with the place one visits and everything to do with what a visitor is hopefully not doing there. This story is instead an attempt to understand this wine as a somewhere and a time, by sound, by feel, through how and what a wine might tell you about its place. Therefore here are road maps for reading it, perhaps for reading difficult texts in general. It will not simplify or sometimes even make coherent. As a wine writer I am assumed, expected, to be a journalist, and insofar as I am this is also my attempt to restore the complexity, long-term nuance, back then, someday, and being lost that journalism denies in its insistence on news-ness and newspeak and disembodied nowness; this is writing in direct opposition to that deceptively simple and easily manipulatable world built on the interplay between words like *facts*, *truth*, and *falsehood* that creates a new kind of illiteracy for readers incapable of reading other forms of writing[49] and other forms of language and thereby incapable of fruitful wonder, curiosity, meaningful inaction. This book is meant to be something you can hold, turn over and over, run a complicated finger along. "A lot rides on the facticity of the map," writes radical cartographer Denis Wood,[50] critically, finding his way, while never leaving the realm of making maps, by thinking narrative and poetry instead. If you are a foreigner in these hills, remember as you read that you are a visitor, a guest who will be treated well. Remember to have respect for your host, which is a little bit me and much more the hills' people and places. That respect will guide you through the moments when you do not understand, a situation familiar to you if you've been a visitor to a place you do not know or whose language you do not speak. The people in this story are your hosts. I, writer, am their guest. As this is a book written overwhelmingly in English, if you are a native—and perhaps only—English speaker, you might find when you begin reading that you wish some words or syntax followed the English you know, but this is a story about Italy, about the occupied territory it was right before then, with present the Venetian republic it had been for a thousand years before. It is a personal invitation, upon which demands must not be placed. If you accept and do the work fitting a respectful reader or visitor, you will, as upon paying deep time and attention to any place you stay, learn. You must not want a false, a tourist's, a consumer's, a worst of all colonizer's taken and distorting familiarity. No *discovery*, no *hidden gems* or *exploration*, and indeed if you listen to this place, any such rich place, there is a story that your listening will make to tell you, how one's patience turns a text decipherable or not.[51] Start here. Move slowly, or at your own pace. I do not mean to confound, but I do know that from such a place you can see better and for yourself. It is an act of diplomacy, my colleague William McIlhenny suggested (diplomatically) about my approach. And that often it is the unconventional, even oppositional, plans that work. One always hopes for good guests.

While the official word for these wines' main grape changed in 2009, when the story does not reach past that year I refer to both wine and grape as *Prosecco* as was said and heard for so long in these hills (even now, the word more widely applied survives, as all old things do, in local daily and unofficial speech). All translations are mine, and I have kept them as smooth as possible while heavily weighting toward what was written or said to make clear that who wrote or spoke did not do so in English, a sleight of hand that has always made me uncomfortable[52] as a reader. I have used Italian words and words in the varied dialects that populate these hills. They are never italicized save for when they are a word that appears *as word*. Same of course for English words. Italics, too, for what are otherwise sounds, memories, echoes, creations through the medium of writing. For those readers who wish for help with Italian or the dialects, there are three tools, different instances: my thoughts in English in the main text, my thoughts in the notes at book's end, and, best, your wonder and the learning that comes of it (supplemented if you wish by dictionary, the internet, friends, other books) when you encounter something you cannot immediately, if ever, consume. There is as much pleasure in making such paths of your own as there is in the drinking of these wines; vocabulary, of sounds and sights and smells far from only of words, increases when you are in its places. My translations are also appearances here of who wrote or spoke, forms that allow the originals their due precedence. There are long sentences and paragraphs written in English, familiar sights in Italian, but mine are correct in English, too, if helpfully unfamiliar, and they are there to help you acclimate as well. I want to avoid the overdetermined immediately familiar sentence that sells, that therefore is written in order to require no work by the reader. Messy words, participatory book. I am not leading you somewhere, this is not a tour, but there are many, many places here for you to wonder into, and out of; feel free to simply listen to the words you see here. There are these wines which like this book require close, careful attention at the end and along the journey and revisiting. The end is an arriving recast as heading there, stopping, distracted to think of something else, movement, movement, movement recast, too. "An assertion," reads one review[53] of a book about the mathematical kind,[54] "often can be proved by many different kinds of arguments. . . . The way it is written may appeal to one kind of intellect or another (or sometimes none at all). But the style of the argument—the writing—is not just about convincing the reader; it is also about the manner in which the reader is convinced. One truth, but in the range of tellings, different aspects, implications, inspirations, and connections are revealed." A reading as inspiring as the book read, in turn inspired by one whose author said later that he hoped his collection might act as a "rust remover of literature" without "boring the reader too much." Same here; with variations on the reviewer's noting that "English is really not well suited to algebra." Fittingly, a lot of my bibliography is made of Italian works, and, just as important a layer, this book's original and spiritual title is *Qui da noi*. With no equivalent in English, this phrase in Italian "means" something like *here where we are*, or perhaps *here in our place*, or *here by us*. It is something that is said by a person in their place, and it was said to me by Luigi Gregoletto, who founded a small prestigious Prosecco house in the mid-twentieth century, and who I was lucky enough to speak with in late 2019, at the start of what would become the pandemic, and who passed away in early 2021. An utterance, that title is also a presence. The original subtitle—*Histories, Geographies, Languages, Topographies, Personalities, and Typologies of Prosecco Superiore*—is a layer of the book as well and is a map, trans-spatial and transtemporal.[55]

I used several tools to order this story: sounds, thoughts in mathematics which I have studied and which is my own contribution to the subject of rational (see chapter 3) that is a big part of this story. Throughout I followed a constant yearning to understand without trampling, a constant worry about who am I writing for and to what ends my writing about people and places. This book is a listening to what I am wondering about that could be applied to any subject, or to the same one at a different time, results differing and true as per complexity and author. Given what my research yielded, and that was also a process of movement through pleasure, what of the many things to be done with that research would I do, and why? What of the many not only stories but ways of telling? "A physicist is just an atom's way of looking at itself,"[56] famously counseled physicist Nils Bohr about a researcher-narrator's role in telling about something. Writing and reading themselves also come with claims to be the only knowledge, an overtaking that is changing right now as oral history work and collections become, perhaps oddly, academic and which is made even more problematic by that who knew and tended to the Prosecco wines of the Marca Trevigiana hills were very much mostly oral cultures. Exclusion in written texts is also of certain writing people, women most (un)noticeably, along with others deemed not publishable or recordable, more names, work, and thoughts that formed Prosecco but very difficult therefore to hear today. I have included what voices of theirs I could: testimonies collected by the Italian oral history society (AISO) from their descendants who grow and/or make the wines today; written testimony from their nineteenth-century allies like Illuminato Checchini (see chapter 7); my imagination. In July 2021, I returned to the hills, in part as a student in AISO's Seconda Ecoscuola di storia orale nel paesaggio del prosecco superiore, and have included words from the first course, in 2020 when I could not, writing in New York City, get there. The testimonies of those interviewed should be regarded as a text, position oral historians today. Do not translate, do not render "publishable," I understood. I have included producers' website language, too, part of wine's mythology and remembering, inseparable from its marketing and no less a true part of telling wine stories. More about Glera: there are two varieties in this zone by that name: Glera Tonda and Glera Lunga, later recognized as a separate variety, as well as at least five other varieties traditionally included in these wines and today allowed in Prosecco Superiore at up to a combined 15 percent. Considered the area's star quality grape, Glera Tonda, commonly *Glera*, is required at a minimum 85 percent.

This book was researched and written from late 2018 to early March 2022. Much of that time was during the pandemic, which brought death, unknowing, fear, borders and libraries closed. As I had first thought of this book as an interaction with the hills, which meant I would be there a lot, my initial visits and what texts I'd brought back with me were where I could stand during the pandemic, which also brought a reliance on books and memory and imagination and an at times almost desperate sense of being away. This at least matched the loss of the Carpenè sources, of the contadini now mostly gone. I hope it helped this version of my first intentions. "Some of the ancient world's best cartography and ethnography was done by scholars who never entered outside the city walls, or perhaps even the Museum[57] grounds to gather data," I read at some point in an helpful if unrelated book, "but instead simply drew together the best and most reliable maps and travel reports from library holdings. Euclid's geometry falls into the same category: more a synthesis than an original work, it nonetheless stands as history's most influential mathematical notebook."[58] Throughout the book a sense of mathematical referring should lead the reader to make connections. Some of the material is imagined, made up. Several of the pathways are. If you are after something verifiable in the sense of someone else has written this, too, perhaps think of this as a highly verifiable story, and once more a no less true part of Prosecco's story.

For some of this land's details, both their existence and where they might have moved into one another, abutted, been named something they are not named now, I referred to the astonishing 2005 publication that is the Benetton Foundation's cataloguing and reproduction of the 120 Kriegskarte maps, the Austrians' frighteningly precise military mapping from 1798 to 1805 (no time wasted) of their new territory that they imagined into a national buffer zone, widened if disorderly border for lightning-quick attacks. Theirs is a re-cataloging of the land into disembodied pieces: roads that are equestrian, pedestrian, country ones, principal, edged with ditches dried or carrying water; structures from castles to little houses; brick or wooden bridges; villages and whether they had official markets; more than a dozen classifications of rivers, streams, canals, and torrents. It is an imagination hard to ignore, and almost entirely devoid of persons, certainly no one's name except for those of the military officers who had compiled each section. I tried to put some of the pieces back together. This book needed to be a defense in more ways than what I had started out. With the UNESCO designation had come a mandate for those who live there to protect both the land and the people shaping it. That means many things, including what substances are used outside where the grapes are growing. The year of the designation, glyphosates, soil-destructive, potentially carcinogenic, and widely used throughout agricultural zones globally, were banned throughout the DOCG zones. The two DOCGS and the DOC are currently at work on a common growing protocol. The climate crisis part of this story, an overt part of every viti- and agricultural story, is not overtly addressed in this book; for some of that, see, for example, Telmo Pievani and Mauro Varotto's *Viaggio nell'Italia dell'Antropocene,* Ugo Mattei and Alessandra Quarta's *L'acqua e il suo diritto,* and Donna J. Haraway's *Staying with the Trouble*. Agriculture is a detailed story of one's relationship with the earth and to their land and as such is filled with words that do not translate. Winemaking, which has always been a technical act of understanding and making, is hard pressed to do so, too. The Prosecco in your glass today is the occasion for these stories to be told, and there is in a great sense far less to explain than there is to hear and think and taste. You could think of this book as among other things a workbook. A map that might not get you there. I think of it as a recording of finding my way through landscape, a wealth of books and conversations, a few films and television shows, how and from where. A story of finding my bearings, a process in writing, a look at why here. I looped back, recasted, resurfaced, rematched. And like winemaking and even more like spumante-making, writing is a medium even if it is generally thought of as in service to all else, as a way to present something else that has been done. It is all the more imagined so by artists who work in other mediums, whose work can be hard to work through for others and for themselves, so that writing becomes the way to clarify what their real work is doing. As a writer I retain the right to use my medium in that primal, causal way.[59] By retracing, or perhaps better setting out to take some steps, here is the information in between. By book's end you will have acclimated. It won't seem strange; a place or a story of wine, or of anything, is not just content, is a language that describes, invokes (a reminder that many words cannot be translated, and no sounds can be and language works in many ways to give meaning). After reading this book, I hope that your own curiosity and some new fundamentals will guide you to these places and to others and that you will give some of these bottles a steady place among your favorite wines. Look also at the guides from Slow Wine and Gambero Rosso for a list of more good ones to drink. "People are hungry for knowledge, for guidance," my colleague Marilyn Krieger said to me. Go.

To handle a subject as broad as Prosecco Superiore, I have chosen (a following in form of the way wine-knowledge certifying bodies like the Institute of Masters of Wine and the Wine & Spirit Education Trust present their study courses by casting each wine into compositions of, say, location, viticulture, viniculture, wine regulations, market, history) what I see as nine essential facets, for each of which a chapter, indispensable for thinking about these sparkling wines. Each is a wading into, a stopping to rest, a willingness to grapple, an ever-deepening curiosity and love for this place. As food historian Michael W. Twitty wrote in his intensely researched, personal, and delicious *The Cooking Gene*, the hope is that if you love something enough it will give up its secrets. Hoping for the same, I occasionally found myself charged with custody of such—and protective. And so the first chapter, "The Force of Liquid and Gas," is a start by wondering about Carpenè; it is an overview and a scene setting; introducing grape varieties, textbooks, belief, maps, and institutions. "Combinations of Half-Molecules" works through the chemistry of Carpenè's time, reflects on CO_2, gas behaviors, pressure, temperature, and on the role codes, chemists, and the contadini played in unification; it is headed by an entry in Giuseppe Boerio's 1867 *Dizionario del dialetto veneziano* that is in a sense a model of and key to this entire book. "Rationalizing Change" is a meditation on *razionale* and related words which come up repeatedly in writing by Carpenè and other scientists and whose exact meaning then troubled me from early on; *empiricism* and synonyms; it is also like chemistry a chapter about change. "Walking Conegliano" is about the city, political and technical center of the hills, and therefore also about the school. "Prosecco's Machines" thinks about Prosecco-making, technology, and the materials that shape the methods. "The First Prosecco Bottle" is the label that first spoke *Prosecco* and *Conegliano*. "Promised Lands" moves through these lands that are especially fitted to the grapes and the method of Prosecco Superiore. "On the Prosecco Road" gets to the first years of the boom and presents some of today's best producers, among the many and growing. With a style shift into the present, "Old Mixed Vineyards of the Twenty-first Century" goes to Asolo by way of one producer, 150 years after Carpenè, converging and just as astutely diverging.

On the last day of my second visit to the hills, I went to one of the zone's most influential wineries, opened in Valdobbiadene just after World War II when the push to Prosecco spumante was if quiet and idealistic a gathering force. There, Elvira Bortolomiol's father, more than ten years now since his passing, is no less larger than life, was the champion of this part of the DOCG. Elvira now heads his winery. She opens her wines for us, in four, to taste. She wears her role as guardianship. We visitors sit at a long table where five glasses are lined up before each of four chairs, and when she twists from its cork a bottle of the wine made in tribute to him, you can sense her pull away, too, a fact and action both of personal grief and of solemn officialness, as she stands straight, shoulders drawn this time, feet securing against the ground. She walks around the table, she pours from that bottle, she begins to tell us who her father was.

1

The Force of Liquid and Gas

Nothing, of course, begins at the time you think it did.

—Lillian Hellman, *An Unfinished Woman: A Memoir*[1]

This is a book about Prosecco Superiore.

And so it begins with a quest by a man named Antonio Carpenè living in what until forty years before he was born had been the hilly inland cultivated by academicians and contadini[2] of the Republic of Venice, ruled since the 1200s from the lagoon city on the Adriatic sea a distance, plumb-line south, half the length of Hadrian's Wall and a half day's walk away. In another chapter, it begins at an earlier time, too, in the crucial years that lead to his days. It starts also twenty-two years after he has died. *And* a half-century later still, when the materials, method, and apparatus his vision requires would first finally exist and then be accessible to all those who will become winemakers in these hills. This wine that Carpenè works tirelessly to invent in the second half of the nineteenth century is a fluid meeting point of all the forces that are becoming the nation of Italy (even official, the date of Italy's unification is one of three, determined by where you are or what you're waiting for) and that carry through its future and into now. Since it is he who first tied the family's name to Prosecco, and I, doing my best to know him better in these pages, do not yet feel invited to call him by his first name, he will begin his place herein as *Carpenè*, if—I hope a less egregious liberty taken—without an honorific. To start, I have poured from a bottle of his Prosecco (it is best that we start out remembering this: 150 years later his wines sell quite well everywhere. By 2021 the number of all producers'—in time many will follow what he has started—Prosecco bottles sold each year is unshakeable even by the pandemic). The glass I have chosen is heavier than I meant, and in it bubbles line up skyward within, gas, liquid medium, searching balance with the air above their fluid colors of stable warming straw, presence and escape together reweaving the pour into something that is made of gold. Why is Prosecco so successful? "A good answer," to quote a recent writer about trees, "must be reinvented many times, from scratch."[3]

Nothing else we make is as much of us as sparkling wine. We are in a sense hardwired for it. The cells that let us taste sourness are also a physical foundation for one of the carbonic anhydrases, an enzyme that in its many incarnations in our bodies regulates a cell's water-holding capacity, balances ph, makes stomach acid, transports carbon dioxide, and it turns out allows us to taste fizziness. Tethered like wavy banners to our sour-sensing taste bud cells, anhydrase 4 tells our brains that we are sensing carbonation, reassures us that we know it well, reminds us how much we like it. This is a book about Italy and its foundational sparkling wine, and so we're going to set off on the right word for Italian wine with this sensation: *spumante*. An elaboration of a meeting place for two very different fluids: carbon dioxide and wine, gaseous and aqueous matters whose differences make it, deliciously, hard for them to remain in the same place and time. An exaltation[4] of a vinous base. Spumante is flavors and textures brought by a measurable presence of dissolved carbon dioxide or carbonic acid, enough of it to make the liquid "fully sparkling," that is bottled with three to six times more pressure than the surrounding air and which mixed with the water that is at least 80 percent of wine becomes the salty acid anidride carbonica[5] Carpenè was after. What I have just poured into a standard wineglass sparkles. Striking and as important as its announcing tartness and creaminess is that drinking it you can taste the grapes it is made of.

Grown here for centuries, the telltale aromas of wine made from the Prosecco grape are as if from an apple tree in full flower, of its fruit when those are a crunchy green and yellow, complicatedly noticeable herbs and something lemonlike so rich and ripe you begin to think of papaya from shades of orange, fruit far from here. And there you should stop. And turn back just far enough that the world before you shifts into something yellowing but with assumptions that are recognizable. The point of this wine, of this book, of the man that it must begin by being about, is that it is from *here*, and that *here* is three collections of hills,[6] Conegliano, Valdobbiadene, and Asolo, each of which you can think about in relation to their easternmost point which is the *city* named Conegliano—scraping up into the hills against the Prealps, local cultural, political, and scientific center and final reaches of the flatlands of Treviso, Terraferma province of the once Venetian empire (the Dogado[7] is said to have drunk three hundred liters of wine per person each year, some of the finest of it from these hills) and its most closely tied, by administration, loyalty, and language though unlike that imperial taker-protector city, this one, small and until 1865 entirely walled, the locus, leader, champion of its dense, mostly impoverished network of cultivated, inhabited lands. In that year, when a part of Conegliano's old wall had been destroyed to make room for the two new small pillars named for the tuscan poet Dante who wrote not in Latin as he was supposed to but in the tuscan dialect that would become Italian, Carpenè is a chemist, is blurring into winemaker, had been a focused soldier for the cause of Italy, unified and nation.

Carpenè is not tall. He verges on slight, more broad-shouldered in presence and squared resolve than by measurable width, and while he is addicted to reading, to writing letters and articles both forms which he publishes steadily in the newly popular, newly lasting, scientific journals, he is equally immersed in the new hands-on science of making things. He says many things loudly, cajoles and cheerleads, is devoted to the new tools and the new nation, and is very often, and I know he meant to be, hilarious. He is, marker of his male academic and revolutionary peers, fully mustachioed. He walks. In 1860, from Conegliano, his home since his father moved the family in 1846 from nearby mountainous Belluno where they had come six years earlier from the boy's birthplace in slightly farther Brugnera di Sacile,[8] he had been a twenty-two-year-old pharmaceutical student at the ancient University of Pavia.[9] As he read and memorized *polveri*, *specie*, *siroppi*, *diaforetici*, and *ipnotici* in his heavy textbooks of farmacopea italiana[10]—"Is a pharmacist a scientist or a merchant?"[11] wondered a chapter of Federigo Kernot's 1871 history of pharmacy, Carpenè's first studies now rapidly changing work, while later, "in the small paesi[12] of our provinces, the pharmacist's workshop is the meeting place of all the most noteworthy characters of the paese"[13]— he also absorbed terms from a smaller and more portable book, written by the republican, revolutionary, spiritual journalist-lawyer Giuseppe Mazzini from Genova who had once wanted to be a novelist. In *Doveri degli uomini*,[14] published in 1860 in London where he had been living in exile, dodging a death sentence dispensed back home for exhorting the Risorgimento's until soon failed attempts to make the nation, Mazzini cried out for Italy for a hundred pages. "Believe the words of a man who for thirty years has studied the affairs of Europe. . . . You will not succeed unless you improve," he wrote. "You will not conquer the exercising of your rights without deserving it, with sacrifice, with activity, with love. Those who speak in the name of well-being, of material happiness, betray you. Improve yourselves and others; this is the first purpose and the supreme hope of every reform, of every social transformation."[15] Carpenè counted on those words. He carried them as he stepped away from textbooks and joined the armies that might make Mazzini's words into bigger and more solid things, tucked *Duties* (growing heavier with his own words by inked steel pen when it was necessary to sit and wait) into the green jacket he wore in the piemontese military, deserted to join the second[16] wave of (mostly[17]) men following Mazzini-protégé Garibaldi, the wrinkling rain-, mud-, and promises-bearing book in the pocket of his Garibaldian-red shirt.

And then it is 1861, and unbelievably there is an Italy. Carpenè went home. The days themselves are bubbling, fluid time of social, personal, and chemical foment, and now that he is back from the fighting he will finish his studies first in pharmaceutical then in 1862 in chemistry[18] while the old Venice waits to be let in. Between a thesis on element aluminum and the second one on compound acetic acid, he chooses the future as a chemist, enlisting in the new science that was harvesting the knowledge of the past century and a half, digging Italy into philosophy and agriculture, writing a new language that can map the ancient into the new, track a fast-unfolding playground of materials. He is watching the decades-long dreams of the small portion, well-educated, literate, of the population who planned and fought in the Italian Risorgimento come true, as a growing mass of infinitesimal stubborn forces rose faster, ever steadier stream, and now France is gone and in 1866 by the annexation of old Venice to Italy Austria will be, too, and Carpenè will marry Teresa Zannoner, and on September 20, 1870, when papal stronghold Rome falls to the Italian army the nation will have its capital city, the whole nation signed into law on February 3, 1871. The particulars of those dreams—for followers of Mazzini that Garibaldi has allowed, helped in fact, the piemontese monarchy to absorb the new nation is, politely, an unwanted form—are to be worked out later. Now is for everything at last possible.

The city of Conegliano is still in its ancient Venetian province once, unusually, ruled directly from Venice: in 1862 Treviso had until sixty-five years earlier been for half a millennium a foundation of the Terraferma and Conegliano the center of a devoted rural element whose citizens' cries of *viva San Marco*! upon the 1,100-year-old empire's dissolution rang in rebellious heartbroken concert with those of their seaside compatriots. Throughout the province and striking in this city of northern hills that runs equally at ease over the last moments of the flatlands, just reachable by train at the Conegliano station opened in 1855 (railroad the technological borderland from the ways of plains to the exigencies of hills and that could transport wine) raced centuries of unyielding Venetian patriotism[19] which might now find a home in Italy. Betrayed by Garibaldi, the hopes of the Mazzinian republicans were to be looked for instead in Italy's cultural, personal, and industry realms. But there they offered glinting possibilities: there had been neither an Italian nation nor an Italian sparkling wine. What each would look like needed to be dreamed, the future crafted now and carried there through making and through writing. "The pharmacist," Kernot drove home, "is an instrument of civilization, in the provinces' paesi, in our villages, in our countrysides."[20]

If anhydrase 4 had been something he could know, Carpenè would have added this: the mechanisms of spumante are no less constitutive of a nation. If the unwanted presence of their armies had been solved by Italy, the French whose wines were becoming what defined quality across the newly solidifying Europe had just found success in almost mystical terms everywhere with the sparkling ones from their northern region Champagne. In 1858 in Paris, the food-famous French chemist Louis Pasteur, anti-materialist and spiritual,[21] had convincingly shown yeasts to be the crucial presence in fermentation processes, putting an end to the competing theory of spontaneous generation and handing studious winemakers definitions and processes by which to make the wines they could imagine. The French sparkling process, ever more fine-tuned since Pasteur's substantiation, was to ferment their grapes (a group of seven varieties of which Chardonnay and Pinots Noir, Meunier, and Blanc are the ones most often used) into a still wine, then distribute that among glass bottles made strong enough[22] to contain and nurture what would happen next. Into each bottle the winemakers introduced enough wine to almost fill, set it to ferment again by adding sugar to attract airborne yeasts, then sealed each bottle to let those organisms run their havoc through the wine once more. Escape impossible this second, enclosed, fermentation, the CO_2 gas created alongside warming ethanol alcohol as the yeasts turned sugars into cellular energy was forced to dissolve into the liquid, making a new substance that is sparkling wine. In this confined state the wines of Champagne remained at times for years, long after the yeasts had done their work and that second embroilment had ended, gathering flavors no grape could supply: bready, briochelike, toasty aromas from dragged-out contact with now-dead yeasts in small quarters. No need to cover up our wines with the flavors of winemaking, Carpenè armed with Pasteur's report must have thought about the star grape variety growing in his hills. For Italy, we need new technology and our ancient grapes. Let the juice of Prosecco not be reacted upon except for fermentation, be limpid and profound, make a long-lasting impression of the mix of Italian ingenuity and Italian lands and grapes. In Treviso, he began to build his own project that was a local inquiry into chemistry and nation: a spumante made of grape varieties that had been long been grown, practically exclusively and at their best, in the province's hills, their intricate flavors and delicate colors observable and verifiable attributes that showed the inhabitants' understanding of their materials. Prosecco would be a wine desirable to an outside market, exportable, enriching, but only as reflection of local taste, which admittedly also needed some reawakening. "The principal articles of the corano enologico, wine made by enologi turchi,"[23] Carpenè wrote of the heavily macerated highly alcoholic wines too many made here, condemning such subpar tastes and wines to match as phenomena foreign to Italy in both spirit and flavor. His Venetian spumante would lead long-lost Italy to itself, earn it overdue pride nationally[24] and on a global market. To get there from here, two scientific roads were necessary: the ways of chemistry and the long path of technology.

How to explain the beauty in practicality that is something Italian because it is something of once-Venice? In 2019, the United Nations Educational, Scientific and Cultural Organization (UNESCO) recognized the hills of Conegliano-Valdobbiadene as one of its 114 world heritage sites (and one of Italy's eight), a cultural landscape shaped by "techniques of land use that guarantee and sustain biological diversity, or embody an exceptional spiritual relationship," testament "to the creative genius, social development and the imaginative and spiritual vitality of humanity" and as "part of our collective identity," an area which must be safeguarded from development and planting pressures carried by the vector that is Prosecco's twenty-first-century success. A "distinctive landscape," the UNESCO description fawns, "where nature and human history have shaped and been shaped by an adapted and specific system for viticulture and land use." The buildings that appear here and there where vineyards are planted are now mostly no longer used,[25] farmhouses—masserie and chiusure made of stone and lumber—with stalls and plots within which the contadini tended to Trevisian life. On some landings immediately at the hills' end, notably different from massive counterparts on the region's southern plains, are Treviso's small industrial buildings—enclosing rectangles in white, yellow, red that are shrines to the nineteenth- and twentieth-century industries of silk, wool, wine, kitchen and bathroom ceramics, hiking shoes. To my twenty-first-century eye they tell their own stories with no disrespect from within the agricultural landscape. In 1867, old military maps in hand,[26] Carpenè walked along those hills.

If geography is said to be in the service of war,[27] in Italy it is also tied to the making of wine. In his jacket, his Mazzini's *Duties* (eighteen years, he remembers, it took to write) is now easily bent crescent-like into a side pocket, one hundred pages survived a battle and so reread that their rose-brown cover wrinkled and crumpled, awaited kinked at the corners, was spotted in uneven depths where suddenly raindrops had landed on springtime walks and soldier downtimes, frayed spine a thumb-width down where he'd held it open solid to follow each word of the man who had begun to dream Italy into being ("I have loved you since my earliest years"[28]) when he was sixteen years old. Mazzini had studied medicine, at first, too. "Credo nell'eterno progresso," he'd written, and Carpenè walks along the beat of those words, too. "I believe in the eternal progress of life under God's creation, in the progress of Thought and of Association, not only in people of the past, but in the people to come."[29]

How much of Venice's civic Catholicism was of Carpenè, too?

Mist filled in where the yellowing light stretched thin into gold and broke. He walked as the path moved slow and steep and rounding into the hills' jagged faces that locked together like wrong puzzle pieces at their edges to form the two chains of hogback hills that make up, in a broad line squished southwest to northeast, the Coneglianese and, west and north in a stretching furrow that hinted in a geographical way at the line from the Adriatic to further west the Tyrrhenian waters onto which faced Rome, the Valdobbiadenese countrysides. Both men wrote heavily, poured out their hearts to reach their compatriots, spoke of destiny. What Italy would become had to do with what Treviso could become, to be drawn from what it already had but was neglecting, raw materials evident to those who knew to look. Above all keep believing. There was evidence now that faith had restored a nation, and so faith could build it into its long-suppressed glory. Disciple by books and walking, Carpenè—and all those who believed along with him—would make through a process of both restoration and invention Italy's Trevisian wine. How much of this chemist's knowledge of materials was studied on pages on paper? How much worked out through while he wrote? How much knowledge from the company of other chemists? How much was from the paths he walked and how much was he able to talk to the contadini he met along those ways? How much in the schooled Italian the borghesia spoke in public? How much in which Trevisian dialect he might have spoken at home? How much in this-is-how-we've-always-done-it which I hope he knew to speak as well?

Forest echoed and repeated around him. While they do so in greater numbers now and over pastures and plots since abandoned, oak and flowering ash grew along the sunless northern faces of the hills and the woods-rich gamut of southern faces was interrupted by those occasional contadini structures and by pastures for the summertime. Farmhouses were also centers of subsistence and tiny production—rows of vines encircling the local silkworm industry's native black mulberry trees and some but far from all trained with fruiting arms draped like grids along the branches of cultivated cherry and maple trees, mixed with rows of wheat and oats for selling and, for eating, corn. "Più vino e meno pellagra,"[30] Carpenè would write later[31] addressing the poverty that also claimed a large part of these hills. "Where wine is not a luxury, life is better, people are sturdier, happier, more humble and in balance," he wrote. "More wine and less granoturco." In using for *mais* (dried, ground, stirred with difficulty into polenta, hated deliverer from desperate hunger) a word that carried *foreign grain*,[32] he hints loudly at something else. Wine as a specifically Italian resource, Italian wine as cure for recent problems that ail the new nation, now finally fermenting once again, necessary turbulence to uproot what has been misgrown so that the real thing, grown right for the better part of Italy's history, can come through again now. In those grapevine rows an explosion of varieties. More than fifty cultivars were planted heartily throughout the Trevisian countryside, and "if we count the ones planted less often, we would have more than a hundred,"[33] many of which could be found growing together in one row, from which to then be harvested at the same moment regardless of the ripening time each variety required—in one site we should plant one variety only, or at least in distinct groups, variety by variety, he insisted, in support of the specialized planting he saw occasionally and which is the way it's almost always done today. "Today, if a wine is thought good by the masses, one must suspect that it's terrible, and vice-versa," he lamented of the outcome. "In these conditions, it is impossible that the worthiness of one who walks along the right path is rewarded. Long is our journey,[34] and if France is jealous of our enological progress that means it has foresight, because it is thinking now of how to defend itself from what could happen—perhaps from here in a century,"[35] he cheered up, temporarily. "Too slow is our progress!" From tiny worlds along the hills, he could see plains south past the Piave river, past the ancient forest on the long, slow Montello seemingly one hill and whose trees still hold up Venice, could picture what from his readings he knows is most egregiously there: a few of the grapevines grown coupled to a walnut strong enough to support them for fifty or more years[36] and which gave hard and lasting wood and fruit that needed no preserving, but also strange flavors to the grapes. Walnuts are rarely part of a wild forest. A walnut is a solitary tree, guarding its space with a substance called juglone, toxic to other trees and stored in its roots, leaves, and bark for releases into the terrain. But this is also a story of flat Treviso from the Po River Valley to the train station at the base of quick and sharply rising Conegliano. "In the hills, the population is more industrious," he writes pridefully when he is back to sitting at his desk. "More intelligent, more affluent, and more dignified; traits that lessen as you descend to the lower parts of the Province."[37] Within the city limits, slopes, upon which vineyards in mixed plantings and some in rows that ran as in the hills in hatch-marks across the direction in which their piece of the land rose. As form gives hints of the process that formed it, read the latter pattern as a practical move against erosion of friable calcareous clay or in begrudging obedience to hard white Biancone and red Scaglia Rossa. And, there are steep coaxing sandstones and more gently rising marls, all sedimentary formations carried and formed by battering rains that now fall evenly and lighter throughout the year, running alpine waters and their offspring downhill torrents to feed the century's many mills. There were northern Neozoic glaciers that long ago creaked and shoved their descent into the plains,[38] recorded the lakes that are now along the top of Treviso's border. There are water runs throughout, and as they have for centuries some still shift their course while others runneth over, destruction or less painful road-blocker. At least fifty that I can count, over a hundred if one were to add canals and streams that run alongside city street, trickle forgotten under low walkers' bridges, with honorifics like *ruio, rio, roggia, torrente, torentello, valle, rosta, acqua*, a trickling *sorgente*. Tradition, I begin to see, is a solution to what is directly before you, devotion and reciprocity in the place one lives.

Now there is a hill for me to climb.

With unification suddenly a fact, the restoration of Italy's wine is a first and urgent task. And chemistry, science emerged from the cocoon of wonder that characterized its beginning in the eighteenth century, is in full effervescence, source of a flood of answers to the questions that had been building up over the past two hundred years and more questions, more rapidly answered. By the later 1800s it was comfortably connecting its practitioners throughout Europe by numerous journals whose pages filled with a common language of symbols and substances, a rationality powered by belief. In this science that over the past century has transformed from holder of wonder to a campaign of curiosity, another element draws Carpenè in: chemistry is just as much a way to understand Italy. It holds out before him this golden grape.

As of other sciences, it makes sense to ask: is Prosecco an invention or a discovery?

In a series of undated sepia photographs I look through on the Carpenè Malvolti winery website it is easy to recognize the hall that you can walk into today. There you can buy wine or join a formal celebration in the historic and decorated hall. But today's place is two rebuildings later. The hills that circle this small city were rich in outposts for Austrian then Nazi Germany's and Mussolini's[39] soldiers and in strategic hiding spots for the mostly[40] Italian partigiani who fought against them. After the first world war and again after the second one, vineyards and knowledge were again rebuilt, and it remains hard to not think about exploded shells, bullets fired, blood of soldiers, villagers, fighters, wounded and worse, returning to land, memory and earth deepening textures of these hills from which to keep an eye on the strategically navigable Piave that ran red, here's inhabitants still tell you, a scarlet which the blue and mud of this great river could not dilute. Here, Prosecco is as good as blood. In one old photograph the winery hall is lined with enormous wooden wine tanks. Along what is today the walkway, in the images a cluster of bottling seats, walnut-dark wood bolted to blackened cast iron and upon each one a bottler sits and seals by hand one bottle at a time, 320 of them per sixteen-hour day. Stretching into a corner that meets ceiling instead of floor, a giant spumante-style bottle—it's not clear what it's made from—crowns a massive wooden barrel. Since he lived before steel, the wine-holding vessels of Carpenè's machines were from necessity vats made of color-, flavors-, tannins-imparting wood. Were they made by Garbellotto, historic cooper of the zone since the eighteenth century and whose large casks made of local Farnia oak harvested from Treviso's forests continue their quiet reign in the winery today?

Until now, the science represented by the icon of a bottle had been the pharmacists'.

In an 1868 assessment of the new nation's wines, Carpenè and professor of agriculture Angelo Vianello—who fought for Venezia in Italy's first War of Independence in 1848, and believed—together complained that the "lack of typologies in our wines is a great ailment for us." In their Viaggio in Italia[41] they traveled through the new nation tasting and researching how its wines were made. The results appear in the 1874 *La Vite ed il vino nella Provincia di Treviso* they wrote together and published as a complement to the 1873 Vienna exposition in which Prosecco wine was first presented to the world, though based on the exposition catalog seemingly not sparkling. Its cover points to an earlier and more practical outcome: author Carpenè is credited as both professor of chemistry and technical director of a new local winemaking cohort, the Società Enologica Trevigiana. Launched the year of their travels, which we now understand were reconnaissance for forming back home in Conegliano a society to make the best of the province's wine with an eye to both export and education, the Trevisian enological society was cofounded by Carpené and abbot Felice

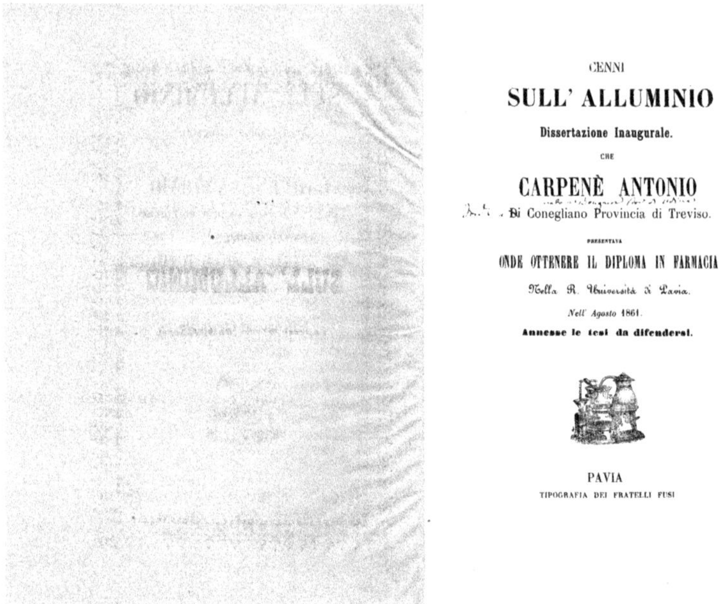
CENNI

SULL' ALLUMINIO

Dissertazione Inaugurale.

CHE

CARPENÈ ANTONIO

Di Conegliano Provincia di Treviso.

PRESENTAVA

ONDE OTTENERE IL DIPLOMA IN FARMACIA

Nella R. Università di Pavia.

Nell' Agosto 1861.

Annesse le tesi da difendersi.

PAVIA

TIPOGRAFIA DEI FRATELLI FUSI

Title page of Antonio Carpenè's 1861 pharmacy thesis with added notations of place
Carpenè Malvolti, Conegliano (TV)

Benedetti (president of Treviso's agrarian consortium founded in 1867 and who a year later led the compilation of the *Ampelografia della Provincia di Treviso*, a very large book covering Treviso's land through ampelography, the new nineteenth-century science of drawing, naming, and cataloging grape varieties via visual markers, that would be published in 1890 with in-depth tables and illustrations of 120 principal wine-grape varieties[42]). With Vianello as president for the twelve years it would exist, this unity of local producers made wines from grapes bought mostly from sharecropping growers who in springtime, se met do[43]; in late summer used a cortelazh[44] to gather gariòta[45] by gariòta; were too often dissuaded from improving what was already planted by the always impending late autumn threat of needing to "*far* San Martino"—leave house and plots if their three-year contracts, too short a time to harvest grapes of any vine they might plant, were not renewed on November 11, day of this patron saint of pilgrims, drunks, and owners of Venetian gathering and drinking places called osterie.

"In every locality and in every year in the same locality you find types that differ one from the other," the duo bemoaned first in that 1868 report. Italy's winemakers, "thinking to do good," create imitations of foreign wines, "instead of studying those techniques and using them to make their own kind." Today, Italy is celebrated for its wealth of native grape varieties and biotypes, most confined to the places they've been grown for centuries and local languages unto themselves, but in days that needed steadying the duo's assessment of such abundance was twofold and worried: an export market is easily confused, and not all cultivars fare deliciously in all places. "Whoever wants to dedicate themselves to selling wine abroad must restrict themselves to making two or at most three types of wine, or better just one," they advised the Società's would-be members. "These winemakers must adopt just one variety of grapes, or at most two. Every viniferous district must take advantage of its own vines, choosing those which have been judged to be the best, that is, most adapted to local conditions." The hills' top grape had been chosen, mostly by the academicians, in recent pre-nation history. During an agri-academic stirring in which academies had added agriculture and in particular the peninsula's ancient splendor of wine-growing to literature and mathematics, the for now first known written instance of the grape appears: in a 1772 edition of the *Giornale d'Italia* dedicated to agriculture, arts, and commerce, Conegliano nobleman Francesco Maria Malvolti now famously wrote, "Who doesn't know how exquisite are our marzemini, bianchetti, prosecchi, moscadelli, malvasie, grossari, and others, which are made in many of these hills,"[46] in official support for the wine's legendary, if mostly for the academics, presence and blaming the growing and vinification habits of his time for the mostly poor-quality wines then being made of them. By the 1870s, Malvolti's great-great-grandson Angelo Malvolti was proudly and carefully growing in his Conegliano vineyards rows of Prosecco Balbi, a Prosecco biotype especially rich in aromatics and rounded acidity named for the equally impassioned Conte Marco Giulio Balbi Valier who had selectively bred it into cultivar existence in his own vineyards in the steep Col San Martino hills just decades earlier.[47]

As science of matter and change, chemistry creates and provides knowledge but also always materials, makes new substances through "metamorphosis and transmutation."[48] Sparkle is both quest and bottle, it is the technology that might keep liquid and gas united as one substance, is how to enhance a nation, deliver its nuances, delicacies, intricacies, and finesse, and among Carpenè's most prized materials are the fruit of those grapevines that grow best where he lives. If it is his duty to learn these hills, he wants to know them in some way that might overlap with the ways the contadini did. "Blessed is he to whom knowledge is passed down," they would say to anyone lucky enough to receive that gift, but *his* understanding has not arrived that way. He compiles. He counts the varieties of wine grapes grown in each of Treviso's comuni, notes the names they are most often called, wonders about which ones tasted the best—the whole unspoken point of this scientific eye—in each place. "The number of comuni that grow a given variety is a very dubious guide for inferring the quantity of special wine which a speculator might calculate,"[49] he concludes in 1873. "In the interest of an industry which began its rise in our Province only four years ago, it is useless to show the quantities of wine, which we could calculate for each of the main varieties." But he includes them anyway.

With *La Vite* came a map composed by engineer Giuseppe Carpenè, brother of the winemaking chemist. This Carta Enologica della Provincia di Treviso (1:145.500) holds the province's ninety-eight comuni and forty-six varieties of their grapes, drawn into wines that are deeply colored either *red* or *yellow*, volumes of each as millimeter-rectangles[50] and numbered by variety, each rectangle a way of letting you know where you are alongside railed roads and pathways for horse riders. By stepping into the winemaking hills with books and maps in hand, Carpenè is breaking apart what had for so long been a continuity. He is recasting Treviso's wine-growing into steps, making it into processes discrete and thereby measurable.

In late 2019, my companion and I drove up the path that stretched toward the towers of Credazzo, solid hillside world on one side from which errant plants occasionally reached our car, vertical end of that world barely reaching the opposite tires. Our plan for if a descending car arrived to face us was only to tell the driver that we were not from here—no visitor could possibly be expected to carry out this drive in reverse. At a sudden widened break, above which the path would no longer fit our little car, we stopped. And walked directly into hillside, 102 footsteps along one interrow walkway to reach the top of a wavy little plot of Glera vines, 117 along another that six rows away rose along an entirely different plane. Flatlands and in the grouped hills dense building blocks that trailed off into the other landscapes, lines of roads and waters, formed a variegated sea before us. Closer, we could see enough to follow along the buildings that like boxcars defined Conegliano's main street, could look north and up to see immediately on a hill that reached higher still those towers. Then tires slipping, crunches and catches along our gravel-led and somehow even more unlikely descent. Down, we headed a little farther west, along the Provincial traffic roads to Valdobbiadene.

When Carpenè walked, there were not yet the asphalt curls at sudden differences between land and sky, where the material edges of where the industrialists have turned landscape into unbearable road for machines run unfinished into the cooler air that, pressure system, follows the hills' great falls back to earth. Nor were there as many vines. And even fewer planted like he wanted them, rows reaching against the blue with nothing else like those which I had walked along. He stepped upon crumbly soils, gray-white breakdown of ancient fossils in long and slowly upward slopes in the east, practically solid sky. Wider spaces, abrupt rises reached their beginning and then kept going until there were spots, any spots, with vines growing, some true vineyards where nothing else was tended, others with trees that fed as well with fruit and berries, light that cadmium yellow, dimming under the great trees' branches, the church bells winds of roar and clang trickling over sounds of springs like wine pouring into the spumante vat Carpenè is building, to receive its dispensation of CO_2. "I coltivatori amano la patria,"[51] he'd written. Stepped onto narrow trails so protectively steep the field violets (flowers the yellow-white of phosphorous) and methyl-yellow clustered radichiella will still be growing there in the twenty-first century when so much else has changed, through the lines of vines growing in tiny waves on the hills' rickety faces, fantastical twists and smoke,[52] to be conducted into sky, air of yeasts and CO_2, gas that touches but by definition moves across the dirt from which they grew. "Agli Agricoltori Trevisani!" enthuses a page that heads *La Vite*, love letter technical and throaty. Read this so that you, too, may see, find faith in the beauty and potential of the land from which you live already, your land which you are dishonoring by growing its true potential badly. Through you, our greatest grapes will be both emissaries and a generous living. "Agli Operai Italiani," Mazzini had opened *his* missive, with punctuation more to the point, more paternal, reassuring, and even more optimistic than his vinous successor who instead now had the immediate honor of calling on his fellow countrymen. Both personal and passionate but only one with an audience now named Italy. Follow me to the promised land. This is what you, the first to reach there, must do.

In the wine-by-the-glass osterie of Venice and Treviso, the red-shirted socialists would continue gathering for decades.[53] In which ones did Carpenè meet the growers? Drink their Prosecco wines, cloudy and determined? What did he drink? "What is a good wine in Italy?" he scoffed[54] in the closest, if by opposition, indication I could find. In solution he built a sarcastic list. A good red wine must be black and form a dark pink foam when swirled forcefully in glass. When you drink it, it must "insolently excite the nervous system of the mouth" causing you to roll your eyes and your entire face to fill with goose bumps. The chills one gets by drinking it is a signal that the wine is good—it has substance. Italian wine must never be perfectly clear, especially if it's a white wine—this means that the wine is adulterated." New wines are preferable to old. "Foreign concepts,"[55] he thundered at list end. "In Italy, a real criterion on the different qualities and merits of our wines, has not yet been made by winemakers, or by merchants, or by consumers." But, once mighty, "glorious in arts and knowledge and . . . inexhaustible in its resources," Italy remains the land of wine: "grapevines[56] are cultivated in almost every region, and in many localities viticulture was less neglected than it is in our days." What is going wrong in the vineyards is easiest to fix. That falls to the viticulturists, and indeed his focus shifts: before *La Vite*, he had written another book with Vianello: 1867's *Nozioni teorico-pratiche di viticoltura e vinificazione*. In 1871, his solo reworking of the former becomes the first edition of his winemaking-only textbook, the *Sunto teorico e pratico di enologia pei vinificatori della provincia trivigiana* to become in 1878 the two-volume *Sunto teorico-pratico di enologia (del Direttore Tecnico della Società Enologica della Provincia di Treviso)*, dedicated *to the reader*, in print until 1901, while he pushes for the nation's first school of enology and viticulture to be built in Conegliano.

So, Direttore Carpenè, thank you for talking with me. When did it become about wine?

Something in the air changes as the bubbles enter. It seems to warm. Thickened it carries the scent of apple blossom and lemon edges.

"Ecco come abbiamo operato."[57]

He lifts the glass and drinks, averaged smooth motion at 10.75 centimeters per second[58] from table to taste as all the producers do.

"Con intelletto d'amore."[59]

Atop his desk and at its feet, all made—like the madie and cassapanche[60] that are furniture in the Treviso fashion and still found today in its countryside and small urban centers—of local walnut[61] trees, aloof but reliable with nourishing and storable food, he has ordered his notebooks into those for future use and those he's busy transforming into books of records taken. Up eleven steps that are more climb than stairway a room that is the highest rung in this compound of winery plus laboratory plus home holds, still walnut, a tabletop like a kitchen one over which a too-large window that can be opened and closed. On it a scale for weighing up to 100 grams, sensitivity in half-milligrams; scales in kilograms, sensitive by grams; a Babo mostimeter for counting grape sugars; a thermometer in centigrade from -10° to 350°, fractions in between; a Guy-Lussac alcometer; microscopes for enlarging 400 to 1,200 times; Bunsen components.[62] Across the room, it's not far, he'll need to cross the tiny expanse repeatedly, a closet that offers his labeled substances and which must be out of breeze reach from where he's exposing them, controlled. On the way there, a Salleron alembic distiller. Another tabletop holds out spouted porcelain bowls, calibrated long-necked flasks of glass, and coverable crucibles for substances that need heat to be dissolved; a copper bagnomaria to hold two liters of water. Cross again and, smaller, another tabletop offers straight and tapered beakers and footed glasses for dissolutions he'll do cold; and there, too, are nesting glass vases that can resist a little heat, all for cajoling materials that precipitate from other ones. Misunderstandings and mismeasurements happen and should from them come fire or poisonous air to breathe, there's the window, and this top floor to protect the rest. Downstairs, a vast and hall-like room holds towers of wine wrapped in staves of oak, smaller tini on concrete stands, barrels stacked on ground which will be filled like the baskets of six bottles giving meaning to the bottling table they crowd.[63] To enter Italy is a chance at resurrection, reassumption of a tale as old as agriculture. In the coming winemaking of quality and curiosity, Prosecco, Venetian pride abandoned by all but a few, will also be new knowledge. Like its complexities in bitter and tart, the wide range of Prosecco's aromatics are Venetian patrimony *and* Italian sensibility, which France does not have. "Italy has a great future with the industry of spumanti wines," he would write bubbling with confidence in the fifth edition, "corrected and expanded with many illustrations and a table in chromolithography,"[64] of his *Sunto* in 1888. But first, Prosecco presents two problems: preserve the wine's primary aromas *and* dissolve within it enough gas to make it sparkling like Champagne.

It was easier to be a soldier than to understand this shy and sensitive grape. Carpenè found one solution—keep the base wine very cold to keep those flavors unchanged, a state that would also let it, transmuting, hold more CO_2—then built a compound apparatus around his spumantizzazione method, a unified system for making up to four hundred bottles of Prosecco per day.[65] In 1882, this patented design earned him the written admiration of fellow agronomist and enologist, founder[66] of the *Giornale vinicolo italiano* Ottavio Ottavi, who wrote, "after the invention of the Carpenè apparatus, we no longer find advisable the production of vini spumanti with the method of Champagne: this method is long, painstaking, expensive, difficult to implement, and almost no one does well with it in Italy. We know all the vini spumanti we've sold during these twenty years, but few can compete with the vini spumanti alla Carpenè or alla Sciampagna"[67]—a toast to technology and wine as national traits, and credit I'd say mainly to the grapes. How to create the gas in the Carpenè system remained a sticking point in conversation about how good the wine was. Champagne's bubble force came from the fermentation of its base wine. Carpenè's from his apparatus's first and second stations, while the base wine awaited its arrival at the fifth and final stop, waited to be carbonated the way seltzer water is today. He describes the method in his *Sunto* patiently, pages of instruments drawn precisely, parts clearly labeled not by name but by a letter which appeared, too, in the surrounding instructional text so that a reader might know exactly what and how to use it. Even assumed knowledge was laid bare and accessible in the hopes for a community of Italian winemakers, a society open upon instruction for anyone to join: "There is nothing that merits explanation in apparatuses 1 [a carbonic-acid gas generator labelled *a, b, c, d*] and 2 [a gasometer labelled *e, f, g*] because they are common for the preparation of carbonic acid, which is obtained with sulfuric acid and soda bicarbonate, to have it pure and inodorous" he writes therein of his five-part spumante method, and then includes non-chemists anyway. "But for those with less practical experience, we'll say that in flask *a* goes the sulfuric acid, in the botticino *c* some soda bicarbonate and water . . . opening the tap *b*, the sulfuric acid enters the keg, decomposes the soda bicarbonate, creates the gas, . . . then by way of *e* enters into the gasometer *f* to fill it." It allowed a winemaker to regulate its release (not more than forty-five bursts per minute) into an ice-sleeved copper boiler ("diligently tin-plated on the inside" and pressure-sealed with a cotton or rubber gasket and "viti di pressione") filled with 120 liters of wine and to be slowly shaken during the process to help the gas dissolve into the liquid, human movement mimicking what molecules of two states do in close contained proximity. To handle the explosive power of CO_2 created by fermenting wine—and at a volume large enough to be an industry—autoclaves in stainless steel would become first a reality then a necessity of the twentieth century, but his sparkling process is among Italy's earliest and while ultimately (soon) the Italian container-based method would turn Italian methods "natural," that is sparkling through re-fermentation, he defends his technological step against the French way of doing things. "We must deplore how in Italy there are, even among instructed people, those who discredit the Italian system of production, asserting that the bubbles of sparkling wines obtained with a non-French system rise rapidly and are quickly exhausted, that sparkling wines made with the French system are much superior, and other stupid assertions repeated even by exposition jurors, absurdities in opposition to the laws of science," he writes from the firm ground of CO_2, unshakable chemical understanding of the gas that makes wine sparkle. "We are forced to ask if there exist two carbonic acid gases, one different from the other, the one that comes from doubling the sugar in fermentation and the other that is separated from bicarbonate sodium."[68] The problem is not just a messy understanding of sparkle, or even these tenacious French insertions into Italian matters. Such conversations must stop "because they make people think we hold the false conviction that what is Italian is inferior, both in genius and in aptitude, to what is foreign. We must have more awareness of ourselves, less servility, and appreciate more, when it merits it, the fruit of our studies, of our intelligence, that has nothing to envy of the foreigner. Here in Italy the maxim—*nemo propheta in patria*—is too true."[69] Further authority added by that Latin phrase which appears in all four Gospels, base of a religion that in Venice was as civic as it was spiritual, and which is an echo of Mazzini's including *God* in his *Duties*, love of whom would guide Italian character in an Italian nation. If Italy was the apparatus, the method that required it was a certain type of conduct among all Italians. It is "the love of the bel paese,"[70] Carpenè implores, that transforms character into nation.

Pasteur's triumphant settling of the debate between fermentation and spontaneous generation comes with words linked to vitalism and to his Catholic faith, but he doesn't bring in France. Mazzini's much-cited *God* was an assumed spiritual presence, ardent difficult faith reserved for the people of the promised land of Italy. For Venetians, catholicism was civic and political, and so rural Treviso was deeply Catholic, too. Was Carpenè? And in which ways? Until just after unification, the province's civil records were held in each parish, an official subdivision of ecclesiastical territory complete with church and priest, and measurement of place that preserved the Roman *pieve*, synonym of *romanità*, and as such of *the people*, updated by Christianity to mean a territory's community of baptized believers, then by invading Lombards to mean a community of their subjects. Where Carpenè walked, woodsy Pieve di Soligo extended south from the Valdobbiadene hills with centurion oaks and olive groves along the north-south run of the river Soligo, water as much as memory and fed by the Lierza that until 1953 powered the seventeenth-century grain mill at nearby Refrontolo, faithfully adhered-to work that sustained body and soul, interrupted only by floods and lack of corn to mill. The torrente Campea still links the Pieve to the newer Farra di Soligo, *farra* the Lombards' toponymic salutation to indicate a place of these invaders' own, west of the river and where the Volpere and the rio Bianco run. Throughout these hills went uncountable rivers and streams which, it might make sense to hear in them, carried, separated, defined, and promised echos and reflections of their Republic's imagination and power, *vineyard*, *God* and church, *Venice* or an *Italy*. Difficult hills and many waters, the religion they'd been carrying might to Carpenè have been visceral and efficient knowledge, occupying body the same way science did, duty and system and philosophy of wine and nation. Italy runs steadily throughout his writings—but along with faith, for him there has to be a *why*. Why else do chemistry like this if not to educate Italians, instruct them, transform them into themselves?

In Pasteur's *Memoire sur la fermentation alcoolique*, illustrations are detailed microscopic views of the discovered ferment, illustrative diagrams of yeast at work, a formerly magical and messy world laid bare, made knowledge shown off to convince, to take the prize. But the illustrations in Carpene's books are something else; they are use and labels, drawn frequently with heads and hands, tools that suppose workers and await them. They are calls to take up *these* arms now, you battalions of newly knowing and best-equipped winemakers of Italy. "It is the story of too many among the Italians of today, drinking as they have been of the old ideas of France,"[71] wrote Mazzini. "A very sad story, but how to interrupt if not by changing the guiding principle from which we depart?" Walked in from the city, in the hills Carpenè bent his feet against hard boot leather, pushing up into the chalky gray ground over which waved stalks of corn and wheat colored like threads of silk, the steady breezes flapped like big soft hands Prosecco's grass-green leaves to show their smooth gray-green undersides, shimmered behind them the yellows and citrus-green rounds of grapes gathered in generous bunches while the whole-body efforts of the bell-ringers of Col San Martino[72] rang out over the Campea's hastening run through the same fine-grained Santa Lucia marl the grapevines grew in. Counter-bought best, his brown leather hat marked him as from the tiled streets of Conegliano, none of its rain- and sun-negotiating porticos out here, built reflections of a Venice that was also a major veneration site for Siracusian patron saint of sight Lucia placed by Dante in Paradise, to cover him from the rain that Prosecco grapes so desired. Returned back down into Conegliano bounding connecter of flatlands and hills almost reachable by the new nationwide construction of iron-railed roads, under cover on an osteria's sidewalk he sat down to his own glass of Prosecco alla Carpenè, wine like Trevisian silk (itself between a liquid and a solid, part crystalline, part disorder[73]) in strength and weave.

"Encouraged by the success . . . and once more happy to see the fruit of my labors used to feed the fount of our national viniculture's progress, inexhaustible richness for Italy, this eminently winemaking nation," the introduction to his fifth-edition *Sunto* reads. *Eminentemente vinicolo*. You can see him taking his bow. Since the edition of 1878, the textbook's pages have included a section so new and fierce it has been titled "Volume Two": "Enochemical Analysis" is devoted entirely to the chemical analysis of wines and showcases instruments and techniques—four moveable wooden stools each impelled by three legs for the painstakingness and time that a person doing analysis, bottling, closure-fixing needed; a wooden box that is a truncated light-blocking pyramid built in mulberry and fitted with a measured glass tube with scaffolding brass support for measuring the color of a wine. It is one, united book, new and discrete section bound together with the original pages. On its cover page is a symbol I have yet to officially decode but it looks like this: a badge upon which an ornate *E* and sword, topped with Venice's lion of San Marco holding an open book, time of peace, to herald Italy's once-more majestic wines, brought by the character proudly sovereign Venice was said to enjoy in its citizens, each a blend of power, wisdom, justice, peace, and religious piety. Of the writing that became volume one in honor of the latest science, thirteen pages[74] are headed "Fabbricazione dei vini spumanti"[75] (an urgency) and include the guide for using the carbonation apparatus he's invented, written with an insider beauty rigorous and inclusive, an ego-bound force no less invitational for that. But the section's first four pages are given to the necessity of, before use, fixing his countrymen's characters. "Spumante has a merit relative to the quality of the [base] wine used," he wrote in unmissable analogy and pragatism. "Italy, with excellent grapes graciously perfumed, at its disposal, has neglected the packaging of its spumanti, when from these it could make a healthy profit." And while any wines could be made sparkling with his system, he encouraged that "for spumante, white wine not many months old is preferable, made without skin contact, pale in color, with a light gracious fragrance that is natural to the grapes." What is indispensable is that they all be "limpidissimo," shine clear and true. Defensible knowledge and understanding are products made of consistent proximity. They are things of the land you have known for generations or have walked long in tribute to, things of the land you love and therefore can see as they are, in good times and in bad. Prosecco spumante made of grapes and technical aspects entirely of Conegliano, grown and invented in the hills that surrounded its scientists and implicated them in land and wine, in the Province of Treviso, in the nation of Italy. We still have a way to go to reach the *excelsior*, he wrote. Lofty principles, so what. The work, the matter of it, was very much on earth.[76]

In 1876, Italy's first winemaking school is built in Conegliano because of him. The place, complete with weather station (from which there are now centuries-spanning decades of daily records) and experimental vineyard on a slope that rises near instantly from a heavy wooden door hooded by stone entryway, is a draw, fold of testers and learners and gatherers, and most importantly in which the future is being made. Prosecco was the solving of a problem, answer to questions found in the strange space between its civic enormity—earn money—and everyday life—drink wine. "Wine is a millennia-old integral part of our history," writes food historian Massimo Montanari. "The Romans diffused it throughout Europe, making Italy one of the cardinal locations of this culture.

Culture because establishing wine-growing practices has never exclusively meant technology, just as the consumption of wine has never exclusively meant food. Fundamental social, symbolic, religious allusions have always accompanied the history of wine."[77] In Carpenè's turn to winemaking questions: What is Italy? Closer, how do we fix our dire economic situation? The questions themselves would not be fully formed until well into the twentieth century, desperate complexity that enriched the possibilities for answers: Veneto, the wars and emigration, the Austrians, the poverty and misery that have become our lands. But our victory is already the terrain, too. We will find and codify the ways to reach it again and again. With chemistry, with rational science, I can help, guide, shape the questions right. In Conegliano where in 1867 Garibaldi's words to Trevisians gathered in the now Italian piazza where grapes had long been sold ran through the little city then followed waterways and slopes up into the hills, Carpenè spoke, too. *Shouted*, says his winery's website. He stood upon chairs and wheelbarrows, taller, louder, "exhorting the contadini to believe in the land, in the quality of their fruit and to invest in the renovation" of their vineyards, "of their rows."[78] Throughout the piazze of the villages in the hills of academic industry-thinking Conegliano's conglomerate lands and Valdobbiadene's morainic earth, wilder and precipitous and merchant-minded, he brought the school to the contadini, taught, pleaded and incited to the cause that had turned from nation into wine. They, but more precisely their children and their grandchildren, would become the school's enologists, viticulturists, agronomists. They will be the citizens of a new Treviso, in their pockets passports from the specialized, divided lands of *-ists*, newcomers to these hills who walk with notebooks, look through study-trained eyes, hearts filled with experiments, even though most of their families have lived here for generations.

Prosecche. As are the nation's dialects, Italian is a gendered language and that final *e* denotes a feminine noun in plural form. Today, Italy's fruit are vastly feminine words and the trees that bear them the masculine form of the same word, but wine grapes are an exception: the vine itself is feminine, and the variety names are largely masculine. In nineteenth-century records, many grape names appear as both genders, inconsistent and swayed by pressure systems of tradition and power-building. *Prosecche* was what Treviso's contadini called a specific collection of grapes, qualified as such so long as from those they could make the lifted crushed-blossom cloudy and slightly fizzy—that is due only to weather, sugar, air—wines that everybody here both made and drank. In oral historian Dino Coltro's book *La Nostra Polenta Quotidiana*,[79] the narrator (or better, to use the term that oral historians do, the *witness*, with in mind that the written records produced from such interviews are to be analyzed as texts of literature)—is an older woman who had been a contadina in a family of day laborers[80] and remembers for him enough to fill fifty pages with what they ate, when, and at whose discretion. "The hallelujah was not celebrated without tagliatelle with beef[81] and full-bodied wine, the wine of the hills, you understand," she says to him. "You went to buy it in the osterie, but not everyone had it, there was many a sign on walls with the words *Here we sell mountain wines*."[82] From those steep places inaccessible to carts, grapes were harvested into side- or back-slung wicker baskets[83] carried to the roads where the new year's empty vats or carts awaited.[84] Stuck with working when and where it was available, rice fields as easily as vineyards, she is at times envious of the mezzadri, the sharecropping contadini tied to land under tenuous contracts with landowners, a system Carpenè and others blamed for low overall quality. Believe in what you have, break these bad habits born of a de-incentivizing system! But the vision of spumante, to be made of the hills' brightest grapes and interpretive machines, did not yet exist for the contadini. Polenta did, dull and thick and yellow pale graying. What wisdom, what understandings that those scientists shared with those contadini went unwritten as the Prosecco Tondo variety and its Prosecco Balbi biotype, the ending *o* of a masculine noun, became official and cataloged, the very most Prosecco of the prosecche, the grape with a mission, quality, and nation and the one we know as the source of these eminent wines today?

Why were you obsessed with this grape? How did you already see this wine that not for years past your death would begin to become itself? What is the Italy you experienced in these hills and in your passion and your imagination? Though the word is just two decades[85] into filling out with meaning, Carpenè is a *scientist*. And so what his work is can be reformulated here as an hypothesis: In Italy's Treviso, Prosecco is our fine grape, and our sparkling wine that we will make of it will best Champagne (France be gone, second in wine as you have been since the Romans brought it to you). Students of Prosecco—for a while, they, too, will all be men—unite: the new spumante-ists born of grape and of machine will be Italian makers of Treviso's finest wine. Was there another destiny in what Italy's first chemist winemakers knew? Facts that came from what they have studied, from as today arguing against doing things their fathers' way, from as today freshly softened stories of how and why their grandfathers made wine, trinity to guide specialists in vineyards and cellars, a devotion to wine as branch of knowledge? What else was redirected in these translations? What was taken from the contadini and turned to market concerns, giving country air to counters and compilers? It is easy to stand against him, intruder!, or step in step with him, visionary swoon, but everywhere in his writings I catch glimpses of a complicated man. But while he dreamed in writing and equipment, the growers—"ancient and primitive rootstock of the contadino culture"[86]—were of a world that would be left out of *Italy* enrobed in Chemistry.

In everyday talk, the English *chemical* has come to carry an utter and destructive split between a nature and a laboratory. Italian *chimica*, while endangered, often avoids this, regularly preserving the noun as *substance with which a chemist is concerned because these are the words of which the world is made*, each a measurable and identifiable element of it. For Carpenè and for decades his fellow chemists, winemaking can be cast simply as a search for substances that make up nature and which therefore are also knowable in lab. "The *assaggi chimici* [emphasis mine] of the principal constitutive substances of musts and wine have made themselves known among many oenologists,"[87] Carpenè wrote in his *Sunto. Chemical tastes* that could describe wine by listing what it was made of. "And this proves that rational vinification methods are rapidly gaining on empiricism, as progress is on stationarity, which is the real cancer of industries." Stationarity: a statistics tale of a collection of changes that does not vary across time.[88] Everywhere there were threats to the new Italy. As destiny and history, the past was both a source of strength and an abyss of famine, disease, invaders, repeated ways whose reasons had long since been forgotten, an unshifting diet, whose only constant was so much uncertainty that it kept the present shifting in search of some rough equilibrium. If Italy was a golden light, it was also an onslaught, but it just might sparkle and shine: concerned with materials and transformation, the chemists saw in the state of its wines an unnatural lack of movement. Some among them might have imagined in the stillness of a wine without bubbles a dreaded stagnancy.

Moving metals into machines, pen into letters and books and articles and patents, Carpenè walked, drew by iron and woods a rational system to make Treviso's viticulture and enology what it should be in glass and national[89] and outside market. *Insight* drove him, or *innovation*, or any word that means a deepening of understanding and not departure or newness. Spumante was always already here. For the Vienna exposition, he had advised producers to send wines that represent, "per mo di dire,"[90] the climate and the terrain "upon which they are made."[91] Pour. *Look*. Here is a thread that moves through cloudening yellow liquid veiling strands of gold, is become laboratory notes. It is as if he were writing with a microscope, instrument he praised as a "divine emanation from human genius."[92] At all places a quality wine must shine, must present its colors as a treatise: golden for the hills where just past them are their mountains to protect them from the worst of cold and wind and rain and below them the plains to gather the worst excesses of air and water. Spumante, abstract, perceptible pleasure, technological proof of Italian land. Like the hills' daffodil sunlight upon their blue-green waters, Prosecco gleamed.

In Conegliano in late February 2020, I page through the late-1800s volumes of the *Rivista di viticoltura e di enologia*, Conegliano's viti- and vinicultural journal published with the school since its founding, in no particular order as I search for sounds of Carpenè's days. At times his words therein are sharpened to a point by concern. "I am extremely worried," he wrote in October 1872, "about the harvests in the Trevigiano which are successful barely one-sixth of the time."[93] Extremely poor-quality grapes, poor in sugar, rich in acids, the average price they took "was 42, maybe 44 lire for a quintal." The price of new wines I can tell he thinks are still cloudy and shoddy: 900 lire for "black wine" in an 8 hectoliter bottle, or 75 lire per hectoliter. Five hundred lire for "white wine" in the same, that is 62 lire per hectoliter. "Here are the solutions to these questions:"

End of letter.

But already I am wrong. What was it I brought with me to this reading? An assumption of revolt, distrust of power? I realize I am missing the rest of the letter, and that even in this introduction, eighty-one-word letter drawn with a quick clear aim, he couldn't keep the passion out. In 1798 as the Austrians prepared to pour into the fallen Venetian territories France would hand them, they saw medieval social and administrative structures made five hundred years earlier, few roads that they might use, town after town "backward and motionless for centuries,"[94] a world they declared unrecognizable to those of the science, recorded, industrial places other parts of Europe were becoming. Under the General Anton von Zach, the first thing they did was make maps, for three years measured, named, defined through triangulation technology first used by the French. First, the territory was subdivided according not to existing geography but to fit Austria's occupying needs. For each new place, an officer was made responsible for reporting it, for plotting it onto paper and to make his section into a list of findable Austrian resources—possible soldiers' quarters, inhabitants, livestock, bridges, waterways, food grown, viticulture—and of avoidable perils—flooding rivers, landslides, hills, sinkholes, mountains, and other hiding places within the earth's surface, bridges built of brick or lumber. There had been no cartographic work before, their work from scratch suggests, leaving out at minimum local map-telling work which had been taken or destroyed by France's retreating soldiers. By 1805, the 120 sheets of these Kriegskarte military maps covered all of the Venetian territory, a scientific knowing of it as rich for war as it was for the new age's hope that all might be knowable, and in the seemingly uncountable things to be found, one osteria Gai pinpointable "beyond the bridge, to the right, [then] 650 steps to the right, a road travelable by carriage"[95] at up to thirty miles in an eight-hour day, half the distance for who rode upon it on a walking horse, twice as fast on a galloping one. A record of questions. What with what didn't fit? How does what Carpenè was doing differ?

I have made my own map of Treviso's rivers. For now it is the ones that cross the territory of the Conegliano-Valdobbiadene DOCG, networks of fiume, torrente, ruio, roggia, rio, val, and valle in electric blue lines, rough and geographical. While I drew them they wiggled out of more precise navigation and became a map to tell the kinds of stories rivers do—that they cross and join and end and cause separations in foods and languages and feast days. What they might know of the hills' territories as they power mills, obey canals, are memorialized in names of locally connected streets. Soon, I will add what runs in the Asolo hills. The Piave river which draws together Conegliano-Valdobbiadene and demarcates the vineyards of Asolo does so largest and as an outward boundary and because of it the other rivers here can be thought of as internal infrastructure: after avoiding each other for most of their very different paths south the fiume Lierza passes into the Soligo three kilometers from where the latter will join the Piave on *its* long run to Venice's sea. A native of the northeast's idyllic Prealps, urban Monticano waters shakily bisect the city of Conegliano, where centuries ago it also changed its course a few streets west. Into it run the Crevada that links by water the city to hilly Refrontolo and Susegana just east and south and the Cervano, first gathering strength from watery threads like the Rio Acqua Grande and Fossa di San Pietro, that had long connected the city to northern economic stronghold Serravalle which in 1866 combined with nearby quiet Ceneda to become Vittorio Veneto though the addition to the name of Italy's first ruler was by local usage after Italian victory in the World War I battle around there, nation but also Venetian. Back in Conegliano, the walled and double-denominated Torrente Ruio[96] will take you from the city's ancient Jewish area where the winemaking school first was, "suffused with that austere but gentle beauty that is characteristic of the Coneglianese hills."[97] In 1883 the school began its already necessary move to a new larger building on farmland (key to the times: *that* construction featured both a stall and an access road) farther west across the city. There today, beards the signal of revolutionaries, a bust of Carpenè complete with his iconic whiskers greets you after a gravel-pathed walk to the school's front double doors. It is the image you see of him again and again: skyward gaze in all his portraits, steady coiled curiosity, even when his eyes are made of stone, and in this one he is noticeably older if unfaded. From the engineer's oenographic map I know that roads suitable for riders or horse-drawn vehicles might weave an even more intricate world than what the rivers make, a thick north-south red line to Conegliano from both the east and Venice, hundreds of little ones to everywhere, already the railed road that like today means trains to and from Venice's mainland and to and from Udine to the east, alterations for human use that preserved the landscape anyway. You could travel "the Province comfortably in carriage in all directions," Carpenè promised in 1874. "The smallest village is not without well-built and graveled roads for reaching neighbors; and now they're even building just to facilitate access to the countrysides." One thousand eight hundred and fifty-four linear kilometers of "very good streets over 2,325 square kilometers of surface."[98]

If spumante is a heightening of the viticulturists' well-expressed base wine, then careful attention must be paid to the grapes from which the latter is constructed. Measure the land instead by traits of the grapes that are grown on it. Today the vineyard sites that give Prosecco grapes their most discrete aromas and textures are identified by the dialect word *Rive* plus the name of the locality they're in. Along pathways in these hills the steepest slopes appear in traditional conversations in that word unsurprisingly tied to the Latin *ripa*—sheer plummeting steepness that exists specifically in relation to rivers. "Something *pendulous*, that is to say, these *rive of rivers*," Innocente Nardi, president of the Conegliano-Valdobbiadene Prosecco producers' consorzio until his term ran out in later 2021 and whose family grows Prosecco in the Rive of Farra di Soligo (vines rooting in calcareous clay soils a near-perpendicular three hundred meters above the sea), told me, casting an image of hill from the top down. Spoken, the word carries a linguistic concept into this hills which unlike other parts of the province hold no significant geographical examples of it: the Piave doesn't have especially formidable rive, he explains. "Therefore in our dialect the concept is deeply changed by our territory, and is identified instead with those especially steep parts where there are vineyards." A landscape where steep land does not point you to nourishing and navigable water but instead promises of a certain difficulty in working and a certain quality of wine. Rive is used in opposition to flatlands' *brolo*, a word itself repurposed from an older meaning of a garden inside monastery walls and since a while here used instead for vineyards planted alongside houses.[99] "I'm of a certain age," says Nardi, and he remembers his grandparents' use of *la riva* to point to the vineyard of a hill particularly and abruptly steep. Through further indications of heady particularities in soil types, shifts of mineral compositions, more or less stony states, and how they face the sun, these old and steeply sloping places result in fragmented Prosecco—with some exceptions including that Verdiso and older vines of ancient Trevisian cultivars, that's primarily what's grown here now—wines that taste very different from each other. Recently, *Rive* has also become a legal term, forty-three names of places each of which can be written on a bottle label if the spumante inside has been made of grapes from only there. But with so much attention for so long paid to the characteristics each site confers, grapes grown in the Rive are also made as base wines that will be deftly blended for integrated tastes of the Conegliano-Valdobbiadene hills. If one place grants grapes of higher acidity, it might be blended with another whose fruit is heavier bodied; fruiter places might require partnership with hillsides that give more austere mineral and blossom flavors. The rivers as clearly related are deeper, older, numerous, and harder to find. They are a map of what is otherwise not there.

In the Rive of Farra di Soligo where Prosecco grapes yield wines of such fascinating austerity that it is necessary always to allow the final wine to retain a few extra grams of sugar but which instead of sweetness make this steep acidity as lush as its lime and beautiful green fruit and herbs can be, Carpenè stumbles. And for the first time almost falls. His next step is of sudden and unbelievable difficulty and while too slowly moving into it he thinks of how the objects in his laboratory, inert wood, yellowing paper, the accumulating ink words in his ever denser laboratory notebooks have all been growing heavier. Of what the wine must taste like, of the beauty in taste of this nation's character and wines, he has written nothing, not moved pen driven by his enjoyment of it, of a life with it; nothing. Like landscape and grape, the wine must have been too obvious for written details. Only strengths and what needed fixing or inventing needed words, news to both the growers and the foreigners. Pasteur's tale of fermentation has made understanding wine as important as the tradition of it. But what from the actual traditions? Of vino da messa and the one del prete, bottling in the 'l calar de luna,[100] starnir le vache, fighting hail with the sora la tempesta? Pleasure and appreciation was something had in person, and no less to the hills' chemists it might have seemed bizarre to have written of such intimate and obvious things. At least three senses have the final word over the final result of winemaking, a physical response that drives all of this, stubbornly resisting rationalization though not for lack of effort as ongoing scientific research into just what a taste of place might be. "I firmly believe that the best of our hills, of Italy's wine is yet to come," he might have said, and then what? Is his work meant as opposition or to overtake? His ideas would provide stability not only of wine but of a nation and especially of these hills. But that stumble comes in the first days of his ataxia, a progressive, trembling hands and interrupted gait, paralysis that in his early forties will take him from his directorships, his winery operations, but not his microscopic attentions or experiments and never from his writing. His walks continue on the pages he will continue publishing until his death in 1902 at the age of sixty-four.

In 1891, Italy three decades in, Carpenè will use his system to make Marsala,[101] the aged sweet wine of Sicily whose producers had lent Garibaldi their agile shipping boats so that his men could land there stealthily in 1860, sparkle, too. "Entrusting the progress and continuity of work to the zeal and usefulness of all producers, would make the nation rich, manager of its own production and consumption," Mazzini had wound down his Duties, adding a triumphant tale of fourteen men who formed a workers' società, and took care of each other until—"a victory after starting a war"—they ("most of whom had not tasted wine for a year") earned their first small salaries, and promptly spent them on a celebratory meal for all, their families united at a common table. "And this, Italian Workers, is your future," he wrote like a parable. "You can hasten it. Conquer the Patria, conquer a popular government that represents its collective life, mission, concept. Organize yourselves into a vast universal League of the People, so that your voice can be the voice of millions and not of a few individuals. You have truth and justice; the Nation will listen." Mustache and resolute shoulders, Carpenè's most defining feature is that he believed like this. His chemistry teacher, also a revolutionary, admired his hard work and curiosity. His machines, iron and bold, are stamped with his name. The wines that are his legacy are more and more inseparable from Italy's becoming, tied most deeply to this light-filled grape that is the color of a mix of straw and daffodil, that is the light of dwindling later-summer green, is the radiance of belief. Exiled philosopher, what need had Mazzini for wine, for chemistry laboratories, for a writing that communicated materials and bottles strong enough to hold the spumante that Carpenè at home already in the new Italy poured all his faith and association-making into? There were new things to read now.

Carpenè is clearheaded. Soon into the Enological Society, he is at odds with many of its members, who "ignoring the present viticultural conditions and therefore exaggerating the value of our raw materials,[102] would like that from this establishment come out, after one year, wines superior to all the others in the world. These pretenses belong to those who rise from ignorance, prejudice, or envy, every time the more intelligent and thoughtful try to build[103] any kind of industry with the sacred scope of developing the nation's resources." In this premonition, a warning of the inferior products of nationalism lying wait in the upcoming century. Critical and nurturing, he remains a patriot and when the Society collapses at the end of the 1870s, from the work he's put into it he builds a new winery on rational Prosecco growing and friendship. Soon to be mayor of Conegliano, Angelo Malvolti backs such faith with both his grapes and finances and together, in the building where the Società had failed, they launch Carpenè Malvolti. Until 1924 the bottles are labeled with the words *Champagne Italiano*, the key to what that meant toward the bottle's base: *Carpenè & Malvolti, Conegliano (Veneto),* for what words were available yet to anyone making such a wine? Materials having not yet caught up to vision, they were likely made alla Carpenè and alla sciampagna. In 1895, the piemontese[104] wine scientist Federico Martinotti had patented[105] his method and apparatus to make sparkling wines of that region's brashly aromatic Moscato Bianco grapes growing for centuries on the steep but wide and sweeping hillsides of its Asti zone. If France had reliably durable, strong machine-made bottles, Italy now had Martinotti's pressure-withstanding tank, at least on paper. It materialized slowly. Interior-enameled cast iron second-fermentation tanks were used until at least the 1950s. The solution in stainless steel would come soon after, material of the early twentieth century, made into autoclaves by the 1960s, accessible in quantity and price beginning in the 1970s. In 1930, Carpenè's grandson adapted the Italian system for *their* wines, setting the way Prosecco spumante is made. The winery-founding friendship was immortalized, too: when Malvolti retired in 1912,[106] the Carpenè family insisted his name remain a part of the winery forever.

In 1890, Malvolti's vineyard included another grape traditional to the Prosecco hills. Yielding lime-green large and ellipsoidal grapes under pale-green squared and pentalobal leaves of even paler underside, the Verdiso variety was grown there, too, with regard for quality and very little solfato potassico[107] applied, noted Carpenè.[108] "It is produced in all of the Marca Trevigiana," Veronelli will write of the variety in 1961. "Especially appreciated is that of Cison di Valmarino, of Farra di Soligo, of Pieve di Soligo, of San Pietro di Feletto and of Vittorio Veneto."[109] Around the same time, De Rosa will write that "the various organoleptic constituents [of spumante], composing a harmonic picture based on the elevated CO_2 content, would immediately show a significant imbalance if the wine were tasted upon elimination of the carbon dioxide it contains,"[110] in his *Tecnica dei Vini Spumanti* still today a reference for Italy's spumante-making and an early testament to what Carpené foresaw. Already in the years immediately after annexation, in Venice newspapers were printed based on rationalism and liberal thought,[111] and so a little later, on that walnut laboratory chair, brought outside to sit supported by the school's already storied grounds, in the brightening cadmium light Carpené will open the new Sunday paper, *L'Ateo*[112] written by another Ligurian living elsewhere, and read.

Late summer afternoon through the windows at the osteria, he pours from a bottle that he has brought. Unclouded. A strong yellow with light-reflecting properties, he observes out loud. The healthful sapidity our growers know our Prosecco to be, the floral and bitter freshness of our best grapes and to make sure of it, we fermented the wine for an hour on the grapes' skins, which will make it sturdier, too. I have made our way to make spumante, and the bubbles are fine and steady and there is no reason to doubt our fine wine of Treviso. The wine is an elusive color he might have described as something like *giallastro pieno* had he gone looking for less practical descriptions. Winemaking is how we enter into conversation with the grapes, with their hills. What UNESCO has recognized is that this is a landscape still in talks with its people, but also with something to say to a few of us who are not of there. The hills and city offer traces for us who have taken the time and ingenuity to find ourselves with

In the Carpenè Malvolti winery in Conegliano, portraits of Antonio and his son Etile dream of revolution and spumantizzazione of Prosecco.
Susan H. Gordon

Prosecco, microscope and notebooks in hand. There is another mill, smaller, still working, and powered by the Soligo river on its run via an ancient roggia before the Lierza reaches it, in the Pieve di Soligo—which it turns out was two Pievi[113] each on either bank of the Soligo from the Middle Ages, ruled separately until Napoleon and the Austrians' takeover[114] and combined into one comune by the Hapsburgs, that kind of map, further bottling local pressure. Here in 1970, the pharmacist Schiratti, known for his "enological passion"[115] and the wine cellar beneath his pharmacy, is a stop for Soldati on his way to see the oste Condo in Col San Martino. Both are Rive designations. A little west, among the top 107ish (even with the twenty-first-century insistence on data, the number corresponding to these faces and twists and amphitheater turns is regularly reported as anywhere from 106 to 108) hectares of the entire Prosecco DOCG zone, the Valdobbiadenese's sharp amphitheatrical friable-gray Cartizze whose vineyard lands today are Italy's most expensive, is the sprawling destination spot Osteria senz'oste, an osteria without a host, and without a dining room. Instead chairs and tables are placed outside, seatings like balance wires throughout the gnarled hill faces. Alongside centurion vines cuttings which are supplying traditional genetic Prosecco to where in these hills they are most suited stand vending machines and an old stall now filled with the local wines and foods for sale, mostly on the honor system.

In 2019, I see, dark candlelight, on a sparsely stood bookshelf in this room on Bond Street that is now (osteria plus food) a restaurant whose famous wine cellar houses one of New York City's best known and most personal Italian lists built on the church-altar-staining red-wine Sagrantino grape from central Italy's Umbria, an old determined and loved book. Though no one reads it today, it is widely known of. This one is the washed vermillion of two centuries back fading, drying, the founding manual of Italy, and I place it speciman-like onto the table where it lies gently rain-dotted and off-center-creased, softly brittle with when I open it curved steep handwriting in black along the margins of its straining yellowed-sunset pages. *Io ci credo* read thickly turned letters by steel pen, heavy hand five pages in. *Antonio Carpenè*, deliberate, elegant, even, thinner first on the inside cover page which announces *Doveri dell'uomo, Giuseppe Mazzini* above *Thought and Action, God and Populace*,[116] and a drawn symbol of the risorgimentali conspirators the Carbonari. I am poured a glass of translucent deeply yellowish sparkling wine, apple blossoms and their green fruit, the savoriness of dismantling anadride carbonica, complications from a second fermentation which *his* wines could not yet have, in this room which was once the receiving room of the entire townhouse, owned by a friend of writer, poet, journalist and essayist Edgar Allan Poe who is regularly rumored to have set one of his most famous stories in this cellar, and who wrote ardently of ciphers ("it may be roundly asserted that human ingenuity cannot concoct [one] which human ingenuity cannot resolve"[117]) and of secret codes.

2

Combinations of Half-Molecules

Ciòpa—ten-cent bread that is equivalent to two "panetti."

—*Dizionario del dialetto veneziano*

$C_6H_{12}O_6 \rightarrow 2\ C_2H_5OH + 2\ CO_2$ (fermentation of grape sugars)

What makes a nation? In 1873, Italy had a problem. Ancient land of Enotria now modern European nation, it was sending representations of itself to world expositions: earnest gestures of artwork, descriptions of educational systems, and, particularly personal, wines. With a few exceptions (the foundation of modern Chianti Classico by tuscan wine-grower-baron once-minister of Italy Bettino Ricasoli who recast it as a local fine wine based on his region's Sangiovese[1] grape had impressed judges[2] as had tentative bottles of Prosecco[3]) that last category was not showing well.

"The climate of Italy is eminently propitious to the cultivation of the vine," declared piemontese chamber of commerce and arts VP Giovanni Boschiero in his report on that year's pageantry in Vienna. "The favorable disposition of many pleasant hills makes it naturally the wine country par excellence." What and who then to account for the discrepancy between historical and geological propensity and shaky present? Too many submissions were easy insights into what was going wrong back home: graying yeasts like tiny storm clouds, still-fermenting liquids in bottle a-tingle with carbonic acid whether it suited them or not, grapes turned to wine saturated with signs of winemakers disregarding rationality and order, said Carpenè and other wine-enchanted chemists, linking their format of empirical knowledge regulated by theory and their laboratories' reliably repeatable transformations of matter to their nation's well-being. You must, they said to each other and to the contadini they could reach, rack, clarify, and at bottling add oxygen-shielding sulfites[4] so that wine shipped could arrive at its destination ready to present Italy's natural excellence, good disposition, and stability.

Early each year, these pleasant hills swelled with the weather, sugar, air that can make slightly fizzy wines of the grapes picked in the past year's final months and quickly (when done right) crushed to juice for fermenting. It is cold. Wines of those Prosecco grapes are in bottles or enclosed cask by now and likely they are still a little sweet, still a little bit a grape juice. Wintertime winds from north of the Terraferma trail Treviso's lace of small rivers and valleyed dips to prime each waiting grapevine, double arms[5] with a craggy cane apiece from last year, each of those with seven to nine buds (only the fourth and on are fertile enough to count on[6]), cocoons from which, eager assumption of the spring, will emerge this vintage's leaves and grapes and tendrils along with buds foundational for the next. Now, those making wine had witnessed, the juice they'd pressed from grapes which from hard green had metamorphosed into translucent and tenderly ripe yellows across the summer months had for a while around that time of crushing bubbled in barrels as, if told in other ways, fermentation turned each molecule of grape-sugar glucose into two apiece of ethanol alcohol and carbon dioxide. Still at last, they now awaited local markets in barrel and occasional bottle. Today, it is common knowledge that the stillness is outcome of one of several things: the end of sugar supply for yeasts, a level of alcohol or temperature that they deem too uncomfortable. Yeasts, responsible agent between juice and wine, have long been known for their (or barring knowledge of them, the bubbling fermenting must's) dislike of heat.[7] Allow them too close to 32°C/90°F and they will die, but in return they are famously adapted to cold and gently freezing temperatures, halting and planning as winter temperatures move in, moving back into action as springtime warmth arrives. In late March and early April across these hills' small vineyards on gentle clay slopes and steeper limestone, buds on Prosecco grapevines begin their break. And in inescapable bottle- or vat-defined quarters, the new wines began to warm, many with sleeping yeasts and grapes' sugars waiting, too. Where fermentation starts again, gaseous CO_2 enters into uncounted and uncontrolled solution with the waiting wine. When it stops at sugar's end, there is Prosecco that is a little fizzy, very cloudy, and (given the pressure of gas and liquid forced together and breakable glass of bottles prepared to hold still wine only) to be presented to exposition or market (both with demands of predictability) in entirely unpredictable form.

The three senses of sight, smell, and taste (the last two closely related in flavor detection but the latter perceiving textures of tannins, acidity, alcohol warmth as well) have the final say over the poured result of any winemaking, physiological responses that drive what we shorthand *taste* and that have stubbornly resisted counting. Today, rationalizing taste is an ongoing grail—in the

United States alone, hundreds of scientific articles chart chemical compounds matched to *honey, fruity-dark, cooked apple, violets, hay-like* via references to branded products such as Welch's grape juice and Sun-Maid raisins—building on smell and taste as molecular sensing.[8] Organisms, we process tiny pieces of the solution that is wine, pieces that are an edited if efficient version of the wine they speak of, defining it. As a category of molecules that can separate from the wine we see, those pieces are bodies specifically processable by us because they are able, warmth and movement, to become volatile, airborne like yeasts, rich in meaning and information, and delight. It is easy to imagine Carpenè's thinking through these bits of taste, possible to work out his immense enjoyment of the materials he made a lifetime of in those writings that instead feature Treviso's ideal hills, the reassertion of its winemaking, the hard work of chemistry and Italy. It seems important to place among his writings more missing words on drinking Prosecco, much of which, he knows, will come later: I want to add augural *amarognolo,*[9] *apple, lemon, floral character sparkling* and *clean.* As Italy formed, its chemists thought and rethought matter, differentiated molecules from atoms or argued against doing so, concentrated knowledge of how and under which conditions matter's particles were able to unite at all while their field rapidly cohered into a classic chemistry momentarily based on indivisible atoms which were already rushing to break apart in the earliest days of the new century. In accordance, Carpenè's Prosecco is its own understanding. It is an equation, and so a story of input and output. It is as all materials are a chemical formula and therefore a formula of structure, composed of measurements of molecules, gas and liquid weighable and of calculable volumes reliable by bottle. Prosecco as Trevisian base wine plus new grape sugar plus new yeast equals precisely calculated tone and sweetness by balance and tradition: for when the Italian method comes, the quantity of residual sugar, grams of it per liter of wine, that every wine answers to even if in quantity of zero and that in sparkling wines, based as they are on grapes with elevated acidity, is an always considered balance component and that in Champagne is a *dosage*[10] which is *addition* to balance and flavor as a final step in bottle, will be built into Prosecco with the second fermentation, and like the wines that changed upon the springtime, no sugar changes at the end of it. Prosecco of yellowish-golden grapes pruinose and a little dotted, smooth green leaves as big as writing hands, on if mostly neglected now vines that had long been cultivated in the places where he walked, on our hilly common ground so far from New York City in the time of the pandemic, on clay and crumbly trails and paper. The school he had wanted into being and the wine he had seen past his own lifetime at some point became both paths and wall to me. Too much of other parts of his life resisted my looking for them. "Carbonic acid, carbon dioxide,"[11] he wrote in his textbook either confusing or, simplifying, collapsing the two compounds, "is a colorless gas with a slightly acidic taste. It is soluble in water and more so in alcohol and wine. At a temper. [sic] of 10 gr. and at normal pressure (0.76 cent.) one liter of carbonic acid will dissolve in a liter of water, or in about half a liter of alcohol, according to its density, that is, its more or less anhydrous state. A liter of wine under those conditions of temperature and pressure can dissolve slightly more than an equal volume of said gas." To know more about Prosecco, to speak further with him here, I have to learn about his chemistry.

Matter is a measured thing. Carpenè knows that it changes in accordance with the world around it, is substance with well-defined properties which human hands guided right can transform into—as with all fairytales the deal is bittersweet—concrete but unwieldy material, complaining composition of substances and states. CO_2 is always produced when wine is, companion of alcohol and changes in flavor compounds in the turning, temporarily agentless sentence, of sugar to cellular energy. For most bottles opened, it is the forgotten part of wine. Spumante restores that term in the equation at the top of this chapter, but it is not a simple task: bottled, a spumante can contain five to six times more carbonic acid by volume, he writes. "The amount it contains is due to the pressure to which it is subjected." Sparkling wine is a temporary balance between liquid, gas, and infinitesimal solids, forces joined together in agreement with the air and temperature outside the bottle and as densely packed and explosive as the truck tire ("the internal tension is equal to that of 6 atmospheres, since in addition to the five of carbonic acid, there is also that of the air that was trapped when the bottle was sealed by a cork."[12]) that does not yet exist but which will, a century later when this wine will make its way into the world outside Treviso, carry millions of bottles to restaurants, shops, homes. Today, a century into industrial chemicals' invasions of everything including agriculture and vineyards, it is easy to forget what chemistry is. From its foundational meaning of distinct substance or compound, sometimes a medicine, used by alchemists and then by chemists, *chemical* has become falsely equivalent to *artificial.* It must mean something that *chimica* still carries some of its original substance and process, as does *chemist* in both languages. "Chemistry can be seen as a sterile and invasive subject, but that's not how it is," Cristiana Tiberio, one of Italy's most celebrated natural winemakers and who holds a degree in classic chemistry and who though she does not make sparkling wine I see as a descendent of Carpenè in the ways she understands and sees her land in central Italy's mountainous Abruzzo,[13] told me when I asked about *her* mix of winemaking and that other state of knowing. "Chemistry is a science that studies phenomena in nature and is a greater part of us than we think. A human being is the result of chemical respiration and we live thanks to this chemical phenomenon, the same way biodynamic preparations[14] are the result of chemical reactions, as is wine made by the spontaneous fermentation of musts," so-called *natural fermentation* in which yeasts present in vineyards and cellars do the work instead of packaged yeasts bought from specializing companies and introduced by winemaking hand. Obsession with knowing and making matter, Carpenè's chemistry is also a responsibility. His 1861 pharmacy thesis on aluminum, 15 pages on the infractible "white metal tending to sky-blue" newly isolated from crystalline rocks like feldspar and mica by the chemists and assuming a wide variety of uses especially in, as it was incapable of being broken up or subdued, food and drink, begins: "Aluminum is never found in its natural state . . . [and] was for many years an object only of scientific curiosity." His degree in chemistry comes the following year, with a twenty-four-page thesis on, new scientists' formulations based on ongoing problems in milk and wine, lactic acid.

That lactic acid existed was well known. It was in muscles of people and animals, in cheese and spoiled milk, in raw vegetables turned tangy and soft, in wine turned sour. The matter was a practical one: knowing what was causing milk and wine to become undrinkable meant making sure it did not happen, applied issues of both health and industry. The process was just now being figured out: the mechanism seemed, based on Pasteur's *Mémoire sur la fermentation appelée lactique* published in 1857, to be a case of certain substances combined with an independent third party. Italy, destined to be but as foreign occupation ended there remained the problem of its "production of wool and cheeses scarce, crude, and not requested, overcome always by foreign competition on our own internal market,"[15] foreigners' goods filling the pantries of Italian homes, France's wines ruling Italian markets. One of several scientists supporting a bacterial theory of fermentation that had been tied up in debate for a century, Pasteur became its prime bearer with the help of Reggio-Emilian Jesuit biologist Lazzaro Spallanzani whose 1765 *Saggio di osservazioni microscopiche* he had built upon[16] when he realized he must clear the scientific landscape of spontaneous generation if his stories of a fermenting organism were to capture all scientists' passions. While other theories assigned the process solely to interactions between substances, in his 1858 *Mémoire sur la fermentation alcoolique*,[17] Pasteur announced that "in the course of these researches, I have the confidence that whoever will judge them impartially will recognize that alcoholic fermentation is an act correlative of life," a final rhetorical blow that severed fermentation from materialist explanations and tied it instead to a tale of life and metabolic processes that cast yeast and us as similar creatures. "Of the organization of globules," he wrote with spiritual and even religious tones, "not of death or of the putrefaction of these globules, no more than it appears as a phenomenon of contact in which the transformation of the sugar would be accomplished in the presence of the ferment without giving it or taking anything from it."[18]

What was Pasteur's ferment, agent of fermentation? Was lactic acid always formed by the same process? Always due to the same cause? Were all lactic acids the same thing, equal to saying all possessed the same composition? Admired by his teacher chemist Tullio Brugnatelli for both his hard work and his curiosity, student Carpenè has built his second thesis not only on what Pasteur has so recently published but also by taking on all the materials and writings available to him, rigorously laying out what other scientists have posited and confidently inserting himself into the international chemistry conversation as it is gathering steam. He is already showing signs of what is about to become the two pillars of his work. In a student's voice already lifted by the breezy, personal tone fed by self-assurance that will characterize his articles, letters, books to come, he writes that, "There has not yet been a sufficient explanation of the phenomena that constitute lactic fermentation; first of all because nothing precise is yet known about the nature of the ferments, and secondly because science has not yet managed to change sugar into lactic acid with direct and purely chemical methods,"[19] while he details the experiments recorded by French, German, Russian chemists. It is a voice in the distinct shape of a man who has already fought for his beliefs, who is watching triumphant Italy form and awaiting *his* land's entry into the nation, who wants to make sure it is formulated right. That voice's service to theory is matched by the man's physical practical work. He moves in a mix of confidence and awe—boiling nitric acid converts lactic acid to acido ossalico, he shares as observation; sulfuric acid develops a large quantity of carbon oxide and "gives origin to a black material of humic nature"[20]—about the laboratory himself, setting up his own experiments alongside the others he describes, acting on lactic acid by applying heat so that it will lose water molecules and change first into an "amorphous material improperly called anhydrous lactic acid" then, hotter, into one he calls[21] only $C^6H^6O^6$, and which because "we believe that lactic acid" has a dibasic structure, "we must consider it a true hydrous lactic acid."[22] He writes *HO* for water, an expression of preatomic chemical formula-writing in which elements were rendered equivalent by how much of them, weighed in gaseous form, reacted with some agreed-upon substance: in alchemy a base, in the day's chemistry often hydrogen. It is thinking, practice, that places material first, molecules as prime players with their components defined in terms of their proportions in the ponderable substances they formed.[23] This mid-nineteenth-century science is a balance of experimentation and another kind of writing and so over and over he makes a variety of stoichiometric[24] calculations that *could* add up to lactic acid: the chemists have accepted the principle of mass conservation, but atomic compositions of compound or simple substances are far from agreed upon, including whether atoms exist at all or what they are. Following Swedish chemist Jöns Jacob Berzelius who created the symbols chemistry we use today but was convinced that the world adhered to a dualistic balance and so (along with referring to *atoms* and *molecules* interchangeably[25]) wrote substance as duos, comma-joined salts each of which was an aggregate result of a chemical reaction between an electro-positive base and a negatively charged acid, he first writes lactic acid in older dualistic notation—"2HO, $C_6H_4O_4$." And then writes it and rewrites it as chemically—therefore physically—unified entities, one molecule bound by a countable chemical process, substance considered through what are ways to write it, a thesis that is a trying to reconcile what he is observing with a rationality on paper. For two years this atomist view had been gathering acceptance, thanks to the joined forces of a piemontese scientist who had died in 1856 and a younger Sicilian chemist who championed his overlooked work on the reaction between hydrogen and oxygen atoms to form the compound of water—and correctly counted these diatomic elements as the method by which the molecules of the new substance was created: $2H_2 + O_2 = 2H_2O$.

How a chemical entity reacts is more accurately a story about how much of it does. As the chemists observed and calculated, the elements they were rationalizing united or resisted or came apart in varied forms, unique material outcomes in terms of how much of them took part: an atom of carbon plus one of oxygen is a substance very different from one of carbon plus two oxygen. As the chemists found more and more elements in the compounds around them, creating terminology and numerical standards that could accurately narrate chemical mechanics became one main concern. Two years before Carpenè wrote his exploration of lactic acid, Sicilian[26] chemist Stanislao Cannizzaro had presented piemontese[27] scientist Amedeo Avogadro's 1811 work over two late summer days in Prussian Karlsruhe at the first International Congress of Chemists, called in search of if not unified meanings for words like *atom, molecule, equivalent weight, atomic, base,*[28] at least further understandings of substances through worrying about such terms. Their combining power and behavior were simply a counting device on the way to molecules, which were matter itself, said Cannizzaro, following Avogadro's notions of so-many molecules equals a substance and holding tight to materialist concerns to cast substances as materials, useable and useful[29] goods. As their smallest indivisible parts, atoms in and of themselves mattered little and could be analogous to the laws of numbers. So long as matter was measurable according to a consistent system, further inquiry into atoms, additional terminology, other kinds of meaning, he wrote, should be left to future chemists who might be concerned with something else. Carpenè is equally happy with this atomic state built on the relationships between the bodies involved. He plays in the laboratory, finding materials ever more possible to manipulate (a word[30] whose first form is a way to mine silver, element known since ancient times and[31] in the nineteenth century transformed by the chemists to mean handling apparatus and reagents), calculate, and categorize thanks to the restructuring atomic order.

Reading the chemistry of this time is thrilling. It feels as if I have been granted looks into a chaos at a point just as everything was organizing into earth, stable, enough, in our solar system, a ground reliable and rationalized. The way back is not an inevitable line but a meandering and episodic trail, made possible by many moments. "In any point along the chronological line," writes Massimo Montanari of origins and the nowness of identities, "identities are an arrival point: well delimitated mental and material spaces characterize them, but [those are] always unstable and mutable, as is fitting of all that is alive."[32] I find this a responsible positioning device, and see reflections of these chemists' work across an urban refosso[33] of chemical understanding and national identities that changed everything between them and us of a subatomic age, survivors of those world wars and fascism,[34] sufferers of industrial pollution, all events just years or decades away from them. "The zone around Conegliano is furrowed by too many ditches[35] and is extremely covered by vegetation,"[36] the Austrian military had complained of the land they occupied in 1798 until, excluding the decade in which the French returned to rule, 1866. From the 2000s, it can be hard to see clearly and cleanly what Carpenè was doing, newly soaked as we are in idyllic images of small farms[37] as *nature* and as of their wines as *natural*. The now of Italy and Carpenè and Prosecco is one response to theories of the past one hundred years, to the knowing of a long time before that, too. It is one ending of piled-up eras, foundation of all that is about to change, bridge from me to Trevisian contadini's[38] older ways (in the century's final days my great-grandparents will leave and carry ghosts of their old ways to Rome and into cities of the northeast United States) still pressing upon the world, a swirling meeting place, chemistry and unifying. So much information is now coming in, organisms that make wine and lactic acid, numbers to tell a disentangling atomic theory, but this chemistry is still tied to earth. A divine microscopic, yes, but it is among tools and concepts that never left the material world, chemistry whose work is to rationalize the walkable places. Powerful as they are, atoms and molecules are still knowledge. Matter is something else and ties together the scientists and the growers of wine grapes. Gather and compare, order and define. Pasteur's ferment which makes lactic acid and still and sparkling wines is an agent. It will drive the → in the equation at the top of this chapter. It is a third party that metabolizes one substance and in its process makes that into other ones without its own molecules entering into such reactions. Questioning and weighing with the same rigorous curiosity with which he entered into conversation with "Pasteur's classic studj on fermentations"[39] in his thesis, Carpenè will a decade later, with his characteristic faith in people if mostly those of a hardworking and -thinking near future, in his textbook ponder that, "The word *fermentation* expresses a complex of phenomena that are constantly occurring in nature, to which belongs also the conversion of must into wine,"[40] as introduction of such metamorphoses to winemaking students. "These phenomena, which worry contemporary scholars, have not had sufficient explanations; for this, in part, they are still outside the domain of science. [They] cannot be considered as correlative to simple chemical reactions, due to the intimate contact between them of substances with different properties. In fermentations, the industrious presence of organized living is shown. The genesis of said beings, their true physiological functions are problems that cannot yet be completely emancipated from the domination of disparate hypotheses, hypotheses on the other hand which increasingly show the progressive refinement of human intelligence." He will follow the line of intimate contact between substances with different properties in his laboratory and in his cantina,[41] synonymous with *technology* its entire ancient existence, where his winemaking involves in the same way grapes grown and must and the wood and iron of his new machines. But first, in 1862 Carpenè writes and rewrites lactic acid's laboratory presence, a chloride which is written both as one German chemist's dualist $C^6H^4Cl^2O^2, Cl$ and as Cannizzaro's "rational formula" $C^6H^4O^2Cl^2$, because he is as sure that he, chemist, will soon get it right as he, soldier, believed that he was helping to make Italy. Six years later, two after the old Venetian empire has tied itself into the nation, he will begin changing into spumante the golden hills' Prosecco.

In uneven moments back to the ancient through both aspiration and self-defining chatter, wine has meant something about being Italy. "In the Rome of the times of the birth of Christ, wine runs as rivers," write Giovanni Negri and Elisabetta Petrini of the moment wine turned both synonym and economy of Italy. "From the Palatino to Ostia, the taverns are counted by the hundreds. Vinum is SPQR. Vinum is Rome. And where Rome advances, advances Vinum."[42] Wine has remained Italian code, patrimony, and cultural and goods export ever since. Which grapes were grown and how, what the winemaking results were like, who drank them, had been following courses that changed unhurriedly and in the nineteenth century their arrival point looked like a Vinum with most working traditions in disarray, a tenuous holding-on through poverty and occupation and economic structures that took the time span of neither grower nor grapevine into real consideration, the former's reasons why and guideposts long made difficult to impossible and now mostly inapplicable, forgotten as misery and emigration spread. Here is a collage of Italy precursor. Lands whose greatest conversation topic was agriculture, viticulture its greatest assumption. Faced with crop failures and famines, in the later eighteenth century academic institutions launched as humanities and arts concerns three hundred years earlier began to set agrarian studies alongside literature and philosophy, fixing practical, local, economic, and scientific worries as equally important parts of education (which shaped character) and instruction (by which came knowledge). In 1753 came the founding of Firenze's Accademia dei Georgofili which still today covers agriculture and economics. 1757 brought the Accademia delle Scienze to Torino in Piemonte. In 1769 by will of the Venetian government, losing its maritime supremacy and turning to the study of agriculture to reclaim its dignity through a superiority this time on land, Conegliano's century-and-a-half-old cultural-studies Accademia degli Aspiranti added agrarian classes and vines, pumpkins and snails among the literary courses.[43] These local fixations on how the sciences and studying them affected fields from classic literature to ones in which rooted wheats, fruit trees, and grapevines linked a fractured Italy to European centers that had been thinking about the same. As unification began, wine—now in addition analyzable by the chemists—was as complex as the nineteenth century was showing everything to be, fermentation no longer even a straight line from sugar to alcohol but a chaos of flavors and textures, a multiheaded trajectory that created glycerin and flavor compounds and promised decipherability and definition, a place clamoring to be read as components of a whole that might be made coherent and solid.

At the same time, chemistry was finding order out of its previous century's separating from alchemy in which a "multiplicity of synonyms for the same thing[44] [had] served not merely to obscure matters for the uninitiated but to increase greatly the degree of symbolic allusion."[45] *Mercury* was also *living silver*; substances like *silver moon*, *vinegar*, and *alcohol* were written into formulaic phrases that wound according to who wrote them or even read them. *H* and *2* which would serve the chemists well were called upon by the alchemists to indicate silver, and at other times lead.[46] This alchemist haze was necessary. It stood as informal patent protection and guard, pseudonyms and ciphers that protected trade secrets and hid the identity of those whose findings and worldviews ran counter to jealous less-flexible powers: "Gold making, if ever perfected, threatened state economies, or it might be used to usher in the millennium as some religious fanatics believed."[47] As chemistry's search for openness and understanding matched the political needs of the nineteenth century's push into nationhood, pre-Italy's alchemic obscuring served the time running up to that: when the French army arrived in 1796, it found its own notions of revolution being refashioned into fresh if covert and competing resolves to make a united Italy, free of foreign rule. Throughout the conquered lands, secret societies plotted different notions of that future: continuity with older dreams of Italy in the early-nineteenth-century Carbonari—whose public work was to burn wood to turn it to carbon, affordable fuel for among other things stoves and thereby revolutionizing cooking, too—from the south; the Filadelfi and Adelfi brotherhoods whose ideas had emigrated from France into the north; Catholic ones in anger at the anticlericalism Napoleon had brought; exiled apostle of Italy Mazzini's Giovine Italia formed out of the Carboneria so that the material of such freedom might be a unified republic of the Italian people, everywhere—membership in all as much about knowledge as it was about values. Intricate organizational structures communicated their methods by codes, hidden references[48]—the Carboneria who would count Mazzini a member took carbon, water, salt as insider symbols in their quest for transformation—and hushedly passed publications, each insisting on its own path to a moral height built upon opposition to tyranny so that mind and body and soul could be devoted to an inevitably, one day, free Italy. Matter was the solid ground which we and they last shared. Matter was where everyone could see it, laboratories held at a safer arm's length in, if it came to it, expendable rooms at the top of buildings with terrible sequestered smells that assaulted the senses and taunted chemists with the ever-present danger of explosions: until gloves, masks, and gowns appeared at century's end, as in battle many were badly injured during their experiments, as winemakers come into too close contact with fermentation's unbreathable CO_2 continue to be today. "The gas bubbles which, rare at the beginning of fermentation, then go to take possession of the entire liquid mass," Carpenè warns in his textbook, "shaking it and upsetting it in all senses, are precisely carbonic acid . . . unable to dissolve [in] the higher temperature of the fermenting liquid [and] heavier than air . . . instead of expanding and mingling with that, as it releases from the must it gathers in the empty parts of the vat, and when it has filled it . . . like a liquid which can no longer understand its vessel, it overflows . . . and settles onto the ground. As those chemical-physiological reactions continue, the must is gradually transformed into wine."[49]

Carpenè's chemistry was a nineteenth-century soldier. Proto-atomist Avogadro had blazed the Risorgimento's soldier scientist path: in 1811 he published the *Essai d'une manière de déterminer les masses relatives des molécules élémentaires des corps, et les proportions selon lesquelles elles entrent dans ces combinaisons,* in the language his region was quite at home with—noting, his to-be-famous foundation for turning substances into numbers, that in equal volumes of any gases constrained by the same pressure and temperature could be counted the same number of a "molecule" unit, whether simple[50] or compound—then in 1821 took part in anti-Austrian uprisings and was barred, temporarily, from teaching when they failed. At this time, Italian chemistry is a self-contained thing,[51] and with Italian states isolated from Europe's chemists his work remained ignored until in 1860—remember, the year of Mazzini's last book, cover ablaze with slogans of Giovine Italia—Cannizzaro, "builder of the institutions of post-unification Italian chemistry"[52] and fighter in 1848's also failed uprisings, joined both Garibaldi's army and Karlsruhe's international gathering of chemists where he made sure that Avogadro's work was at last accepted. "I believe," Cannizzaro began his abbreviated lecture now that he was this close to both a nation *and* Europe's scientists, "that the progresses of science, made in recent years, have confirmed the hypothesis of Avogadro"[53] and then, secret societies turned to rationality and access, handed out take-home pamphlets complete with tables of elements identified by both name and—supporting his predecessor's proposition that some of them occurred in nature as pairs of their smallest pieces on paper, a paradigm-shifting observation which he stopped short of calling *atom*—a number based on how many hydrogen "half-molecules" they weighed based on calculations of observed and measured volumes.

Revolutions can look like battles, or petitions, or open breaking of unjust laws. They can also look like the making of new institutions. Carpenè's "whole idea of reform and change is a professional group," Italianist Robert Lumley pointed out to me as I wrote this book. "It is precisely the scientists, the technicians, the professors, the ones with specialized knowledge. They are the focus of attention in the formation of the new." Fittingly proud members of brave and resolute battalions,[54] chemists were applying tools of "generalization and systematization to chemical practice [allowing them to enter] an ideal, imagined world that provided a guide to the behavior of real substances,"[55] and transformed the way they viewed their chaotic subjects. If alchemy's reliance on a symbolic language was meant only "for practitioners of the art,"[56] then chemistry's codes were a light angled onto matter made readable in a new world tired of secrets and of the conditions that had made them necessary. Chemists were members of a newly public and usable science which held the imaginations of literate Italians, who saw their vocation[57] as the way to make steady long-precarious Italy. As they wrote up matter and how the natural world worked, their calculations joined other numbers that formed the nation: key dates are reiterated in cities and towns throughout Italy. In Conegliano, via XX Settembre—on that day in 1870 the Italian army overran papal forces and at Porta Pia entered the Rome of Pope Pius IX who had called nationmaking a "pestilence of socialism, communism, and secret societies," to make the city the nation's capital—is the historic center's narrow portico-lined throughway from which I can walk from the municipal building ("In memory and honor of the men and the women who 150 years ago united l'Italia, forming the nation that today we love," reads one of the building's several stone plaques, dedicated by "the citizens of Conegliano" in 2011) to the restaurants and bars they cover, into the wide open of the Cima museum dedicated to the city's great Rinascimento painter who created an entire archive in paint of the Marca Trevigiana[58] hills. Via XXIV Maggio for the day Italy entered World War I (and when each year the contadini brought their sheep and cows back up to pasture in the hills[59]) from which after the battle of Vittorio Veneto it would emerge victorious (piazza 4 Novembre[60] as was renamed the piazza delle Pecorelle,[61] where livestock was brought for selling in one of the city's several blurrings with its countryside); via XXVIII Aprile from when at the end of World War II the partigiani captured a fleeing Benito Mussolini and his companion and executed them.

Like their pamphlets, much of the written codes the risorgimentali relied on for a one-day free Italy indirectly excluded the contadini, who far outnumbered them and had instead for centuries—in the Trevigiano from at least the Serenissima's earliest presence in the 1300s—been building ongoing cultures of orality and landscape rooted in passed-down knowledge and preservation. In opposition to experimentation and academic and individual voices and bylines, *these* citizens "appropriate[d] actuality in recurrent, formulaic agglomerates, communally generated and shared." Their knowledge was of "elements . . . formulary to an extent radically greater than what highly literate cultures can ordinarily tolerate,"[62] spoken in a variety of local dialects (sounds which almost no one wrote and few read though those who did read out loud to nightly gatherings of family and neighbors at omnipresent livestock stalls, which in small towns might also face out onto the communal piazza) and with a legitimacy and daily reality derived from centuries-old laws and loyalties. Unification was an occasion to impose[63] the tuscan dialect as everyone's new language, to demand payment of new taxes and service in the Italian military, itself used to brutally enforce Italy in the face of "the indifference or outright hostility of much of the Italian population . . . the existence of deep-rooted regional division, the absence of strong economic, cultural, and linguistic bonds." If there was any common ground[64] on which Italy rested, if about 70 percent of Italians were involved in agriculture—though Italy's mythical fertile zones have always in reality been breaks within a great density of rocky, hard-to-farm slopes and mountains—then awaiting success throughout the nation, in mixed plots of wheat and vines and vegetables, in tiny vineyards of quality-granting hills and merely small ones of generally lesser flatlands, in the large commercial farms[65] of the south, were Italy's wines, which with more recent exceptions of academic noblemen planting for pleasure and research had been mostly in the hands and methods of the contadini.[66] To make quality, durable and dependable wine could be unifying and income-generating work for the chemists to study, teach, and guide. Students of the natural world they might, through finally a sensible alliance with the contadini, turn Vinum once more into an industry that looked abroad, a patriotic tool that would also banish those final traces of tyrannical presence on Italian tables. The chemists' symbols would be communication for everyone involved in one of the new nation's oldest symbols of itself, arranged into formulas and books read to oneself or out loud that could teach winemakers to stay, to send instead tasteable bottled emissaries to Paris, to Zurich, across an ocean to Philadelphia.[67]

Chemistry is a language, composed of simplifying information-rich codes. It is a literacy in the natural world around you, not, approaching misstep, for mastery and lordship but for understanding and listening[68] and taking part. As symbols filled with meaning for rational participation, *C, H, 2, +, 3, 8* allow deep communal thinking about substances and their processes. The chemists relied on such symbols and formulas, albeit analytical and abstracted ones read and written in books on desks, as the contadini relied on ones of more outdoors observations: March, usual month of Prosecco bud-break, was made of three days of water to one of sun; June was filled with rainy threats to vineyards, and se pióve de San Barnaba, l'ùa bianca la va ia e se pióve da matina e sera va ia la bianca e anca la nera.[69] While both sets were capable of complicated combinations, the chemists' were also specifically for granting access, open to all for seeing and producing, work that was both theoretical and hands-on. "Metà a scola e metà fora,"[70] said the contadini in apparent agreement. To make Italy, Italians could make both intellectual and economic use of these codes, chemistry[71] allied to agriculture and viticulture, revolutionary intellectuals bound to—if lamenting their version of *empiricism* and calling on them to adopt *rational* methods—their now compatriots who had long known Italy's lands. Inquiry and change powered scientists' interactions with the sites that contadini wine-growers navigated instead by oral evaluatory "units which discourage and even defy analysis," but the formulas by which the latter did so (in a twist on the chemists', theirs also worked to "bind . . . in familiar, unbreakable bonds" but *not* to produce new knowledge[72]) were a meeting place where rational systems could act as mystical key to better and more reliable things. "Away from print," writes Jesuit priest and historian Walter J. Ong perhaps encouragingly about dictionaries of "modern European vernaculars" printed beginning in the seventeenth century, "all was chaos."[73] In the 1869 *Dizionario Generale di Chimica*, complied by a group of Italian chemists, sixteen pages on "Chemical affinity" which read more as instruction than definition offer that "combination becomes more stable the more the combining bodies are of different and heterogeneous nature." In the ways it is possible to read Carpenè's inclusive and encouraging textbook, the codes and values of the Italian chemists might, taught, be a newly shared language between them and the contadini. Based on principles of ordering,[74] chemistry was both practical knowledge and a tool for understanding what Italian wine was made of, was a bridge between literate and oral Italians. As were the economic possibilities Prosecco offered in this growing world. "Italian spumante does not have the fame of the centuries,"[75] Carpenè wrote, underlining his entire project with a resoundingly intended *yet*. Even if a wine is good, *sublime*, he pauses, "the bag of a rich man will only open exceptionally for an Italian product and will open somewhat for bottles of French, German, or Austro-Hungarian wine, if they have been known for many years." But, in his moment, "if the judgment is impartial," a French sparkling wine set against an Italian one "of the same price" does not always win ("it is understood that we do not intend to say that there is no more progress to be made"[76]). As better and exportable wines promised both higher incomes and more affordable quality spumante, Italy's concern was not only one of making. It was possible to reach also the level of consumption of sparkling wines abroad, dignified measure of income and taste: while in Italy a sparkling wine like Champagne has been "a privilege of the rich to be consumed at banquets and solemn occasions, today it begins to democratize and its consumption will be increasingly extended to every class of society, thanks to the mild price at which it can be offered." Italians will have a spumante that satisfies fully, "will be able in the future to consume from 15 to 25 million bottles a year and to export it as well." The role of the degreed chemist was not only to witness but also to then put in its correct place what he (unfortunate but accurate pronoun) has seen. Chemists with their formulaic and one-to-one codes and their proven, patriotic, participation in the physical world could bring clarity to Italy and its wines.

While Carpenè is working to set such production, symbols and names for substances, not to mention the substances themselves, are still being gathered. Some names were primitive, tying a substance back to alchemy and chemistry's common days—*oro*, *argento*, *solfo*, and *ferro* (*gold*, *silver*, *sulfur*, and *iron*) were elements known since ancient times—while a newer element might be named for physical or chemical characteristics or the circumstances of its finding—

bromine for its smell; *iodine* for in Greek its vapor's purple,[77] potassium[78] for the wood ashes from which it was first extracted.[79] Until recently, this growing collection had been organized by equivalent weight—how much of a sample size reacted with a set amount of oxygen, a system derived from alchemy's considerations of how much of a metal reacted with a given base. The need for an organizing principle beholden only to chemistry's own considerations had been growing stronger all century as, leaving behind the long-lived ancient base of four elements—earth, sky, water, air—substances vying for element status began adding up: when Berzelius hit upon his system of symbols in 1814, he listed 47 of them and assigned to each a first-letter identity based upon its Latin name.[80] His written system also holds information about how each element was classified: one letter for the gases, one for metals whose names hold no first letter in common, two for those that do.[81] A number lay behind them all, said Cannizzaro about such lists, and it was to be found among them—it was simply the element that was lightest so that all others could be a version of it in weight.[82] A half-molecule of hydrogen, but what was most important was that it be both something that related what materials were made of and part of a self-referential system. If there might be values each held on their own, what did that matter? The interactions would go on as supposed in a referential framework, every Italian a calculable portion of the reaction whose output included his or her ("The emancipation of women must be continually coupled for you all with the *emancipation of the worker* and it will give your work the consecration of a universal truth,"[83] Mazzini had ended his *Duties*) nation. What need for its own number did an element (whether atom, molecule simple or compound), a substance whose entire meaning was wrapped up in reactions and behavior with other elements (same wavy definition), have? What mattered about oxygen was how much of it might join how much of carbon, how much might make other elements with how much of hydrogen, how much could be ignored by silver, copper, or gold. Rapport between each other, spirit of combination and association, would liberate the elements from scales which were not theirs, would keep them matter, straw-yellow wine to pour, blue-green rivers of water over calcareous clay and rocky river beds. On the eve of an Italian economy,[84] there is an Italian chemistry. Avogadro's realization of equal numbers of molecules in all gases under the same temperature and pressure is the basis of understanding atom behavior within molecules, of Cannizzaro's insistence on materials, of Carpenè's practical, quality sparkling and sellable, interpretations.

To overlay for a moment what every wine writer today knows easily onto the type of difficulty Carpenè's spumante is: a solution of alcohol, acids, the phenolics that bring flavor, color, and tannins totaling less than 20 percent in water, wine transformed to sparkling has a further composition: adding CO_2 by any method results in both another element in solution and a chemical reaction: $H_2O + CO_2 = H_2CO_3$, a compound of elements hydrogen and carbon and oxygen, "formed in small amounts when its anhydride, carbon dioxide (CO_2), dissolves in water."[85] But that resulting carbonic acid, salty and savory, is at pains to exist at all when in an open bottle or a glass just poured the wine's pressure drops to that of air—at which point some of the solution reverts to water and CO_2 which bubbles up, as you can see. "Reflecting in fact on the properties of carbonic acid and in particular on its solubility in wine in different conditions of temperature and pressure," wrote Giuseppe Cappelleri in a wine-focused look at late-nineteenth-century chemistry,[86] "Carpenè could project his system[87] basing it principally on the fact that for water as for wine, the coefficient of absorption for carbonic acid increases with the lowering of the temperature to which the liquids are subjected. With wine, due to the presence of salt and alcohol, the solubility increases further, and on this latter consideration was based the construction of the Carpenè apparatus with which wine was made spumante by keeping it at a pressure of 1 atmosphere at a temperature of -6°C. To avoid possible cloudiness, vini da spumantizzare were first stabilized at a temperature of 6, 7, degrees below zero," which allowed the spumante maker to separate larger solids from the liquid, clarity and, given that solids add to the pressure, stability. But if the underlying counting was incorrect, the materials being observed could run awry. For the foundation of Carpenè's rapidly rationalizing chemistry to hold a place in the new nation, he needed Cannizzaro's work to stand on Avogadro's correct numbering of matter: "In my third lesson," Cannizzaro wrote in his historically cajoling and convincing forty-page *Sunto*, "I show that all the new studies, from Gay-Lussac to Clausius, confirm the hypothesis of Avogadro and Ampère that the distances of the molecules, so long as they stay in an aeriform state, don't depend on their nature, nor on their mass, nor on the number of atoms they contain, but only on the temperature in which they are and on the pressure that they are enduring."[88]

In his 1861 *Trattato di Chimica Generale* textbook which Carpenè must have used, with a way of turning any interesting object whether idea or substance over and over matched by that of his student, Brugnatelli echoes Cannizzaro: the numbers of equivalent weights themselves tell you nothing in absolute, he explains, but the ponderable relationships with which these bodies enter into such combinations, rules, and order that form a rational whole. And though the names of what he termed the simple bodies lacked an underlying law, the opposite was true for the compound bodies, compositions whose names formed a scientific language—"chemistry has a means as simple as it is clear to denote not only the constitution of bodies but also the reactions that they produce; it is a type of symbolic writing with which one can abbreviate the longest descriptions of chemical reactions, and expose them,"[89] as promised rendering knowable the elements involved, via the quantity in which each one takes part, and displaying the chemical laws to which they were held. Oxides clued you in to some inorganic body combined with oxygen and came in three classes including the dualistic ones still hanging on. Acids, the result of a chemical rearranging between a simple body and oxygen and which changed the ending of the former's name to *-ico* or *-oso*,[90] depending on which share of oxygen it gained. Salts, products of an acid combined with a base and whose names began with that of the acid but then building upon what had been changed before, *-ico* endings transformed back to *-ato*, *-oso* turned to *-ipo*.

Sharer of the kinds of words and symbols that once hid and protected, manipulator and interpreter of materials, unlike making literature and philosophy and a lot like making wine, by the second half of the 1800s chemistry required solid ground, places to fill with tools and writing and machines and for Carpenè winemaking equipment, too. "Chemistry," writes Italian chemistry historian Luigi Cerruti, "is a system constituted by the *knowledge* acquired in the field; by the *scientists* who produce this knowledge; by the *instruments*, *establishments*, *institutions* etc. that let the scientists carry out their work, by a *documentation* in which the knowledge produced is deposited and preserved; by *specialized means of communication*, journals, magazines, books etc., that let the knowledge circulate within the system; by the *substances* and *materials* that are used/researched/produced by the chemists."[91] The field's fundamental concern is the production of both knowledge and substances, and at its epistemological base an "opposition between what is perceptible and what isn't."[92] Cerruti's analysis includes an important observation of movement between macroscopic and microscopic realms, and the ways in which chemistry's differ from its "sister science" physics': while both chemists and physicists manipulate macroscopic substances to work events at a microscopic level which then add to their respective discipline's knowledge, the substances that the former produces alongside the knowledge return chemists' work to the macroscopic. This material production that keeps apace with the field's resultant knowledge makes of chemistry a particular kind of social act, involving language, too. First, the word *material* has a stronger affinity to *goods* than it does to *substance*, by which is meant "a body that possesses well-defined physical and chemical properties,"[93] such as composition, solubility, density. "With 'material' we refer to a body constituted of one or more substances, endowed with chemical, physical, and mechanical *properties*, meant for some specific function,"[94] writes Cerruti. *Thing* appears several times in *material*'s entry in the *Oxford English Dictionary*. For those after nationhood, knowledge and the ability to transform materials and place, some power to make them lasting, their roles in countering economic, medical, and vinous misery, were intoxicating tools.

"It debases us, to speak frankly, that this fertile, happy, civilized region in our Enotria was forced to track down true viticulturists and true enologists for their methods and sciences in the land of beer," Carpenè wrote in 1873 in a mix of frustration and unwavering faith. "This does not mean we can give up hope that Italy will become a teacher to other nations in this important branch of the agricultural industry."[95] But winemaker and wine-grower have the wine they deserve,[96] and with education and instruction Italy would make its own[97] again. Which grape varieties grew and tasted best were well known by then.[98] What needed to be looked at closer were the causes of why the wines weren't up to speed, and so Carpenè's chemistry is also a technology. He makes tools to improve his winemaking and offers copies for sale to all who wished to come along: the pages of the journal of Conegliano's school regularly hold advertisements for his temperature-controlled Enoterma Carpenè, buyable for 150 lire and accompanied by offers for instruction. Moving with those longtime rural Venetian assumptions of participation, mutual support, and reciprocity,[99] Carpene's chemistry is a public affair, too. It is a science of professional associations, collaborating with scientists local and abroad, counting in the growers and the contadini in order to produce *and* drink good Italian wine. At home, "even to those not familiar with the Trevisian campi,[100] it is obvious that as in other parts of the Veneto, the hilly part is almost completely covered with white cultivars,"[101] he wrote in 1881. And it is obvious to everyone that although "the consumption of white wines was once greater than that of reds, today the opposite is true and not even here can white wines be considered true table wines."[102] He attributes this to consumers' changed tastes, but also "very much to the population's nervous conditions, for which they prefer tonic red wines to stimulating white ones,"[103] and suggests adding the "red coloring material enocianina,"[104] derived from red-grape skins, and thus "converting white wines to reds," not only to give them a "tonic and less stimulating character" but also to preserve and sell them "like those in the form of vini spumante [which are] eminently hygienic wines, for the carbonic acid gas they contain."[105] *Consumers* is important, too, chemistry an analysis also of both cognitive procedures and methods of production[106]: to bring richness to our nation, Carpenè writes, look not to "the hundreds of bottles costing 8–12 lire but the millions of bottles of clear, good spumante, with a flavor, let us understand each other, that is Italian, at about 2 lire per bottle, at a price that is accessible to most of the population, to the largest number of people—as it is the masses who in Italy, as overseas, weigh the most on the drinking scale."[107] Counting, he knows also that cold liquid can hold greater amounts of a gas and that CO_2's dissolving is exothermic, a heat-creating process in which, uncooled, a system's chemical equilibrium[108] would tend toward dissolution as the produced heat increased, molecules moving faster and faster until the gaseous ones broke free of their liquid compatriots—the cost of his *Sunto Teorico-Pratico* in which he shares this, clear, good, and undeniably with flavor, is 2 lire. The chemistry of spumante is a unification mechanism, and accordingly he builds his apparatus.

The huge variety of this century's laboratory technologies has been preserved in large books filled with hundreds of illustrations on how to assemble a particular apparatus, how to work it, and how to fine-tune it for the results you want, writes Cerruti. "Most of these technologies had a brief life, replaced by others that were faster, or more precise, or less complicated or less costly."[109] By the first decades of the twentieth century, Carpenè's apparatus would be

BREVETTATO SISTEMA CARPENÈ PER LA FABBRICAZIONE DEI VINI SPUMANTI

Patent with recipe for CO_2: Antonio Carpenè's apparatus and guidance for making Prosecco sparkle.
Lit. P. Salussolia, Sunto Teorico-Pratico di Enologia, Vol I. *Ermanno Loescher, 1888.*

replaced by Martinotti's method, which the new Carpenè generations will manipulate for the needs of delicate Prosecco, character well envisioned but still being figured out, but the illustration of the former remains an icon. Each of its five steps has a physical structure that rises above ground on wooden legs, each is a chemical process cast in iron and Treviso's prized chestnut trees or maybe of the Farnia oak from its fabled Montello forest from which were cut the piles that hold up Venice and from which were built the boats that gave it power. It is a drawing not only of Carpenè's quest for spumante but also of the hands-on ingenuity of nineteenth-century chemists who if they were to follow the insights that came to them demanding form usually needed tools that did not yet exist. In his illustration of the apparatus's fifth and final phase is, 5°, a vat for holding the solution-abetting refrigerant material. Drafted with a steady hand, centered in the vat is 4°, the cauldron which holds the solution in its unified and sparkling state. As among Pasteur's ferments are also the acetic-acid enemies of quality wine, the cauldron to hold the latter is made of antibacterial copper, tin plated or silvered on the inside. The one-person sit-down station he's made for locking bottles of spumante with pressure-fighting wide corks then tied in place with coarse string is stamped with *Carpenè, Conegliano*, and, and I don't know if it's to keep himself modest and competitive, *Italie*, or if it is instead a mistake, a getting used to this new name that is still spoken and written *Talja* throughout the lands that were once Venice's. A world made solid, rational and lasting, measurable, presented widely by illustrations of such instruments, underscored in place by the objects themselves, undeniably heavily present and manipulable.

"Under a free nation like ours, science's light disseminates everywhere," Carpenè wrote in 1875. "We cannot deny that even the majority of agricoltori has been persuaded that it is necessary to beat a path different from the one traced to them through empiricism. Now in fact a great number of agricoltori are convinced of the importance of imparting a technical instruction to their children. Now in fact, the youth frequents technical and agronomic institutes more than in the past, and to convince ourselves of this it is enough to turn an eye to the statistics lately made public by the two Ministries of Instruction and of Agriculture."[110] Cannizzaro had in *his* Sunto fifteen years earlier presented rationalized substances by observing that, "once the value of numbers as they appear in the preceding table [of thirty-four substances from oxygen to water to alcohol and ether with both their total and their component weights expressed in multiples of a half-molecule of hydrogen, set as equal to 1] is made familiar to the young people, it is easy to lead them to discover the laws that result from comparing them."[111] On those pages are substances and materials written into their natural order, which is a system of simple arithmetic. Hydrogen, diatomic, equals 2. Alcohol has an equivalent weight of 46, substance and number composed as follows: hydrogen (6), oxygen (16), and carbon (24). For a valiant winemaking chemist, sparkling wine was a technical feat that added to the local grapes agreed to be the best, built systems that a person might work of weights and counters, mechanical machines in gruff metals that could provide the cool and gentle that wine and especially spumante requires.

The word *fermentation* expresses a complex of phenomena that operate incessantly in nature, Carpenè wrote in his *Sunto*. But without sufficient explanation, they were—*for now*, chemistry, continual movement between what we know and what we don't yet, promised—"outside the realm of science." If with a microscope you examine a drop of must in full fermentation, he continued, you'll see the saccharomyces, "so-called germs in the must," globules usually oval, sometimes spherical, varying in size from 1/150 of a millimeter to 1/400 of a millimeter, isolated or in unified groups according to the temperature. Come from grapes via the *industrious presence of organized living beings*, wine is for one brief moment as much mystical substance as it is material good, and to make it sparkle is a question of what is at this point *sufficiently* explained in scientific terms: rationally, fermentation for spumantizzazione was not necessary. In his textbook, Carpenè calculates that wine at "–6° with 1 atmosphere of pressure dissolves 5 liters of gas, making it as spumante as ordinary Champagne," wrote his colleague, chemist and distiller Enrico Comboni in 1882.[112] "It was by applying this property that Carpenè was able to create, it can be said, the industry of spumante in Italia." By the following year, on Carpenè's bottles are labels on which the coat of arms of the province of Treviso, castled heraldic image[113] circumscribed by words like official boundaries, plus its ancient name *Tarvisium* borne on a silver label, and at the bottom the words *Carpenè Malvolti*, the Società Enologica that now that the school was here could dedicate itself to business only and no longer to the necessary public education[114] had been converted by the two men from a debt- and disagreement-filled multimember venture. It is "the love of the bel paese"[115] that makes character and Italy, Carpenè had written. The distance between molecules of a substance in a gaseous state is set not by any property they have but by the pressure and temperature they find themselves in together, Cannizzaro had joined forces with Avogadro and rationality in order to explain. Under the same temperature and pressure, a given volume of any gas is made of the same number of constituent molecules: a story of combined differing but related forces, of the unyielding boundaries of a wine tank in local wood though soon to be made, once the material exists, in stainless steel and of a bottle made of glass strong enough against the outside, that define spumante, that are the nation of Italy.

Lines to make Italy are entangled into the Scuola di Viticoltura e di Enologia di Conegliano.
Susan H. Gordon

Not much of Carpenè's tools or equipment remain. Most of his letters and personal objects are gone, too. Two world wars are coming soon, and when they do the winery, like many parts of the city, will be destroyed each time. Twice as with the city's ruined parts it will be rebuilt in near exact likeness so that still it stands today. Spumante, uniting of wine and gas by those who rationalized the materials and mechanics of the earth, worked tanks and bottles against the air that exerts the correct pressure to allow water to exist, containers of another atmosphere entirely. Even through the dim lighting here, Prosecco gleams ("once the cork is removed and the wine is exposed to normal air pressure, it can no longer hold, at this pressure, so much gas and so it forms that simmering that characterizes sparkling wine and that will last until it has lost five liters of carbonic acid, that is to say, 9.40 grams"[116]), pale emerald and straw; then apple blossoms, their green fruit, lime, and candied sage, complicated by robust lemon and yellow apple flavors from chemical transactions induced by the second fermentation by which Carpenè's spumante was not yet made and by the primitive sapidity of dazedly reforming carbonic acid. In 1902 on 23 March, Carpenè died in Conegliano, the day cold and clear and brilliant in the hills where he had held his maps and from which the mountains fluttered, promises north above the Valdobbiadene hills; Prosecco's vines would have begun bud-break, he leaving when it was all about to begin again, on the twenty-fifth anniversary of the dedication of his school which after that year's harvest will host an Esposizione internazionale of winemaking equipment.[117] Measuring Italy by its chemistry, the end of its nineteenth century can be said to have come[118] on 25 April 1906, the day before the start of the Sixth International Congress of Applied Chemistry held in Italy's capital, Rome, and the first time Italy's community of scientists has hosted other nations' chemists. The Congress's name lets you in on a significant change. Carpenè's field has been split in two: *applied* chemistry to be used by the conquering industrialists, *pure* chemistry as literacy of the natural world. There was now knowledge of the subatomic. English physicist J. J. Thomson and German chemist Richard Abegg had published the first models of atoms based on electrons, carrying chemistry into territories that could not be walked. As the chemists arrived in Rome, an IRT train in Manhattan took its first run as official New York City subway 9 miles north along Broadway near where I live now and the grapevine-devastating phylloxera from the United States which had come to the Treviso province first within the old Venetian empire was found in nineteen of its ninety comuni,[119] while the Trevisian hills kept on with their slow and ancient shift into countable focus, golden and ponderable spumante from the cantina Carpenè Malvolti that in a, finally, arrival point in twenty more years will be labeled for the first time *Prosecco. Vino Spumante Pregiato,*[120] *from the hills of Conegliano.*

3

Rationalizing Change

> Consider that the railways also lend themselves well to importing grapes.
>
> —Antonio Carpenè, *Rivista di Viticoltura ed Enologia Italiana*

> $6{,}022\times10^{23}$ (Avogadro's number)

In chemistry's rational analysis, results are expressed *in such a way* as to indicate not only the percentages of the individual elements present but also the form of their combination.[1] A movement from substances into chemical writing. Strade[2] into *strade ferrate*,[3] into *ferrovie*—iron ways made for traveling along; new main roads for direct or usual routes between a wider landscape made of towns and cities; vestiges through woods, in the *vie* names of Roman roads, and long countryside trails;[4] lines whose material the English *railroad* skips—that run to and from Conegliano's train station. Woods, oak trees (for the Carboneria's charcoal making, unlike pines fast growing and quick lasting and fast burning, oaks burn best and damaged they do not die but sprout back from roots to form new trees[5]) into railway ties; the forests of Treviso's Montello hill on the right bank of the Piave and of the Cansiglio above Conegliano past Vittorio once carried tree by tree for building Venice out of water will now trace the Terraferma with the addition of Fe. A multiplicity of synonyms are the names of old grape varieties as well, but one name can point to different cultivars hence a written and drawn sorting by sight, taste, and feel which ampelographers, themselves new, have started. In Italy, there is a Trevisian[6] ampelography published in 1870 with each grape afforded a page for its image and the opposite one for its "analytic description" headed with variety's names written into a space *commonly named,*[7] *nicknamed,*[8] *corresponding to,*[9] this often supplemented with a handwritten *maybe;*[10] an Italian[11] one in 1897; a universal one in descriptions and not drawings in 1877:[12] "the greater ease of literary and commercial communications provided by the press and railways have greatly favored the exchange of vines and the examination of the same," wrote Piedmontese nobleman ampelographer Giuseppe di Rovasenda, to whom Carpenè sent Trevisian specimens on a train that also carried letters, drawings, journals, in the latter,[13] "so that it is not unfounded to assume that the ensuing ampelographic movement will bring serious improvements in viticulture." Wine traveled this way, too: a price list at the Carpenè Malvolti winery, stamped *Conegliano May 1895*, notes that purchases will be sent by railway, COD.[14] Grape names from indicators of places, local dialects, and shared customs into *synonyms* and *biotypes* of one cultivar, still tied to place sometimes; into ampelographic tables; into details on maps of *giallo* and *rosso*. Experimental vineyards to learn and record grapevines (Rovasenda's includes grapevines grown from cuttings sent by Carpenè). Replace promiscuous multispecied vigneti[15] with specialized ones lined up with only the varieties best for growing there. Lines with direction left and right, past and forward. From *càneva*[16] where the wine waited, hopefully in cool humidity, to *cantina;*[17] from fewer vats and casks to more of them, to a new notion of space and method in making wine, wherein grape selection and an industry. The transformation —> is changing[18] their cantine into stabilimenti enologiche.[19] Elements into substances that are either simple, almost *atom*, or compound, nearing *molecule*, in which case a string of symbols; according to their role in chemical reactions; defined by weighing and as countable; into agreed-upon symbols across languages; Avogadro's number that suggests six hundred sextillion molecules of any substance is a material that we can see and use. Prosecche and what families know into classrooms of teachers, textbooks, and equipment so students can look and try. Orality to written.

In May 1875, *L'Ateo,*[20] journal for a liberal and rational world and against God and priests,[21] is launched in Venice by a Ligurian, but in 1868 the founders of the Società Enologica della Provincia di Treviso are a chemist and an abbot. In Treviso, it has been the priests who create the rural lending and saving casse rurali as they move between state and contadini, charged with telling the latest laws to a population which instead sees them as protectors and *their* messengers when new rules break the much older, inviolable ones by which they live. "Before 1866 almost all the comunal authorities fixed the harvest time and no agricoltore could harvest before that date," Carpenè wrote. "Only since gaining independence has the Authority left the date open, in the name of liberty.[22] Now, the agricoltori of every comune should gather and agree among themselves on when to set the most appropriate periods of harvest.[23] Once educated growers set the day, what is gathered is key. It is necessary to drink wine, and the line that runs there from growing and making must be uninterrupted and

Unstraight lines like remnants of trees in Prosecco Superiore vineyards
Susan H. Gordon

made of only grapes. "It is up to the agricoltore[24] to war against poisons, so-called liquors, that sicken poor people, those lower beings, made in the image and likeness of God, until they are the most disgusting beasts, the unhappiest of animals,"[25] writes Carpenè. In 1888, *naturale* appears twelve times in the fifth edition of his *Sunto*, in which technical and rational winemaking are explained by what nature gives: this word makes it rational work for a winemaker in a disappointing harvest to produce a second wine[26] from vinacce[27] reconstituted with water for if the criteria for ordinary good wine is grapes' charming fragrance, then that fragrance is legitimacy. "For example, with the deposits one has from the best Barolo, from the best Valpolicella, from Prosecco,"[28] or other wines rich in grace and perfume, of vineyards in "charmed places and terrain," winemakers can improve upon grape or place or year to lend "inferior or ordinary" grapes the genio of those blessed ones. Vinous grace made of measurable substances, rational, movable so that "the viscous deposits of a good Prosecco wine, endowed with a very strong and graceful fragrance, poured in quantities of 10 liters per hectoliter into a weak, lightly fragrant wine, will lend it the merits of Prosecco itself."[29] Four men, in a ten-hour-long day, crush from five hundred to six hundred miriagrammi of grapes.[30] The processes of producing in two ways, "that for vino giallo and the other for black, have passed, since time immemorial, from father to son," Carpenè writes in search of Italian lines.

"The usual method of cultivating vines handed down by us for centuries is that of filari[31] sustained by living trees with festoons running from tree to tree. Between one row and another, are interposed large spaces, which are cultivated to cereals," he describes in 1873.[32] "The foreigner who sees this cultivation for the first time, could find it charming, and could consider that vines supported this high up, do not cause damage to the crops below and therefore one procures more from the terrain than he could otherwise. But he is fooling himself, the large amount of evaporation produced by the vines, added to that of the live tree, causes serious damage to the cereals below, in this climate of dry summers."[33] Education can help Italians recover what they knew well before these unfortunate times, but it is not enough: "the impulse given by science, in these times, to perfecting every industry, induced some agricoltori" to make some changes, but theirs is only a partial correction, challenged as they are by the "terrible construction of the winemaking cellars, the outdated wine equipment, which little by little" must be replaced with new ones, "a forma e costruzione più razionale."[34] The objects of technology have always been necessary partners to the wine-growers. As has sudden need, the most pressing of which came to Treviso in Carpenè's final years, "diseases [which] forced the experts to develop research, experimentation, and thus transform traditional empiricism into applied science, giving life to modern viticulture."[35] In the second half of the nineteenth century, from the United States through France and into Italy first came fungal oidium called powdery mildew that in the spring can stop grapes growing, in the summer strike grapes to the ground and break the skins of those still upon the vine, throughout the growing season destroy leaves that generate the plant's sugar stores and protect its berries from sunburn. Then, peronospora, or downy mildew, that lies in wait throughout the winter and in the springtime fatally damages green leaves and stems, turns unripe berries forever brown and yielding. The solution is the element of copper. Last and devastating, phylloxera, an insect that feeds on the roots of grapevines, damage which varieties of *Vitis vinifera*—the grapevine species which counts Prosecco and thousands of other varieties including all of the world's best known and quality cultivars—cannot survive. *Filare* changes meaning, from trees down into rows.

There were roads for walking, trees and forests, rivers. Some of the area's Bisol family will become historical Prosecco producers but first their name is known for the iron they worked by battiferro, machine to do so when the metal was red hot, run by a waterwheel powered by the Teva stream in Vidor in the Valdobbiadene hills, past the railroad-tie trees. The white-wine grapes of the hills of Conegliano and Valdobbiadene, which "to tell the truth" do not have excessive acidity like the reds do, wrote Carpenè in praise, were for a while rationalized literally and equally: a 1937 report on the area's "typical white wines"[36] notes that the wine called Prosecco and drunk by the hills' inhabitants for centuries was made of one-third Prosecco Tondo, one-third Verdiso, and one-third Bianchetta grapes. The vitigni[37] most grown in the province are, Carpenè writes in a two-column table, Verdiso, Bianchetto, Pignolo, Prosecco lined up under the heading *Bianchi* and seven red grape varieties under *Rossi*—"all of which with the exception of Prosecco, Bianchetto, and Pignolo, must be considered ordinary varieties, due to their nature and also for the almost savage system in which they are trained."[38] A multitude of grapes is a road map to terrain of languages and wind and slope. Rational tables that mix up the grapes' varieties under white and red wines,[39] a century later the Strada del vino bianco will run in Treviso's north; the Strada del vino rosso to represent its south.[40] At night the contadini gather alle stalle,[41] relax, talk, trade information, share food and wine and listen—"Mi scrivo par le stale, e no par i gabineti de le dame," wrote Illuminato Checchini known also as Paron Stefano Massarioto[42] in a 1895 edition of Trevisian Catholic weekly *Vita del Popolo.* One or two vats for vinification. Grapes are crushed by the feet of half-undressed men in a brént[43] made of wood, mostly chestnut and sometimes oak, of larch and fir and also poplar, with open topsides larger than their closed bottoms and in which the móst[44] was left to ferment. Which is by air (for a while an element counted among others such as gold, iron, oxygen, ether, rust, carbon). The golden cloudiness of these hillside wines is for families, sold to the ones nearby who did not make it and poured in the osterie where men went to talk and trade, sing and play at cards. Muted polenta is set against the green-and-brown-complicated yellow of the wheat which must be sold to others. Harvest dates and prompts varied wildly from year to year, wines widely better or worse. Grapes and must, in other words, became bubble and noise and heat and finally wine, maybe a little sparkling come the spring. The prosecche vineyards of the hills of Conegliano, Valdobbiadene, Vittorio, and Asolo (pushed into being by the Piave glacier at the beginning, coldest part of the current ice age) do not easily rot but they are being chased (even in the Coneglianese) by oidium, peronospora, and bad weather into only the slopes of Pieve di Soligo, Soligo, Solighetto, Farra, Follina, and Col San Martino. Earth portioned by food-and-wine-growing campo which is 54 are (1.83 of those in an hectare) and almost never of just one cultivar; liquids sold by mastelletto; polenta, seeds, flour by the sacco or the staio; goods weighed by a large libbra grossa (0.516 kg) or a slight libbra sottile (0.301 kg), or stretched an arm's length of cloth (a 0.676-meter braccio di panno) greater than an arm's length of silk (a 0.634-meter braccio da seta). Bubbles of gas that slowly overtake a fermenting must, shaking and upsetting it in all senses. Knowledge passed down through communal voice and memory; authority; unreliability due to outside forces. "If experience dictates a mix of two or more grapes, then do it, every year weighing and maintaining the proportions found preferable by the methods in preceding years" is a bridge to "separate the different qualities of grapes to make different wines," more vats and of different sizes, practiced and written attention to what the grapes from each site taste like, "you all surely know these better than I do,"[45] blending. Turn over the tini and re-wood so that the bottom is now of greater diameter, lesser opening at the top—crushing the grapes by foot is "not bad," wrote Carpenè: "As we are against advising expense upon expense to the winemakers, we approve the system in use here, with the condition that cleanliness is monitored, the men touch the grapes only with well-cleaned feet, that they don't enter the tini semi-naked so that their sweat drips into the must."[46]

Now there is an Italy. The impossibility of plowing transversely the land between the rows and of giving a special fertilizer to the vine.[47] "If we want to gain honor with our wines, especially abroad, we must improve our viticulture, because with good grapes in normal vintages no corrections whatsoever are needed,"[48] Carpenè is sure. Faith has turned to vision, placed before him strong bottles set on tables, with corks popped in triumph, stacked in cellars and shops into reasonable and succinct pyramids,[49] impossible that others do not see these scenes as well. Even the bottles for spumante are not what they are for wines that do not sparkle. "An infallible method for making wine that lasts," he lays the tracks, "is to produce them with diligence and with the simple attentions described in this meager little book."[50] He is among the first of the wine scientists to hold what would soon be official *cattedre ambulanti*,[51] to travel through the hills to where the grapes were grown, to hundreds of small plots on haphazard slopes in rainy springtimes and resplendent summers to teach chemistry and solid ground to the growers, together retraining the savage systems. But "a real rational and more widespread reform in agriculture is not possible to see with the current agricoltori, because it would be like expecting to straighten the old trunks of trees and removing from them all of the bad vices acquired from the bad system of raising them used so far." The greatest obstacle is the present: it is not enough that the contadini have the information, they must be trained in ways to think or they will not use the information well and so his work is for the young, who will make Prosecco, who Italy is for. A straight tree is of course no longer a tree.[52] It is another plant trained and educated with a purpose bent to human need. It is a vineyard in an unshared plot, vine rows aligned to other rows along measured respect of these hills' unwieldy but, spend enough time, learnable curves and planes. It is an educating palo secco used in place of a tree, it is many of them, in a vigna whose particular lines can be planted with new purpose.

"Wine doesn't contain only products exclusively from alcoholic fermentation but it contains dissolved others that did not exist in the must, regardless of more or less prolonged maceration of the solid parts of the grape and thanks to the chemical combinations that have occurred between the atmospheric oxygen and the heterogenous compounds of the must."[53] Did he miss that Italy's spumantizzazione, quicker and more deliberate than Champagne's, might be spirit and material come together, that this transformation was place between what could and could not be rationalized? Perhaps for him the knowledge of CO_2 and therefore the power to include it was vision enough, matched to the ethereal fragrance of the grapes Malvolti grew, like the concrete existence of Italy something still beyond the realm of science. In any case the technology that was the natural and logical extension of his work still awaited the invention of that stainless steel. Around 1140, from the church had come the notion that machines might be providential[54] and now they are the rest of the way to the future: "*La buona cantina fa il buon vino,*" he wrote, "and this is an old proverb but true."[55] Through calculations he can measure what has not happened yet, bring Prosecco into weighable being through theoretical work on paper and then by practical manipulation in the winery. Since, hasty ferment, grapes must be crushed as soon as possible after they are harvested, "so that the order does not fail, one must first gather baskets, ladders [old vines grown up high!], scissors, and vats and calculate the necessary number of workers to end harvest in the pre-established time."[56] Enough grapes become a wine: like Avogadro's number fermentation brings the unit of a material onto human scale. Then, good winemaking can cure disease and poverty, and economic reasoning is the implacable enemy of even deeply rooted habits.[57]

A grape press for exerting pressure equivalent to what men's feet do in "ordinary pressing,"[58] mechanical and made of iron and wood. By law only metric measurement is recognized[59], so wet and dry volumes must be told in hectoliters. A winemaking textbook that begins with a table—*NAME OF THE WINE, Acidity per liter considered Tartaric Acid*—of France, of Treviso's plains and hills. It is known now that there are differing proportions of constitutive components in musts of different years (1868 Refosco[60] with 9,050 grams per liter; Refosco 1875 with 7,122[61]). There have to be other ways to make a still wine into sparkling because how the ferment (composed presence of germs that are 1/150 mm to 1/400 mm[62] big and remain themselves unchanged) produces ($NaCO_3 + H_2SO_4 + H_2O \rightarrow NaHSO_4 + CO_2 + 2H_2O$) carbon dioxide as it transforms must into a wine, wine into spumante, is known. CO_2 is known. Not much else is known about the germs, but a mechanism is understood—"the alcohol slows and impedes the development of the ferments and as it increases by the decomposition of the sugar, the energy of the fermentation diminishes until it arrests, not just for the increase of alcohol but for the simultaneous lowering of temperature and for the depletion of nutrients."[63] Calcium sulfite, acids sulfurous and boric will stop the ferment in its tracks, too. New acids formed via fermentation, "not only acetic, enantic, lactic, but also caprico and caprilico, and on and on enough to send a chemist lost in a labyrinth of compounds many of which constitute an unknown for science."[64] The *Prosecco* grapes Carpenè experiments with for making quality wine in the hills are *gentil; lungo; dal peccol rosso; Balbi,* Trevisian best and recognizable by its varied-sized berries. Due either to greater diligence or to the calcareous-clay soils providing something necessary, he finds that here "Prosecco makes good wines, and truly excellent ones in the best years."[65] Rationality ("how far we still are from giving our agriculture a physiognomy uniform and national."[66]) is how the chemists hold conversation, discussion, argument, how they convince each other, how they open communication to nonchemists.

He writes a table to show his apparatus at work; "it is composed[67]

> 1° of a generator[68] of carbonic acid gas, *a, b, c, d;*
> 2° of a gazometro, e, f, g;
> 3° of a pompa aspirante e premente, l, m, n;
> 4° of a cauldron made of tin-plated copper;
> 5° of a tino for keeping the refrigerant material."

Upon fermentation of the must the sugar found in grapes forms the following products: alcohol, carbonic acid, glycerine, succinic acid, etc.,[69] the chemist winemakers at the Scuola in Conegliano teach the children of the agricoltori. Some of the graduates will teach, others will make, and the knowledge in the textbooks and the experimental vineyards will be passed on. Choose the varieties that yield the highest quality grapes where you are. If you have more than one, plant each variety together in separate blocks, in rows of vines spaced evenly apart and a distance of a meter between each row. Fix them onto pali secchi at a height that lets them gather heat from the ground in cooler places or just a little higher toward the sun if the site has a warmer climate. Pick the grapes only when fully mature so that the wine will be a lasting balance of acidity and flavors. During harvest gather only the best bunches. At the winery, select only the best grapes for gently pressing into must. The flavors of wines grown in the places that will become official *Rive* are already stories of the differences in this territory's discrete parts, components of a Prosecco that helps you think of it also as a continuous system. The flow of winemaking from fermentation to drinking is now a series of discrete and distinct stages, little fields of expertise. Italian statistics a picture of the social body in a rationalized form, following a natural order, ready to receive the stimulus of a rationalizing administrative system.[70] The rationalizing of the Montello woods, the end of the boscaioli. Per. Compare. Opposition. Ratio. Chemist and winemaker, but he is also steeped in grapes, with, for example, Rovasenda. *Razionale*, says Treccani's dictionary[71] is results of reason, founded on reasoning conducted with scientific rigor or on a scientific basis; in chemistry's rational analysis results are expressed in a way that indicates not only the percentage of single elements present but also their form of combination; therefore in the analysis "of a clay, of a cement, of a glass," of a wine, results can be reported simply by indicating the components' percentages "(calcium oxide, aluminum, silica, etc.)," or by indicating the amount of "calcium silicate, calcium aluminate, etc. that they form." Chemical formula, formula of structure. In geography, a rational horizon is in opposition to a physical horizon; in math, the rigor of the demonstration of procedure in opposition to an intuitive understanding. Looking for the people of Carpenè's books, I find three instances of *agricoltore* in the third and fifth editions of the winemaking textbook, one in its more viticultural precedent *La Vite*; 3 and 4 respectively of plural *agricoltori*, 10 mentions in *La Vite*. *Contadino* appears 2, 1, 6; *contadini* 1, 1, 17; no instances anywhere of *contadina*. 12, 11, 0 for *enologo*; nothing in plural *enologhi*. No instances for *viticoltore* single or plural, except for one in *La Vite*, for *il rinomato v. francese Guyot*. *Vinificatore*: 4, 6, 3 (textbook volume one is of course titled "Vinificazione"); 10, 15, 10 for *vinificatori*. *Enochimico*, 5 instances, all in the textbook's fifth edition. No *coltivatori*, even in the singular, except for a page-and-a-half section of *La Vite*, so titled and mentioned three additional times. Then words for work and world: *analisi*: 11, 42 (the second volume is titled "Analisi enochimica"), 7. *Razionale*: 4, 2, 1; while 1, 1, 0 for *razionali*. 1, 1, 0 for *empiricamente*, same for *empirismo*, just 1, in the textbook's third edition, for *empirico*. New groups, new movements.

Would it be better if I said that like this? See Table 3.1.

"It is truly comforting to think that many winemakers today no longer work blindly, as the empiricist does," writes Carpenè, "but instead, within the limits of possibility, want to make sense of things and phenomena for themselves, to rule them, direct them, modify them and achieve, guided by scientific data, results in greater agreement with the needs of *winemaking*." *Enotecnica* is the word he uses.[72]

Table 3.1. Instances of words for people and work in three of Carpenè's books

Term	*Sunto, Ed. 3, 1878*	*Sunto Ed. 5, 1888*	*La Vite, 1874*
agricoltore	3	3	1
agricoltori	3	4	10
contadino	2	1	6
contadini	1	1	17
contadina	0	0	0
enologo	12	11	0
enologhi	0	0	0
viticoltore/i	0	0	1[a]
vinificatore	4	6	3
vinificatori	10	15	10
enochimico	0	5	0
coltivatore	0	0	0
coltivatori	0	0	4
analisi	11	42	7
razionale	4	2	1
razionali	1	1	0
naturale	9	12	2
naturali	2	5	4
empiricamente	1	1	0
empirismo	1	1	0
empirico	0	1	0

[a] "il rinomato v. francese Guyot," 37.

4

Walking Conegliano

> Il Prosecco da alcuni credesi proveniente dal paesetto omonimo del Friuli orientale; da altri invece si reputa che il Prosecco, che diremo coneglianese, sia ben differente da quello friulano.
>
> —Italy's Ministero d'Agricoltura, Industria e Commercio in 1879

In the historical center of the ancient boundary city draped along the slopes of the Giano hill on which the medieval Conegliano castle, along with the incorrectly named and misfit neoclassical Teatro Accademia one of the few things to be built in the nineteenth century is a smooth stone slope. Sideline to via XX Settembre's roadway, the path upon which to move from the street's northside portico up to Conegliano's municipio includes the marble gradient that then replaced twelve stairs so that hill and city inhabitants could have full accessibility to and from their city hall (built one hundred years earlier in expansion of the small fifteenth-century hall still tucked behind it). "North of Venice," wrote Hugh Johnson in 1974, "Conegliano is well known for rather sweet white wine called by the name of the Prosecco grape."

When I walk there in late February 2020 (the pandemic is arriving quickly but otherwise only with schools and museums just closed, group gatherings banned, and official advisements to not hug anyone), topped by a likeness in relief against a lush branch of laurel and inscribed in 1905 a large stone plaque at the loggia of the municipio announces Mazzini with ceremony and apology. *GIUSEPPE MAZZINI / Agitatore e profeta / suscitò la terza Italia,* the top three lines read in conciliatory tone; *e traverso i secoli / si ricongiunse / con la pura anima a Socrate / con l'austero intelletto a Dante.* (As you read, behind him, from by the walnut brawny municipal double doors stares a bust of Garibaldi.) *Nel 1 centenario della sua nascita / i cittadini / auspice l'alleanza scolastica*[1] signed atop a nine-petaled giglio bottonato, a tuscan flourish like a multiplied fleur d'elise, held fast at each side by a large leafed curl, tracing and recording pedigree, establishing his precedence, preordained. Walk along this street then out onto the ones that became more important after him and following a more or less straight line that changes names as you keep going, to arrive at the school that Carpenè so wanted to be built here. It is an institution recognized as much for the technical training it gives as for how long it has been doing so. Its first class graduated in 1878, but from a building a direct six-minute walk from the Teatro and just past the ancient entry to the city. Since 1924 the school has instead held fast at this, its second, location, for which you'll have turned just before passing its now unmarked first place then pass close by the train station by way of a car-filled avenue whose wide sidewalk with a section for bikes and running is a remnant of what the city provided Carpenè and fellow city dwellers then: the viale dei Passeggi, decorated avenue for where the coneglianesi walked. Both sites places of classrooms and laboratories, experimental vineyards planted with varieties to see which did what, training and pruning systems tried out for passing on that knowledge, recorded weather and what

Walk from the Municipio loggia down the nineteenth-century ramp along via XX Settembre in Conegliano.
Susan H. Gordon

the grapevines did in it, was for storage of the materials that chemistry and its questions needed.

Point of arrival, departure, and where trains might change track and destination, when the Conegliano railway station opens in 1855[2] it makes another city street necessary. Both the occasion and the vehicle, it, too, a series of links, demand new infrastructure and here they redefine a space that until then had been firmly on the city's outside. Conegliano has won this station in a competition with other Prealpine cities, and its victory has mapped it into the iron Venice-to-Vienna line, leaving out other centers of the upper Marca Trevigiana countryside like Vittorio and competition-loser Sacile where Carpenè was born in Udine. Its win turns nearby Treviso,[3] Serravalle, Belluno, and Feltre peripheral: before then, local historian and guide Francesca Toninato told me when I returned to the hills for a week in the summer of 2021, those places had been scrigni, *coffers* of industriousness, of traffic, of moving goods, of persons, and of thinking. The necessary viale Stazione will be a perpendicular guide for a short walk to the old contrada Refosso, axis across which flatness and hills reflect and transform each other and where the city's flatness ends at the first step of a broad staircase into ancient Conegliano's still steepening center. But first, today this crossing is a face-off between via Vittorio Emanuele II which moves toward the old city walls to the west and in 1866 has been renamed for the king of Italy who passed along here and the eastward half now via Mazzini which leads to the Monticano river (arrived here "from the little mountainous zone called *al confine*,"[4] which I am placing at a two-hour walk north, and along the way fed by the ruscelli Favarro, Piovesella, and Cervada to bring it to eight feet deep, banks at the level of its water, but from Conegliano banks are twelve feet high[5]) and brings you close to the Carpenè Malvolti winery. Split between guiding spirit and what happened next, but soon, stable and celebratory, there will be no need for the names that made Italy to be in agreement. From 1920 to 1922, at the new street's most imposing corner will be built the Palazzo della Cassa di Risparmio della Marca Trivigiana, an exuberant and optimistic building ornate and curving equally toward viale and refosso to house the Savings Cassa—there is no accurate rendition in English of this bank-ish moniker for a credit, financial, social assistance institution—of the province's entire countryside, from its flatter lands down across the Piave to the northern painstakingly hilly Alta Marca. The station and the streets it occasions act as ties from ancient city to nation. The new avenue recasts the iron-drawn station from proximity to part of Conegliano, star city fusing ancient and recognizable, a new way to get from Italy into Austria that changes just outside the lines of its ancient perimeter into a larger wanted place. I am not the first to see that this city, ambitious since before it insisted on being part of Venice in the fourteenth century, pusher of liberation from under France and Austria and of unification into Italy, small tenacious seat of the forming class of scientists who see no mismatch between visionary pursuit and countryside economic practicality, finds its own image reflected in new nation's possibilities and steadfast, methodical, and from the heart throws itself in. Conegliano is loosening its ancient self to become more seeable by us across that moat in which we can first catch sight of our days right now. Only in 1947 was the contrada del Refosso made half viale Mazzini, spirit and hope and heart again necessary after Italy's two wars and the nation's fascism, but "at the beginning of the century when [it] is still the center of an agricultural hinterland,"[6] the city that Carpenè saw was still like his science and the wine he dreamed. That grand staircase, the salita delle Pescherie Vecchie, had been constructed in 1838, the year he was born (soon after World War II it will take its current name of Gradinata degli Alpini). The first train to arrive at the Conegliano station—on railway time, notes the 1889 *Baedeker* guide to northern Italy, which "is that of the meridian of Rome, 52 min. ahead of that of Greenwich and 40 min. before Paris"—is named Pordenone[7] for the northern Venetian comune and can carry people from city to town, bring to them what is grown by the contadini near each place, move everything to stations' predestined places and travel to anywhere far more quickly than before. Hotels; shops for buying vinegar, socks, hats, silk, and the seeds for starting a silkworm colony, by the century's last days defined the city; a wealth of bread-makers; photographers; hardware shops stocked in iron tools; stores that sold fish, measures, pianos, or olive oils; pharmacists; coppersmiths; the pizzicagnoli of preserved foods; enological machines and tools; the bimonthly *Nuova Rassegna di Viticoltura ed Enologia.*[8]

Down the stone ramp onto the portico's shined and pitmarked stone ground. A few doors down in a southeast-facing site built in the fourteenth century is the Città di Venezia restaurant, founded in 1846 when it was licensed to sell wine by the glass and lodging, and a well-cellared seafood restaurant since, though it has passed through several hands. The current restauranteur, Nicola, has had it since 2004. What was its name in Carpenè's time? Of course he ate here.

Where the Teatro is had been the fontego (fondaco) delle Biade, where grains were stored by Venice, which during crises sold to its citizens at low-enough prices. "From the Persian word for magazzino," Toninato says, repository whose destruction and replacement by the Teatro "destroyed the past of the Borgo," and marked the beginning of the walled city plus the lands, providing and protected, around it becoming the comune of Conegliano it is today. Patiently behind the Teatro's west side is a two-story building: portici, frescos (seeded also throughout the hills) in a mix of geometric and representational, Venice's eastern-styled pointing-upward windows. In the city, the names of buildings and streets can come from those who designed them as well as who they did so for, and here, the Casa Sbarra, Conegliano's first recorded academy was founded in 1603 by notary, poet, curious agrarian Pultio Sbarra[9] in his own home, its last class held far later in 1812 when occupying France would not allow it to go on. It took fifteen years for Sbarra's Accademia Agraria degli Aspiranti di

The view and way from the Conegliano train station up into the older city include the unifying staircase.
Susan H. Gordon

Conegliano, tree-shaded mirror to reflect the sun's rays so that a fire might be lit in these hills,[10] to fall when Venice fell. On the side of his home that does not face down onto the station the hills are right away and from it, through rows of vineyards (in at least the early twentieth century the views of the flatter lands from here showed those planted in the unusual and local pattern "a raggi," the Bellussera pattern in which vines' trained canes together formed a plot-wide diamond-hatch pattern of intersecting rays and trained on maple trees[11]) for ages the city's first immediate steps, one could like from the station walk easily into countryside from city street and back, but instead of down into flat and wider place up, up into the fast and many particulars, mapped by both blessed transmission of knowledge and maps of cultivated plots that were precise not from measurement tools—in the eighteenth-century mapping of the Marca headed by local surveyor Francesco Vidor, "the austereness of the localities, the steepnesses of the terracing,[12] the ground depressions," of the hills made them the last to be counted, step by step, lead drop in hand—but from inside ties.[13] A ten-minute walk east to where the Monticano runs. There, ancient bridges crossed before the war came and beneath them women, just women, wash textiles and rinse fruit along the river's brief banks made of beach and small bluffs, men bringing first their sheep then (when in the first rise of industrialization the demands of city tastes changed some of what the country did) their more stationary cows to market every Friday in the Foro Boario (soon to be divided into the piazzale San Martino,[14] a garden, and various buildings for the new business-minded city) on the Monticano's left bank or, once the Teatro had claimed market space on the river's right, lead them *there* to drink. The turning of the area's work into an industria and soon into something else seems a sure thing, too: the Banca di credito agricolo ed industriale di Conegliano is opened in 1895.[15]

In the second half of the nineteenth century what was this city doing? What was it trying to become? Through it is moving what? The train brings and leaves and brings and leaves, closer now the world to these hills. About six thousand people live here, in this small, nested with an echo of Venice's famous love of secrets, noble, industrious, and resolutely academic urban center fitted on the inside, too, with spaces for walking, though they are dedicated so—the statue-lined concourse by where the school would move, an intersection minutes east of Carpenè's winery where upon a post the sign *Viale riservato ai soli pedoni* made the nature of these streets' meeting point clear. In this city among hill towns people are organizing; forming communal banks with lending desks, cooperative latterie to make quality and lasting cheese of all the milks the shepherds and the contadini need to sell; reimagining older streets with the new names Italy has brought; while in the laboratories and classrooms and in the osterie where the hills' wines were poured a few worried about the poor state of much of the hills' agriculture after all those years of poverty and stress and Venice slowly crumbling until it is lost to foreign rulers. "What matters is that agricultural innovations, for many reasons and for bad reasons, are so slow and difficult, that we can be perfectly calm," wrote Vianello, considerately? sardonically? in 1872. "This improvement will not run on a railroad."[16]

"The social aspect of the cassa [which found] its corollary in the osteria,"[17] writes Casellato, made such nineteenth-century institutions "impervious" to outside control, whether by the Roman Church or the former republic lands' own Catholic-minded, antimaterialist, and protectionist supporters of coffers and casse, or by the attempts of industrialists, who very much favored materiality, to manage the monies of those who worked for them. Where Carpenè might not have spent much time, or perhaps he kept a secret well, are these wine-filled corollaries. Each one of the "numerosissime"[18] osterie that have always characterized Conegliano and still do is a place that is "refreshment, womb, refuge,"[19] each which through extended contact takes a person's discomfort unto itself, each a vital molecule of this little city and the Prosecco hills. "In the city the osteria has always substituted for the piazza,"[20] that most public of places, meetings intentional and not, place where deals are made, the voices of the people unite, their ideas gather communal force, empowering. In Conegliano the osterie remaking the piazza were central pieces of an urban composition, "scattered parcels in the shade of the portici."[21] Sometimes there was lodging, sometimes there was food—"the few osterie with a kitchen,"[22] writes Marca Trevigiana food historian Nazzareno Acquistucci about those public showcases of the hills' feast-days cooking, Trevisian "contaminated by the cuisine of the alpine area of Belluno, a sort of border cuisine" that the cities would soon turn *Italian*—but wine was star and essential promise of such shaded plots, *osteria* a word made of oste, *host*, and therefore also if with no need for it to be said, *guest*. Inside, card games and singing were understood actions well into the 1900s, as was the near absence of any women patrons.[23]

For the decades bridging the nineteenth and twentieth centuries, in the osterie of Venice and those of its closely linked piece of Terraferma, proud, self-assured, and stubbornly ambitious Conegliano (in the sixteenth century, this "little podesteria at the feet of Belluno's Prealps and included in the dominion of the Republic of Venice aspired to promotion to [Venetian] civitas and to administrative separation from [the city of] Treviso,"[24] it bears reminding) the red-shirted gathered,[25] as on the battlefield most but not all of them, men. A source of food, energy, exchange, and intoxication, wine was something searched out, and so the places devoted to it glass by glass naturally became centers of sociability. And more. When at the second half of the 1800s, the banche di credito, casse rurali, and latterie were born, the osteria served as meeting room for all their functions, indoor public-facing places for planning, organizing. The casse peota,[26] Venetian territories' traditional mutual gathering and lending associations, were also to be found in osterie, "often the seat of one or more casse,"[27] writes Casellato. In the osterie, business deals were made, sealed with an unbreakable handshake; emigration made necessary by larger forces but possible by more intimate ones as in the osterie entire groups of exasperated, desperate, angry families decided, coordinated, and made their arrangements—"to shouts of *porca Italia, andiamo via*," Gasparini illustrates for me in 2021 when we meet in the cantina of one of the hills' storied twentieth-century stops—to leave for Argentina or Brasil. A host poured four kinds of wine, he adds: "the rosso, the bianco, the bono, the tristo."

Although a keystone of public living—a notary and access to one of the few telephones would soon join wine as basic city and town needs, all to be found at the same place—an osteria was also a center of a very masculine way of life, concur Gasparini and Enrico Canel, restauranteur-chef-owner of that celebrated stop in the town of Col San Martino on the right bank of the Raboso river as that works its way into the nearby Piave. A woman could enter to retrieve her husband; might stay and drink if she was a foreigner; was respected and confided in if she was, and many women were, the host herself. An osteria was open seven days a week, from around 6 a.m. to something like 2 a.m.: another of its most famous in the mid-twentieth century was along the bus routes popular with workers, many of whom took a drink before boarding their morning bus;[28] at another the local doctor would arrive at opening to pick up his[29] list of patients, who had called the osteria to schedule their appointment the day before. There are hints of another function of this public place in historian Piero Brunello's introduction[30] of *Ribelli, questanti e banditi: proteste contadine in Veneto e in Friuli 1814–1866*: "The protagonists of the stories reconstructed in this book are comizianti di osterie,"[31] which with its flavor of *rally-attending* gives fresh meaning to a common nineteenth-century criticism that workers and contadini spent too much time and money in the osterie. "Allowing the poor to live between the folds of society, declared the new dominant class, means encouraging idleness and the despicable habit of wasting one's own assets at the osteria," Brunello writes. But there were no "other places outside of the osteria in which artisans and workers could gather; place of free conversation, the osteria was therefore considered a place of subversion, the antiparish per eccellenza."[32] The doorways of osterie included one of several symbols to signal their presence to the outside. The most common marker was a fir-tree branch, used for ages by the contadini to announce where they were selling what they had grown, "particularly in regard to wine"[33]—"se 'ndéa par la strada revèsa," Luciano Cecchinel from the upper Valdobbiadenese hills and among Italy's landscape poets wrote in the 1980s, "par rame vèce che no le véa màntega / a dir de i concòi dà de fèr."[34] I imagine in the night the stalls to be such places, too, but where fewer paths crossed and whose presence of women and children might have made them more ignorable by writers.

Today wine is a thing that people drink for how it tastes. "Una volta," explains Gasparini of these hills, you drank because you were thirsty, or to drown your sorrows; you drank from sadness, to celebrate, to forget, wine a thing thought of as for all of life. Today, the glass is wrong, he notes. The vessel you drank from in Venetian lands was impossible to swirl, as one does today to aerate and help along a wine's volatile aromatic compounds when tasting. Who drank in the osterie of the lands of Venice did so from a short and heavy glass into which one great gulp of wine was poured—Venice's famous ombra or ombretta. Wrong, too, if for the opposite direction in time and quality, was the vessel that held the wine before it sold. Fresh and subtle Prosecco grown in its optimal place of slope and rain and sun trained sharp and golden warm on all the hills' faces except for ever-shadowed verso north: Carpenè's time is the point of its first imaginings and so for years to come all wines including this delicate and yellowish envisioning might still be held in wooden botte, absorbing chewy and further bittering tannins from earthy walnut or from sweet chestnut, heavying more still in flavor and color from the oxygen that breached such wooden barriers while in the city sparkling and purpose continued to find their way to each other. Carpenè is building machinery and future in Conegliano, but stainless steel and Italian sparkling wine methods would not come until early, trickling, then well into the twentieth century a steady stream.

Served from barrel and not nonreactive, rational, and labeled bottles, would local knowledge, wisdom of quickly filled and emptied ombrette, know by drinking a Prosecco which hillsides it had grown in? If wine tastes of a place, can perhaps bear the imprint of an understanding maker, then did it also do so while agronomists and contadini slowly tangled and untangled over better growing methods and into practiced, plotted viticulture? In a postcard-size book[35] of Conegliano's osterie and streets told through brief memories and, well, postcards, an osteria is also a place of measuring: six hundred years ago, the city's laws dictated that any oste who sells wine in this city must have official measures, stamped with their corresponding numbers and comparable to the Comune's, and must "use these to sell and to measure justly." But in the laboratories there are other growing aspects to measuring a wine: in Carpenè's *Sunto*, a wooden box that is a truncated light-blocking pyramid built in mulberry and fitted with a glass tube notched in numbers and with scaffolding brass support is for determining more about the beverage's color. Those wines he had criticized, wines of popular taste, loaded with color and tannins? It was well known that the oste would ask the contadini to leave the wine on the skins for a long time,[36] says Gasparini. From there an oste might "adjust" their stocks, adding water when pouring for guests so that a more acceptable color and strength and lightness might be observed in glass. Where did Carpenè drink that he did not write a reason for the preference for deep and tannic hues? The people who frequented the caffès were different from those of the osterie, writes coneglianese author Luciano Caniato. Unlike those of the "basse osterie,"[37] they had money, culture, and time, and they had prospects; some, in sign or imitation of leisure and nobility, were well-known eccentrics. The caffè is for those who stay, the ones who do not emigrate, but it is also for the forming borghesia, new ruling class of businessmen, associated or not with agriculture; in the first few decades of the twentieth century arriving ever more quickly it will be a place also for, new, the men in black shirts. Might Carpenè, well-versed in the tricks of the wine trade—he had written for example against a practice of mislabeling local wine with an easier-to-sell French origin—have known of this practice also? Did he not mention it from a desire to protect? The osterie would be safer bets for those against the black shirts.

"Conegliano has always been, in addition to the head place of an agricultural system, a city of markets. Fortress for protecting earnings and grains,"[38] writes Caniato. The Italian word he uses, *luogo*, does more work than *place*, carrying also human presence, physical location, community, name, legitimacy. This city before the great waiting eastern Alps had long been also a border stronghold where cattle, wine, cheese, armies, and winds wound through on the paths adapting to them in movements from close-worked Prealps to fertile flowering plains south. From Conegliano, the continuities of streets and hills west and north remain best navigated by those knowledgeable and hardy, traits granted to who had been living here for generations and who over centuries had transformed administrative ties to Venice into a powerful sense of luogo and personal identity, a subject city both independent and respectful. These habits had stayed even as, economically tied, too, Conegliano degraded as the greater power's did, its stress and pride both heightening after Venice ends.

Old inclinations turn into lines for a new road map. As an Italy grows louder and moves closer into view and then finally in 1866 arrives, Conegliano finds itself anew: equipped with railroad station, wine, and in another decade the nation's first winemaking school, the city begins to shape a population 75 percent contadini and 50 percent "illiterate" into a locality that is defined by nation, which some like Carpenè are busy shaping, too. Running and refashioned nature, a railway system is also a national space. Conegliano's conduit between station and city, self-consciously first called viale Ferrovia, is soon absorbed into the growing city and renamed viale Giosuè Carducci to honor the national poet of tuscan origin as the city's cultural and market center shifts into its new role by erasing the formal boundaries between old center and what had been its first passages outside. We can watch the adjustment reflected in both its citizens' temperament and the architectural structures in accordance with which they move. This straight wide street immediately outside where the old city walls once ran was now a center of movement, of materials and a mix of people citizens and foreigners. At the ascent via the great staircase, the Teatro, opened in 1869, but before the twenty-four years it took it to be built, this ground was open, free, and urban space. Here hills and center met, gathered, talked, drank and bought and sold by the warehouse of grains and the monthly livestock market. If not Conegliano's only concept of piazza, it served nonetheless as "luogo d'incontro,"[39] legitimate encounters in an "antica piazza cittadina" held in "the embrace of the city with its unmistakable greenness," where the space had long celebrated the city's sweep into the hills, cut when the Teatro is constructed. We are back on via XX Settembre, which runs from the abiding Neptune fountain (ancient sculpture rebuilt into fountain here the year Carpenè is born and that will be one of the few structures to dodge 1917's devastation of the city) to a darker monument installed by the fascists in twisted tribute to Italy's World War I victory and which was until 1870 named the Contrada Grande. Its new name reflects the day Rome's walls are breached and *that* city is taken as the new nation's capital—part of a change of local toponomastics that is a reflection of the coneglianesi's enthusiasm for the nation,[40] writes Caniato.

And we are also back before the adjusted space of city hall. Its loggia is Conegliano's own official shaded parcel, another refreshment, womb, refuge with a host of plaques and busts devoted to those who helped to make the nation. Little and written monuments they are less memorials and far more symbols of concrete and ongoing presence. There are those words to Mazzini. There is an ode to Conegliano's Maddalena of its noble Montalban family: risorgimentale "who dedicated her life to Italian Unification." Here is the history that is ours now, together they seem to say, and to that end there is also, prototype if measured by its subject's place in time and evidence of the city's insistence on having long been here otherwise, there is a position for Giovanni Battista Cima, who died in 1518 at the age of fifty-eight and was painter of these hills with brushed details of buildings such as the city's old Convento di San Francesco and the pointy bulbous cordate-shaped *Cymbalaria muralis* and *Umbilicus veneris* like small vivacious grape leaves that grew along its walls. To the new nation of Italy is applied prideful *Patria*, but more warmly and more colloquially and more purposefully once the fascists have come and gone, is used the term *Paese*. It is a word that means *local* and *ours* just as much as *nation*. From the latin *pagus*, or village, it still indicates this and is used also to point to a *land*, or any "cultivated and inhabited territory, identified by *its own* [emphasis mine] characteristics of physical and anthropic geography," defines Oxford Languages in its role as Google's dictionary. It is a word of ground and of connection and inseparable from a rural cast. If *paese* were a substance, one could describe its composition as indissoluble.

As the nation solidifies, Giambattista Cima is made an icon of the city, secular patron saint whose nickname has always been *Cima da Conegliano*: "the assumption of Cima to character-symbol was made concrete also through the baptism of the ancient medieval piazza that from then on (1893) assumed the name of Piazza Cima," writes Caniato. *In honor of his native city, in the works the world admires, happy and superb* reads the artist's plaque at the Municipio. "Spurious assignment on the inside of a Center that, just a little later, will have a toponymy that is entirely risorgimentale,"[41] Caniato comments. To do his work, Cima had left his city, though he would travel back and forth from his new workspace in Venice. For twenty-five years too much away he had stared back up into his hills more seeable from here to him than they were to those who were not from them. His imagined point of view, recorded in his paintings, would be among the earliest landscape settings for the Madonna and child. Into that landscape reproduced in paint he set also San Giovanni baptizing the child now adult and immersed in details apparitional, remembered, and built from, now that it was left to the artist only to climb to the top of the other city's San Marco tower, wonder and longing as he gazed north across the flatlands the best he could in the direction of the very much missed city from which he came.

Though today he is no less a figure of Conegliano than Cima is, Carpenè is not of here. He was born east across the Livenza (river that now generally takes the feminine article *la* but is sometimes indicated by masculine *il*) in Brugnera di Sacile (counted then in the Terraferma province of Udine and now as two places in Friuli's Pordenone) in 1838. Months later, his father moved the family to northernmost Veneto's Belluno and then in 1846 across the Piave winding immediately south, to arrive here and to stay. On the inside cover page of his pharmaceutical thesis, the first word of the official, printed line "Di[42] Conegliano Provincia di Treviso" following his name is crossed out in recognizably steeled black ink and whoever has done this has added an entire line above the original, now redirected: "Carpenè Antonio / *nato a Brugnera Prov. di Udine / Dom.*[43] *a* Conegliano Provincia di Treviso." Cima, in his renditions of the divine, has painted grapevines well rooted and reaching outward laden with greencast yellow grapes which now that there is ampelography appear to be Bianchetta Trevigiana,[44] knowingly trained on overhead pergola and which can still be found here and with bright-leafed Verdisos stand out in the old vineyards of mixed plantings against Prosecco.[45] From ancient agricultural and pastoral river-rich Godega Sant'Urbano (production center of wine, corn, and silkworm cocoons; middle, southwest to northeast, point between Sacile and Conegliano, equidistant northwest to Vittorio, on the Conegliano-Pordenone[46] postal road) abbot Benedetti was another point upon which the redirected winemaking of the hills converged: in 1866, he had organized one hundred men against Austrian advancement, then followed up Italian victory with expositions and infrastructure for agriculture, from mulberries to ampelography. Armed with him, notes historian Mario Ulliana,[47] within Treviso the city of Conegliano was at the avanguardia. "Carpenè and the scientists were avanguardisti,"[48] echoes Gasparini in a different conversation.[49] Conegliano "is the most opportune of the Province for its central location among the hills which produce the best grapes,"[50] positioned politician and rural economist Antonio Caccianiga in 1874. In notation of the scope of the Enological association Benedetti and Carpenè have founded, it has a technical director, and that is Carpenè.

Without a Conegliano past, Carpenè is instead looking straight into its future. Technical and concerned with land, its people, their raw materials, he is, notes the 1932-launched Italian encyclopedia Treccani, in 1870s and '80s a *precursor*[51] of the early twentieth-century practice of cattedra ambulante, pervasive *ambulatory professorships* that led the chemists, the agronomists, and their lessons to the contadini, and which he saw as not a solution but an emergency measure until there was a school.[52] His is a long-sightedness so enriched by action that it makes him a catalyst for what in one hundred years Prosecco will at last be: in Conegliano he is defining a wine not yet possible to exist but that is, only see, natural outcome of the intersection of these hills and Italy. *Catalyst*, formulating ideas destined for complete confirmation in the future: that the word, substance which, echo, "when present in small amounts increases the rate of a chemical reaction or process but which is chemically unchanged by the reaction,"[53] corresponds to processes observed by chemistry, the science of change, is no surprise. As is not the initial, avanguardista, role it holds: a catalyst is a "substance present in a preliminary stage of a reaction."[54] Turn the focus to wine, and find a world of precursors: in glass, flavors have arrived from somewhere, fermentation a catalyst (as are the wooden vessels and in-bottling sparkling Prosecco is seeking to avoid) for other chemical transformations, from what was already there, creating or releasing flavors from non-expressive precursors in grapes where they had been kept quiet by the sugars they were tied to. What substances from the hills, to the grapes, to what you could taste, quench, drown, mourn, celebrate, forget, plan, and organize in glass? Carpenè's science does not yet know chemical analyses for flavors, but assumptions are already there. Did he wonder, how do we measure the taste of these hills? These grapes? Work for wine which tastes enough the same each year? As chemistry's international community and language of forms generalizes and unites, Conegliano is changing now from agriculture's center into the larger zone's emotional, cultural, and technical core, retaining in deference senses of a "stillness indebted to a rural spirit engulfed in green."[55]

Walk out of the city and into something of the greater paesaggio. Unlike *landscape* as the larger area in which you find yourself, *paesaggio* is specific both to point of view ("a part of territory that is embraced by the gaze from a determined point"[56]) and to—with that root of Latin *pagensis* as the adjective form of *paese* and therefore location as a concept that excludes the observer in all but the occasion on which it is mentioned—the practice of a quietly assumed usefulness. It was not something one took in, strolled, and preserved as is. It was terrain for and made of growing, for and of feeding livestock, wherein grew the trees and plants that are necessary to a life, and across which goods, firewood, and wine were carried from where they had been made. That landscape is made by humans as much as by nonhumans is at its most undeniable here. The entire zone is as much monument as it is landscape, its declaration not backdrop but an active work, whose formation one might watch for seeing "humans joining other living beings in shaping worlds."[57] Even the city dwellers, though they might not need to turn to it, knew the paesaggio's function, usefulness and determining right back. There as in the city the church bells interrupt you everywhere. "Our Province has an extremely rich distinct viability," Carpenè wrote in 1873. In the second half of the ninteenth century, the city of these rivers, trees, and hills had begun to murmur in its own words. Then, half of its inhabitants did not read or write, notes Caniato, despite Conegliano's great number of elementary schools. Carpenè walks in the manner of a city, chemist and creator of wines, his movements useful, to set off thinking and for, observing, recording and making use of; why else would the Società choose its Stabilimento within reach of Conegliano's public strolling places? The province would represent Italy through a concentration on these hills that was possible only from this city. Before the Società was founded, notes Caccianiga six years after it was, "the wines of the province were unknown." And in prophetic warning and guidance: "Abundance becomes good fortune or danger according to human intelligence and activity. For the quality of its terrain, Treviso could become Italy's Bordò[58] and Conegliano has the virtue of being the first to see it."[59]

The writings on the bad state of Veneto's wines then are numerosissimi. Searches for an explanation most often follow a systemic line from the end of Venice and its weakening throughout the decades just before, as sadly[60] its agrarian publications and many academies end, too: conquering France sets the better-known and -priced wines of Piemonte and Modena against other offerings from Italy's north including those of Venice as, like with grains, the zone's wines suffer from the loss of the republic's protectionism.[61] Agrarian studies are economic and rational ones, and they are also moral. A 1771 session of the Pubblica Accademia di Conegliano—three years earlier its members had added Science and Agriculture to Belle Lettere—worried about "in which way are the habits of cultivating in our hills defective, and which is the most acconcia[62] and least wasteful manner to correct them, and so have a more abundant gain?" The following year, that academy reported on what its worries bore: "a certain grade of impurity and therefore durability of the sweet wines of our Hills derives from the quality of their places in some, and from the layers of Terrain, or instead from the unskillfulness in production and which in this case is the way of rendering them more pure and lasting for the objectives of their greater healthfulness and of more profitable and secure commerce."[63] During those difficult two centuries before unification, "the interest the cultured classes took in agricultural problems found expression,"[64] writes Ulliana, "in themes adopted within its many Academies." Starting in the 1500s, such forums appear and disappear throughout the Italian lands. Dozens are registered in Venice's lands: "in Ceneda the Vigilanti, in Treviso the Solleciti, the Perseveranti, the Infaticabili, the Ingenui," a list that reads also as the times' desired traits for researchers or scholars: *vigilant*, *solicitous*, *persevering*, *tireless*, and *guileless*. Conegliano, too has nineteenth-century academies. One with a name lost to history, though written records of its sessions remain. Then a Collegio of the Incamminati—it is correct enough to translate this as *walkers setting off*—and ten years later in the new century, Sbarra's Accademia degli Aspiranti (as member of which Francesco Maria Malvolti wrote of Prosecco). Avant-gardist, Carpenè is also the carrier of older arguments:[65] with its academic past still in mind, Conegliano's task was now to teach an agrarian future, and through that to teach a little mathematics and design and literature as well. Throughout, its roads are flanked not just by trees but by full greenery, color of cultivated plots and forest that were as utilized, measured, as official as the city's named and renamed streets. So what of a city whose culture is agrarian? And of its noble wines that made well again and future might make the paesaggio of which it is both center and edges anew? In questa Piazza[66] where the grapes were sold, Carpenè shouted. Words of the rows in the specialized, grapevine-only vineyards the scientists and experimenters and believers hoped to see. Pushed out loud to them and in writing to ministers and scientists for the building of, right here, the nation's first school of enology and viticulture. "It is indisputable that after the acquisition of independence, the majority of agricoltori, even in the Venetian region, want progress," Carpenè wrote in 1873 to the Ministero d'Agricoltura Industria e Commerco in a letter published publicly[67] as "On the conditions of the wine industry of the Veneto," and followed by "On the necessity of the institution of practical-theoretical viticultural-vinicultural schools,"[68] in 1874 to the editor of the *Gazzetta di Venezia* as 1875's First Italian Enological Congress in Piemonte at which he would help argue for the school's creation drew near. "The Comizj Agrarj, the Expositions, the conferences, the congresses, the press and so on help ["ajutano," he wrote in the first one (offered at the Congress as indispensable reading for the cause[69]), the *j* a Venetian aspect where the Italian *i* would be] the agricoltore to perfect the art of the campi,"[70] Conegliano's most common agricultural size—the hills along which those small and measured sites (still today in this enormously successful and densely planted zone most growers own fewer than three hectares) are set as constant feature in the imagination of their growers and their scientists. The 1892 edition of the *Annuario d'Italia*, a 1,800-page compilation that each year presented the places, materials, customs, and measurements the new nation was made of,[71] noted that Conegliano has grains, mulberry trees, and an abundance of grapes that gave highly prized wines. Commerce, its compilers wrote, and industry are thriving, led by silk making and the production of spumanti wines, "known and appreciated in all of Italy and in the principal cities abroad." There is an important factory for straw casings of bottles and distilling machines. And "several years ago was opened a school of viticulture and of enologia, which is very well attended and gives excellent results." "Italy, the land of wine, does not have an Institute like the one that other more or less wine-producing nations have," Carpenè had cajoled seventeen years before. "We are asking for it not only because we need it but also for the dignity of the nation." If the request is granted, he had promised in a letter to the province's leaders, the school's grounds will include a depot of viticolo-enotechniche machines, a meteorological observatory dedicated to agriculture, "and if a way to publish a special scientific-practical periodical is not found, we will at least provide for publishing the most important results through an Annuario."[72]

Outside the osterie, onto the crowded open street, *Giuseppe Garibaldi / il giorno 5 marzo 1867 / da questo verone al popolo / parlava*. These words in marble on a plaque affixed to the building,[73] or more likely at least one replication in postwar replacement of, from which he did. The dedication sits beneath a window ledge at the first story of a brick edifice rising two stories over rounded porticos against layers of peeling paint. Carved in prized stone on this plain building. Was whatever version of this building fresher then? Had it been painted for the occasion? The word used for where he stood and spoke from is a literary term for a small section projected from a building and parapet-bounded, and something closer to a little balcony juts out from the building to the right, larger windows set in stone and ornate frames of columns and arches, and though far shorter a third and crowning story. In New York streets in 2020 hundreds come mostly in support and some in protest, and I hear them before I see them. It is daytime, well-lit afternoon, and dozens of panicky and jumpy uniformed police on foot shadow-walk the marchers one parallel block away but they do it peacefully. The videos you see of police attacking are all happening at night. Perhaps for this time Garibaldi had chosen this plainer place to make a point but it is a political point only and he is not like Carpenè, the needs of the new chemists to make order of the innumerable atoms, to create a taxonomic scheme, to show off the intellectual aspects of their profession, to have a place for the growers, too, for the waters gathering to come, and on this day there is little in this piazza that is of use for the contadini in the hills. In 1867 Garibaldi is no longer carrying the dreams of Mazzini who will die in Firenze five years later and be immortalized by Carducci as "the last of the ancient Italians and the first of the modern ones" and whose last book Carpenè instead still holds, dusty and corner-bent, a gentle tear on the spine from where it rubs against the heavy seam in the pocket of his walking coat, deepening that separation. *The land, the soils of our Italy are sacred*, Carpenè had added by sloped hand into the margins of its pages while with linear precision he marked what he yearned to write into his own words: Italy is a Moral Law. La Patria. The Nation as One, he underlined. *Italians, we love what we have, what we are growing and what we are neglecting*, he might, too, have included. A *scientist can understand what we are doing and what we must do. If I know, I can begin to help what must be done and we will know together.*

Though this will not be known well enough to teach it until after the century ends, the higher acidity of a liquid injected with carbon dioxide is due to the carbonic acid formed under a pressure that causes ions of hydrogen to dissociate from the H_2O. This "system . . . initially at equilibrium with all species at low concentrations compared with water, and then subject to an instantaneous injection of gaseous carbon dioxide"[74] is a process of electrical charges and movement. Of suddenness and weighed equalities. Of, simply, balance. "A chi che no bef vin Dio ghe cioe anca l'aqua,"[75] warns a Venetian proverb included in a 2012 examination of Cecchinel's work, the examination itself beautifully titled something like *The Steep Word.* "Important for the city but also rather dangerous for its torrential character,"[76] the unstable waters of the Monticano were a constant threat to Conegliano, but they might also, rarely, slacken. The rains that are a feature of these hills replenish their many rivers and creeks, their small occasional lake, are as important as steepness to these wines' claims to greatness, but can also come as fluids torrential and impulsive, courses disobeying what has been built against them in stone and wood and lives. In the mid-twentieth century, three Monticano floods came, one per decade (in 1966 destroying a great part of what pieces at the Carpenè Malvolti winery had survived the destruction from the two world wars). In 1868, the river's waters had been increasingly lessening: it turned only two water mills along its run from Vittorio to Godega Sant'Urbano's nourishing Livenza. Four such mulini sat alimented along the nearby Crevada as it ran into the Piave. But in the city, a river is three things, passages from and into hills, movement and power determining the land as that is used by its people. Containment by promenade and bridges, rivers had interior city uses, too: those housewives washing, sheep and then the stiller cows drinking, currents that turn the water wheels of the battiferri and cornmeal-giving mills. A river is a sight seen from those promenades and bridges that follow it along, skipping over it at times in steel and concrete sidelines along which to walk at slow pace, for entertainment and, though it is a guiding and directional place of precise address, too, without a precise end. In the warm weather of 2020 I know that the river is just a block away and that I used to run there. I know it is there, too, by the steady stream of people walking to and from it and to get to the streets emptied by the pandemic you'd have to walk two blocks east, against the torrent of people coming and going downtown along the Hudson. Ambulance sirens all the time, but muffled from blocks away and only once did I see one stop close to here. The river pathways are too crowded for me to run and already it's too much to take, who is being hurt the most, who we are losing forever, fast compressing deadly consequences. "Drio 'l sas rucià su la lasta slisada," wrote Cecchinel in the dialect he listened to around him then used his poems to replicate its sounds: "andé che gnesuni pi romài riva / . . . de là de i càrpen e i rore spauridi."[77] Into local lines, "e la so botegheta drio la piazha / coi color che lustréa come ferai," Pieve di Soligo poet Andrea Zanzotto, who died in Conegliano (and in a 1972 Rai television documentary[78] spoke with its star, Giuseppe Mazzotti, Treviso's chronicler, essayist on local art and cultures and mountaineering, along with singer Toti Dal Monti and legendary twentieth-century Marca Trevigiana oste Walter Balliana about Prosecco), mixed quotable Italian: "'Rivendita carbone e giornali': / negra pedo de 'n camin, co' igrun / de garbon, e in medo—che canpanele a sonai!" as church bells reached outside the city and into the stalls that come tender nighttime would with wine and stories provide revitalization and escape. "Ma su le òlte galive e le stale," wrote Cecchinel of the ones near him, "de quela nòt tèndra de Tóvena / veja débole tante àgreme dale / da su sora i pra jerti de Nàdega."[79] Planted in the hills, Prosecco cultivars find advantage there in part because they are a grapevine that is particularly in need of reliable water.

In July 2021, I walked up the steep and stepped Cal Zetera road and into the Valdobbiadenese forests of once-managed chestnut and oak and carpino. Today the paths look like forest in the ways we imagine the woods to be a thing of nature but look closely: beneath uncontrolled and shaky softly stemmed patches of low-reaching green and thin-trunked crowded trees are structures people here abandoned not so long ago. Of wood and stone there are stairs embedded into ground, and you can still use them on sharply rising roads that, now that walking is for pleasure here, can easily be mistaken for intended hiking trails. Here and there are the more deeply stacked terraced embankments built

Just past the Monticano river in Conegliano, the Carpenè Malvolti winery is what its co-founder started here in the Società Enologica di Treviso.
Susan H. Gordon

of only soil and grass: the zone's traditional ciglioni that hold vineyards stable against those steep, enriching positions. This trail is from 1880; it once curved off. By now they will be ruins, but then roads official and navigable only on foot made up the forests, too, and those who used them grew and used and maintained the trees and the greens and herbs that could grow around them. From these utilized hills one might walk into and from the city. These are ways that, only see and be ambitious, give Conegliano its strength and reason. "It would be like bringing water to the sea to want to talk to you, dear Signor Carpenè, of the utmost usefulness for Veneto's provinces to found and begin a similar school,"[80] wrote his colleague Giovanni Battista Cerletti shortly before that path would be constructed, too. On 9 July 1876, at last a royal decree created the Scuola di Viticoltura e d'Enologia—in Conegliano. For its first year, Carpenè and Cerletti will lead it. It will be supported by the Venetian provinces, the comune of Conegliano, and the nation (respectively, in 1893: 12,000 Lire from Treviso, 1,500 from Vicenza, 480 from Rovigo, 450 from Belluno, 600 from Udine; 3,320; 27670.50)[81] to cover in addition to equipment and salaries, in 1879 a four-thousand-lire annual rent for the school, cellar, and adjoining land of at least five hectares. "The merit of having Conegliano be the site of the nation's first school goes to Carpenè and his communication to the State of the conditions of the wine industry in Veneto," wrote Caccianiga. "He was convinced that Conegliano, more than any other center of Italy, deserved the fortune of hosting an enological school, because—he said—'it had been favored by nature for its joyful position, for its climate and terrain especially fruitful to the grapevine.'" From his tours in the hills, the website of Carpenè Malvolti proudly shares, was born the relationship of trust and esteem that still binds those "Viticulturists and the Family" today. "You would be doing something extremely useful by researching if among the spoilage ferments of wine and beer there are any upon which sulfuric acid has no effect at all," Pasteur had ended his reply to Carpenè's wondering about the compound's derivatives the year the school is founded. That same year, in Italy arrive also the telephone; the phonograph; a machine for writing; four-stroke engines; chain-driven bicycles; widespread availability of the portable ink-powered steel pen; and the *Corriere della Sera* which in the early twentieth century will become the nation's first daily newspaper, thrive in the fascist regime, reshape after World War II, and is still, in its new form, one of the nation's main newspapers today.

What instruction was needed shaped two distinct courses at the school—one in echo of the accademie to draw students who had a strong liberal education and wanted to add scientific and technical knowledge; the other, this time, for the sons of small growers, who wanted to with that knowledge's fundamentals learn to grow quality wine grapes. Both courses were for shaping winemaking into a new, professional, field, its practitioners into a new community: viticulture and enology (complete with chimico-fisiologico laboratory ready to either teach from scratch or complement what was learned elsewhere) were theory which could be worked out by practical instruction in vineyards and cellars, be raised by classes in literature, mathematics, drawing instruction, natural sciences. How to earn a living and by what paths to reach a larger and farther-away market needed to be addressed, too. In 1882, an ancient Greek problem known as squaring a circle (given a circle, with a compass and straight edge construct a square with the exact same area within a finite number of moves) and long thought to be undoable would at last be proven so by writing the two shapes into numerical form: the π that defined a circle's area was shown to be a transcendental[82] number and therefore could never be a solution[83] to any polynomial with rational coefficients as obviously the area of a square must always be. Even unresolved. the stated problem had been a metaphor for an impossible task. "The problem of obtaining high earnings from capital invested in the earth is one of the most arduous," Vianello had written ten years earlier in articles that summed up his and Carpenè's work to win over propertied contadini by way of an experiment they called "Two professors stipended by six hectares of vineyard."[84] Together they had built two small vineyards, six hectares, in Spercenigo in the Biancade flatlands just below the province's capital and homonymous city Treviso: thirteen vineyards, 32,925 vines[85] for trying out how a wine-grower might make good wine and earn a living. "And for me, though it was not as insoluble as the squaring of a circle," he lamented with an edge of hope, "it was however far more arduous than usual."[86]

The knowing from which to produce Italy's wines must come from thinking of its cultivated lands. Understand the forests and fields, the masserie and chiusure and the small pastures and shelters on the hills, and from Conegliano into them you can, though it requires more imagination today, still walk. The six hectares of the two professors' studii sperimentali are part of a twenty-six-hectare holding with woods, so that the owner can provide himself the pali and vimini a specialized vineyard requires. Softly linear leaves of the willow trees, which must grow near water, vimini were common ties for artfully training vines. Now they might do so onto lines drawn along stakes fashioned from trees growing on the farmer's property, dry wood driven into the ground but that no longer competed with the nutritional needs of vines that once more easily if less deliciously and lasting draped on trees rooted into the same earth. In this way the vines could be planted in plots of land, alone, with boundaries that protected the plants' ability to provide quality grapes. The cultivation of viti a vigna[87] is a remapping, and a new way to measure out the land. Within it, a vineyard is a re-walking. Vianello's article includes careful yields of wine, affirmedly necessary gelsi[88] for the local silk industry's sorted counting of wholesaled bozzoli, the laborious *Bombyx mori* moth larvae's silkspun cocoons.[89] He offers a new word—credited to France, he writes—for the human worker of this new place, name of the traveler over this new map: who will tend a vigna's grapevines is a term the wine trade uses today as if reclaiming though what we are closer to is staking it—*vignaiuolo*,[90] and, presumably, *vignaiuola*. Are the scientists the ones to bring the word to Italy? Or has he missed some language from Italy's vineyards whether promiscuous or specialized? The place can now be talked about in more universal terms, too: in the article is an informal table credited to Carpenè and showing the "mechanical analysis" of the texture of the land, in the process recasting it as *terrain*—about 70 percent sand (composition mostly siliceous-clayey and about 15 percent calcareous), the rest clay. A chemical analysis (even in this rationalized walkthrough, Carpenè's glee is evident in the many-worded list he makes) yields carbonates, acids, an oxide, substances "not attacked by phosphoric, sulfuric, chloric acids," organic materials and water, while the chemical breakdown of the caranto (what the Venetian lagoon is made of; an agglomeration of clay and silt that is consolidated only in contact with the atmosphere) is mostly calcium carbonate, then the same unattacked substances, rusted iron from ancient volcanoes. Unspoken here are the rivers and the gradual slope of ground, ending curves, the lines of this city to its hills, at their basin the grand bank building that is coming soon, and the school which will open in four years. "Therefore six hectares of vineyard can pay a professor of agrarian studies and an enologist! And some despair of the economic future of our nation!"[91] Vianello concludes in abundant optimism. "The little owner must be convinced of this, is the vehicle for our nation, for our identity." Conegliano's winemaking school will be the first of the nation's many to come. Those who entered its theoretical course with sufficient instruction in literature and natural sciences and little interest in the practical part could finish the professional-technical course in one year, and go on to teach and guide the new wineries sure to follow. For the practical students concerned with wine's raw materials, the head growers of the vines, *capo vignaiolo* offers the school's proposal, would be paid a part of what they grew.

Walk, with the municipio at your back along the old city's main street shadowing viale Mazzini below, and three minutes later pass the city gates once named for the little Ruio river which runs past the Jewish cemetery across it less than one kilometer away. "After 960 steps outside Conegliano, accompanied by the little Torrent Rujo," reads the Kriegskarte report on the pathways of what it sectioned off as *Conegliano*. Keep going and the Ruscello Ferrera will unite with the torrent; turn off and at 1,020 steps see the Torrente Crevada, very dangerous when there is abundant rain.[92] The year before much of the old Serenissima became a part of Italy—when the Porta Dante, two new brief towers a street's width apart, replaced a section of the city's ancient wall in six-hundred-year celebration of the Italian poet—that Trevisian assuredness was already there in stone, further reinforced by Conegliano's own ancient loyalty to Venice and republic. Immediately through the Porta is Conegliano's old ghetto, the only place as Venice had required since the 1600s the city's ancient Jewish population[93] was allowed to live and work—until the neighborhood and city were made contiguous in 1797, outside and inside united. On this corner of the tiny quarter there is another apologetic plaque, dated 1998: *To the comunità ebraica that in these contrade—for centuries could live and work faithful to the tradition of the fathers. Dedicated, in perpetual memory, by the city of Conegliano.* A few steps from here was the school's first site, in an old rectory complete with vegetable garden and whose rent was paid by the comune of Conegliano, scientific knowledge of wine backed by the faith of its fewer than eight thousand residents.[94] It is a building that already existed, and in a sense still does: remember that as each world war destroyed large pieces of this borderland city, shattered buildings were replaced by new versions of what they had looked like before the bombs hit. In the first days of 1877, its repurposed doors opened to the institute's first thirty-five students, welcoming them into rooms and halls in which the hills might find respite and careful attention, a within the school equal somehow to within the city. Carpenè is president of its administrative board, and he is also noted as representative of the Ministero. Per the technical drawings of the school's expansion and remodeling published in 1878 in its new dedicated *Rivista di Viticoltura ed Enologia Italiana*,[95] bounded by the via del Ghetto and "colline in considerevole pendenza"[96] the R. Scuola di Viticoltura e d'Enologia is a composition upon three floors. A meteorological station is adjacent to the building, part of the school's footprint as is the small experimental vineyard climbing those mapped hills. Both this floor plan and a 1902 postcard locate the vineyard from a white-pebbly pathway that begins at the school's doors then soon transposes into trellised way up into the rows of grapevines. From the beginning the observatory of agrarian meteorology measured weather that mattered to both plants and scientists. Every day, the data it provided was collected and posted in the school's hall so that students would develop "a precise concept of evaluating the different factors of climatology."[97]

Steps inside from city streets, chemistry and physics together claimed their own amphitheater, theory to the practical one next door where agronomy, viticulture, and enology would hold their own presentations to the students. Close by, a *tinaia*—tuscan and academic word for the room in which wines held in wooden tini might be changed, matured, raised into ongoing greatness—and an experimental wine cellar side by side; classrooms for natural sciences and mathematics, laboratories dedicated to the divine microscope. On the first floor up, libraries, reading rooms, chemical labs, a professor's studio, and a museum of agronomy—until now a field mostly just longed-for—to cast the new enterprises of nation and Conegliano's wines as historically and scientifically legitimate. In that vertical movement that even a small city claims, both kitchen and laboratory awaited their practitioners on top floors, which here also held the storeroom of chemistry materials; below, classrooms into which spring's blue-yellow sun rays or gray-yellow light of falling snow might pool over desks and paper, "the softly spiritual light of the Veneto's hills at dusk,"[98] and soon there would be electric lights in the short evenings; the living quarters of the dedicated director. The hills could be momentarily interrupted here: the school's collections of minerals, models of crystals, botanical collection, samples of terrains, birds, insects both useful and harmful in the vineyard[99] allowed rational teaching of observation skills and created on the inside a fitful, partial, reading of the out.

How the students move about the school is determined by those two courses of study. The first one has solidified into the *Superiore* program that will prepare students—intended young people with a degree of maturity and "good general culture" who want an immersion into scientific and technical knowledge—to become directors of rationally organized wineries. Soon after the school opens, chemistry both "general" and "applied" and the study of Italian, French, and German "foreign languages" will supplement, along with drafting instruction, a wider focus on natural science and physical chemistry. A good part of the budget, fed mostly by the tuition of forty lire for each of the course's three years, difficult by almost every standard so that several of these pupils are offered admission at half the cost, went to the laboratories of chemistry and microscopes, "intended to teach the students about materials" and that would then allow them to continue such research even after graduation. (Steeped as they were in the study of ferment, some will go on to become directors of the latterie; the Marca's first one will come in 1882.) The second course is now officially the *Inferiore* course. It will require a little more than a year and offer an education that is more hands-on, aimed instead at those more "modestly prepared, children of small agricoltori," who wanted training for work "that was above all practical."[100] To *these* students went other instruments, termometro, gleucometro, acidimetro, an alambicco for determining how much alcohol a wine held dissolved in it. Many of the pupils in both courses stayed at the school, some kept at the expense of twenty lire annual room and board paid for by wealthy families or "moral"[101] entities—some, reads an 1880 report, can be twelve to fourteen years old if they can read and write "correctly"; preferred are the figli of coltivatori and small landowners[102] and of the eleven such students enrolled in 1879, eight came from contadini families, three from business owners, and three from "parents in other conditions." The upper course, notes the report, was supported for example by provincial funds from Rovigo, Avellino, Catanzaro, and Istria, an Armenian, a member of the local noble Collalto family. While, "the students of the lower course are largely sent and maintained at the expense of landowners who count on making use of their employees later on; thus 3 sons of settlers are maintained at the School by Mr. Count Collalto, 2 by the Honor. Counts Papadopoli brothers."[103] Students of the second course who did not room at the school, could be admitted without tuition, and were paid fifty lire a year for work in the vineyard or the cantina.[104] "The moral, and with time material, advantages that the Provincia di Treviso," will earn through financing the school should be push enough toward its success, Carpenè had written in 1875. "Everyone understands that the Institute will not be limited to a Comune or a Provincia, but will embrace the Venetian Provinces and the neighboring ones and for some studies will pass even those borders."[105] In the 1879–1880 school year, an increase in students shows that widening reach: more from Treviso, more from other Venetian provinces, from Lombardia, Piemonte, central and southern Italy, and, from abroad, Austrian, Swiss, Armenian provinces. That year, the teorico-pratico course included: "the Italian language, arithmetic, basics of accounting, history, and geography, and more widely and emphasized, basics of botany, zoology, geology, meteorology, of chemistry and of physics, insofar as they refer to viticulture and vinification."[106] In a few years, the report looked forward to, "when the effects of mandatory schooling" take hold, the general classes could be replaced by more developed specialization. The region, Carpenè advocated, where Italy's first such school will rise will receive the greatest part of its benefits and the "slow agricultural revolution that will follow it, will have as its ultimate goal to ensure and grow, if not create, a conspicuous source of wealth."[107]

In a rented podere of 6.94 hectares a kilometer and a half away, at the same time the experimentally minded school added "Pinot Franc of Burgundy, Cabernet of Bordeaux, Gamay of Beaujolais, Raboso veronese, Riesling italiaco, Lambrusco di Montova" to a little local Verdiso already there and trained on pali secchi[108] in the "spalliera" way, akin what was often called by its French term Guyot in reference to the man who had recently popularized it—a vine's trunk grown vertically some fair distance from the earth so that the year's bud-carrying canes could be positioned roughly perpendicular to it, keeping the summer's fruit out of reach of earth's humidity, tied along the lines drawn from stake to stake—as one of the zone's vigneto a raggi. Carpenè's Società Enologica has lent its own 1.3-hectare vineyard, where cultivars of the Venetian provinces including red Refosco and Raboso share space with golden Prosecco, for observation and student practice and will let the students use their winemaking cellar, fitted to the new industry ideals, too. "From Conegliano," Carpenè defended it against those who did not believe, "like raggi di una ruota the beneficial lights of science applied to the most beautiful and lucrative among the products of the earth will spread evenly throughout the Veneto Region and beyond."[109] Though a workshop where the botti were made is an initial part of the school, it will too quickly grow into an important part of the local economy and be handed over instead to Fratelli De Carli fu Pietro (a distributer also of lime for use in vineyards against the dampness-loving peronospora microbe), which will remain an apprentice center for the school. (The tools used by the area's ancient coopers "are the same still used today in Conegliano," writes history-minded Trevisian chef Giuseppe Maffioli, "whose factories are perhaps the first in the world, for the quality of the product, total production capacity and production range, with barrels from 1 to 150,000 liters. In Conegliano, there are also barrels for family use, 10-15-25-40-50-100-150 liters upwards."[110]) "You young pupils," Carpenè had welcomed the future that first day: "I need only to remind you that if your fathers knew to fully reach the purpose of the social life of their generation which was to render the patria free and independent, to you is entrusted a purpose no less lofty and noble to develop one of the most promising sources of national wealth, which is that of attending to the economic future of our paese which will make us as strong on the inside as on the outside."[111] The second location will be built from 1921 to 1924 after the destruction by World War I. It will be farther afield, adjacent to that rented plot for growing vines and seeing what happened on the steepening hills Carpenè had long been pointing to and within strolling distance from the panificio on whose outside wall was painted its latin name, *pistoria*, oven for baking bread, on viale Umberto 1[112] (though that name will be changed again) where the city walkers gathered, talked, and stretched and breathed deeply, idling and often empty-handed. On this second and current site, Italy's first winemaking school will include wine sold in bottles and by glass in the small and joyous hall of the new Bottega del Vino, all construction, decorative details, and botti by Conegliano makers, all glassware from Venice's esteemed Murano, and where on the school's fiftieth anniversary in 1927 a celebratory list of the nation's best fine wines, chosen by a Conegliano jury, offers of the hills of Conegliano prized Prosecco and Verdiso.[113] Here one could meet once more for the wines and songs that make it into something like an osteria, inside on its walls painted, in Italian, *If there were not wine, what would become of you, what would become of us*, Italy's word on this new making of its most ancient industry.

Center of a land poor and stressed but that bent sharply toward the sun, Conegliano was among Italy's proudest cities. When the school is opened, already one could arrive here conveniently not just by railroad. "Seventy years ago,"[114] Carpenè had written in 1873, "the roads that reached the hills were a continual succession of arduous ascents and descents," and "you did not need to be very old to remember that in many, many parts of the plain the best way to get oneself from one village to another was to go by one's own legs, jumping ditches, to pass from one countryside, *paese*, to the other, in order to dodge the so-called streets, in which a pair of strong oxen were sufficiently loaded by an empty cart." But now, he promised, you could travel "the Province comfortably in carriage in all directions. The smallest village is not without well-built and graveled roads for reaching neighbors; and now they're even building just to facilitate access to the countrysides." One thousand eight hundred and fifty-four linear kilometers of "very good streets over 2,325 square kilometers of surface," 1,590 of which built and maintained for the most part by the comuni through which they ran. Well-equipped his province was, too, in workers: 149 inhabitants per square kilometer—"almost as much as Belgium, more than many of the low countries, more than 2/5 more than the Italian average." If Italy had Treviso's density, he notes, the new nation would have ten times the population it did. "The famous French viticulturist Guyot finds that a bigger vinifera production in France is very difficult due to the scarcity of the population," he writes. "We definitely cannot complain of such a lack." These are numbers that in opposition to the Kriegskarte maps are stories of ties from and usefulness between inhabitants, surface, streets. Where they meet in the school, *viticulture* and *enologia* are for giving their winemaking its own slant within the Marca's general agricultural concerns. The two words are materials for an industry that never meant to leave the land, from whence wine drew quality and character via those who knew such places well. A century and a half later, the school's one defect, Gasparini says to me, is that it for years it turned out great enologists, "grandi tecnici who could make a divine wine from apples," but it "has not formed vignaioli, it has not formed contadini." As today's backlash to the twentieth century's great misstep of industrialization gathers speed, many Prosecco makers here think of their bottles as ways to tell the stories of their part among the hills. "Sure, you use what technology offers you," Gasparini adds. "But you must depart from this beast that is the grapevine, and every vine, every ground, has its own perfume." In 1866, already the hills' most expressive creatures were besieged by that oidio and peronospora, and phylloxera which had landed first in France in 1860 will be found in Treviso forty years later and, according to the agricultural minister's increasingly haggard lists[115] of infections by place, in every one of the province's ninety comuni by 1936.[116] In 1908, more of the city's streets will be renamed.

The city's Commissione per la nomenclatura delle pubbliche vie di Conegliano will add many of the great coneglianesi to the Italian names. Where the school first opened, the via del Ghetto is now renamed for eighteenth-century agrarian and literary member of the Accademia degli Aspiranti into via Pietro Caronelli.[117] Across the Monticano the street upon which the rebuilt winery still sits has been cast as via Antonio Carpenè. Both are smooth roads made of cleared brushed earth diluted with darkening gravel to hold wide pathways firm, for those gathering, for walkers and for horses and some bicycles, street foundations waiting for the electricity-based carriages that are on their way. Not very far away in the countryside and towns, as you still see today, there are roads also clearly marked by being made of larger and rough white stones, sharp answer to the hills' softening yellow light. No less a network for being a discontinuous one, they are a collection of local passages and not like railroads at all. They are small and close by, what need is there to name them, and the word for each of them unofficially and everywhere is a *strada bianca*. Outside in New York City's emptier streets, the air is bad: thousands of people overflow from our hospital rooms as everything in the city is shut down and at least thirty thousand people here will die in the first year of the disease that a coronavirus brings and which from the daily published lists of names and numbers we try to track what is happening and to who. People travel to work on crowded buses and subways which run much less frequently now that downtown has stopped using them. On June 8, 2020, the first phase of the city's reopening will bring every construction project back to life in the same moment, NYPD and tourist helicopters overlapping with building trucks' deep grind as if someone had opened the door to where all of it had been waiting, idling, that sudden and that total. The bodega that almost no one who lives or visits here goes into will reopen at that same moment for the returned construction workers. Near us store windows on the small streets that the real estate and tourist industry had taken are still boarded up and joyfully painted and written on in color, freed and giving for a while, quieted with hope and pride and a welcome sense of *finally*.

5

Prosecco's Machines

> Tra gli scopi principali della Stazione vi è lo studio ed il perfezionamento degli apparecchi e delle macchine enologiche.
>
> —Federico Martinotti, *Annuario della Regia Stazione Enologica, Asti*

> Vedar che se un povero diavolo va a robar un toco de legno i lo mete in preson.
>
> —Illuminato Checchini aka Paron Stefano Massarioto, *Lunare*

In 1881 at the international competition of apparatuses and enological machines[1] in Conegliano, the technology of Carpenè's system for spumantizzazione was awarded a gold medal.

His apparatus was equally method, a series of processes embodied—in wood, iron, copper, CO_2, base wine, cold and borders, a grand finale in glass and isobaric conditions—and well ordered. A set of steps in change or action, instruments, and outcomes, ordered by existence and level of sparkling, so that CO_2 is the underlying and therefore least element, from its creation at station one through to the final, spumantifying station 5 at which the wine enters the machine's steps (both action and change) for the first time. If uniquely among the sciences chemistry produces always both materials and knowledge, what knowledge of the materials by which Prosecco is produced? The golden apparatus has been immortalized in ink, only: the illustration with a written walk-through inside Carpenè's *Sunto.* In advertisements in the *Nuova Rassegna* (as the school's *Rivista* was recast in 1887 when Cerletti abandoned editorship of the journal that for ten years had been "the echo of [the] classrooms, the emanation of its studies, the manifestation of experiences in the laboratories, on the terrains, in the cellars; the certificate of life and of work in the institute,"[2] and moved to Rome as secretary of the new nationwide Società generale dei viticoltori italiani). Heftier, yellow-macerated, aromatic wine of families, sometimes the Prosecco (made to be still, yeasts' process, sugars, made to sparkle, air in the hills slowing into cooler temperatures then in the new year's first warm days, station 5, when poured salty carbonic acid decomposing as it left this cursory pressure of carbon dioxide dissolved in water) produced and drank by the growers who owned land, known and lauded locally, held indeterminate fizziness to tell about the hills' snowy winters: "now that the warm season is beginning, many vinificatori are nervous because a certain quantity of wine, especially white, of the past harvest is threatening to change again," Carpenè wrote in early 1872. "There are in fact some this year that did not undergo complete fermentation in the autumn[3] and that therefore remained dolcigni."[4] Today wines like this are made again, if by more controlled actions, softer and fuller and less explosive than spumante: they are instead, typology by level of CO_2, labeled *frizzante*, akin to *fizzy*. By the 1930s, as co-reported[5] by agronomist Giovanni Dalmasso[6] (director in 1925 of, experimentation bedrock of the new winemaking, search and search again, the research-centered Stazione sperimentale per la viticoltura opened adjacent to the school's second site in 1923 and which is now the Conegliano seat of the national CREA[7]), this style would be a robust, hyperlocal on purpose wine specifically of the Valdobbiadene hills, 10 percent of whose productions were called *amabile* and made of Prosecco grapes allowed to dry, develop, concentrate, and sweeten further[8] before pressing. "The amabile typology is generally sold in the month of March in bottles or damigiane," wrote Dalmasso and Giuseppe Dell'Olio. "Subsequently, part of the residual sugar enters a slow fermentation producing, among other things, carbon dioxide, which will give the frizzante, and sometime spumeggiante, character to this wine."[9] To make it, the must is placed into "botte-tini" when the weather has already turned a little cold, "and therefore the fermentation proceeds with difficulty. The wine is always poured off[10] the solid parts of the must before the end of fermentation, with a little residual sugar left of 6–8%." While the still, dry[11] typologies were of must fermented to the end of its sugars, and the wine poured off (movement akin to decanting a bottle if for other reasons) the grape solids up to three times to remove any chance of stubborn sugar or yeasts, the wine made to be sweetish-frizzante was poured off only once at winter's end ("Siamo all'epoca dei travasi,"[12] Carpenè had introduced in 1878), then often followed by a new technology—Dutch burlap sacks designed as gross-gauge filters to remove the threatening sugar and yeast particles inherent to amabile frizzante wines. Important consequence of grape sugars and cold winters and an equally specific chemical outcome—and technique by the early twentieth century—they were nevertheless a far less technical feat than Carpenè's quest after an elated Prosecco, a wine of deeply embedded anidride carbonica and each liter of which is today composed of 11 percent alcohol, 4.5

grams of total acidity, fresh flavors created during fermentations, a wine tensed with a defined amount of pressure, harmonic, pleasantly fruity, characteristic of these hills.[13] There are two types of spumante wines, "which are wines oversaturated with carbonic acid gas," professor F. A. Sannino wrote in 1907 as a new production option between slow container rest and quick CO_2 addition was stirring into being: "natural spumanti and artificial spumanti." The first are those that count on fermentation, of the grape must like the contadini's or of the base wine like Champagne's, for their gas, "which develops in the vessel made for the preparation, conservation, and transport *(bottle)*," the second on introducing the gas, obtained any other way, into the wines "by means of special apparatuses."[14] The solution to sparkling Prosecco would turn out to be an apparatus and accompanying method that could produce a natural spumante deferring to the hills' delicate grapes—by reimagining that vessel.

"All machinery is derived from nature,"[15] claimed Roman engineer Vitruvius in the first century BC. To those looking in the second half of the nineteenth century, the hills' Prosecco grapes gave hints toward the technologies that would make them most themselves. At expositions like the wines of Venezia in 1887, the new tecnici mixed with the wine-obsessed noblemen[16] many of whom were scientists and winemakers themselves, with experimental vineyards that counted on alliances with contadini growers as well as other scientists (in a premonition of the twenty-first-century mix of *digital* and *natural* to counter the industrialization coming in the twentieth century): look, too, to the partnership between the noble Malvolti family—researching quality among Prosecco types and growing techniques—and the distinguished scientist which included the growers reached in the piazza and the vineyards. Along with his sense of national mission, Carpenè had carried notions of materials. So that they might articulate aromatic profile across nuances of growing sites, extend historical and frizzante presence into technical skill and deliciousness, be an economic weapon against poverty and poor crops and nothing but polenta, transportable across the new boundaries between home pride and export, these grapes' rational form was spumante, and for realizing the machines that would best serve that winemaking, first came wood. For its initial decades, the spumante that Carpenè Malvolti made and labeled with the revolutionary and prototypical *Champagne italiano* was of wine fermented in vats built of staves (Trevisian barrels mostly of chestnut[17] trees, Carpenè and Vianello wrote in *La Vite*, and oak). Then, along his apparatus heated gasometers cast in iron created acido carbonico, to be run by a valve under control of the winemaker's hand into wooden cooling tanks to force the bubble-making gas into that base wine and, tightly fitting lids, keep it there. "Artificial, or saturation, spumanti improve noticeably if made with wines that before saturation have pleasing aromas such as those of the white wines of young and somewhat sweet Prosecco,"[18] wrote Sannino, and in Carpenè's apparatus, the motion of man corresponds to nature, also by process. In accordance with the motion of natural spumanti, the machine "has stirrers . . . to slowly stir the wine so that the gas is absorbed more easily. The shaft is set in motion by turning the handlebars during the time that the gas is passed from the gasometer to the boiler by means of the pump. No more than 45 plunger strokes per minute [and as] the gas is pumped into the wine, the stirrer is kept constantly in motion by means of the handle to keep the wine agitated so that the gas dissolves more easily."[19] The hills' grapes and common gas, the first materials were clear. The one that could give shape to the machines capable of making the promised wine did not yet exist, but soon. In the meantime, there was an iron will.

It is important to know this: *spumante* is a word about pressure. By European Union regulations, wine labeled this way must[20] at 68°F or 20°C degrees in a sealed container have least 3.5 atm,[21] the kind of power that yearns passionately for equilibrium, that can push bottles to explode. Imagine the tension of a solution of H_2O liquid and CO_2 gas with in suspension solid particles of color and flavor and acidity, against the air that surrounds an open bottle or a glass poured into—given an outside, the compromise is bubbles. To be made, contained, shipped, or seen sparkling, spumante requires vessels of fitting shape and strength. Frizzante wines linger at less than half the pressure, fewer than half the bubbles by number and far less pushback than spumante. While frizzante wines can be made to happen, they can also for the most part just happen. This is never true about spumante. Taking the utmost of what he knew these grapes were made of, how they turned to wine, how the wine might be transformed to sparkling, Carpenè made a wine that is a story about the closest technology he could make in anticipation of the wine he could not stop seeing. Or, it is a tale of the materials that he and the other chemists had at hand, which ones were coming into being in obedience to other needs but that might also serve other national demands. In his illustration, inside that vat built in staves of local trees to carry a refrigerant mixture of ice and water and a little salt at the final step of the Carpenè method, the cauldron labeled 4° that holds the base wine, then upon apparatus operation the new spumante, is made of copper, plated[22] on the inside, against bacteria and reactions like corrosion from contact with vinous substances but also with a mechanical concern in mind. Against a pressure allied with corrosive acidity and bacteria-friendly sugars, a winemaker needed a material that was strong, unbreakable, nonreactive, and from which could be made a vessel of continuous, unstaved surface. For these delicate wines such a material would also need to be free of flavors and aromas that might themselves take part. To serve a winemaking nation, even one whose uniting in such a mission might render new spumante-makers willing to share cost and use, it could not be very expensive. It was already known that most of these qualities were to be found among the metals, but what combinations would produce such a thing was just starting to be worked out. "In the Middle Ages," writes Harold Cobb in his history of the material which would soon arrive, "alchemists tried to turn lead into gold, but no thought had ever been given to the possibility of turning iron into a rustproof noble metal, nor were any rewards offered for such an accomplishment. Iron rusts, and that is the nature of it."[23]

In the tinaia,[24] wine's fermenting makes the forests around you come into clearer focus. At Colbertaldo in the southeastern folds of the Valdobbiadenese, "you can still walk through a small wood which was first mentioned in historical records around 600 years ago," Franco Adami—third generation of, with first stirrings in 1920, the historic Prosecco house Adami—told me,[25] "when it was known as the Bosco di Gica." Since 1775, in Conegliano the cooper Garbellotto Botti & Legnami has been making oak tini a few minutes' walk from the Monticano river, both low-stretching warehouse building and what's built inside adorned by their name in whimsical letters the color of new red wine that float upward as if while being brushed in ink. Garbellotto's solidity has always been of what they make, which is wood outfitted with today's other ubiquitous material in winemaking: fronting small doorways cut into the base of a "single highly prized central stave" that crowns botti circumscribing ten to five hundred hectoliters of wine via round or oval boundaries are flashes of the non-rusting iron-based stainless metal used to seal it; the gleam of a small reflective assaggiavino tap by which to dispense a slight amount of wine to taste as it ages within the wood; shiny, strong, and smooth valves onto which fit hoses for travasamento. Walk north along the Monticano for about a mile to see at Carpenè Malvolti many Garbellotti and, filled with aging brandy and grappa, one-hundred-hectoliter[26] botti by Garolla in the Padovan plains. Then follow a ten-mile line straight southwest along the line of the Prosecco-planted and softer Coneglianese hills formed by "clayey Miocene rocks which give the landscape its rolling character"[27]: walk through the reddish ferretti soils that are the final iron signals from long-ago volcanic times and over crumbly limestone conglomerate soils dragged and dropped by glaciers in the coldest days of our current ice age, down across the Piave river and up onto the long series of hills that like their forest are together named Montello.[28] There, across six thousand hectares 280 meters above sea level on a single and more rolling hill[29] than the wine-growing ones that sweep that left- and right-bank arc from its northeast to its immediate west, are the legendary oak trees to make Venice stand on water, the woods that gave Venice timber via the sacred Piave that crosses the Marca Trevigiana.[30] When Austrian occupation of the fallen Venetian Republic ended by two-day-long plebiscite in late[31] 1866, the bosco of Montello accounted for more than one-quarter of the entire Venetian forest owned by the state and slated for public use, writes Brunello, but it was also the site of ancient traditions and economic rights that had been steadily eroded for at least the past quarter-century.[32] As public land, the forest had for centuries' worth of changes belonged both formally and informally to those who lived within the comune it belonged to, the so-called comunisti who insisted that what belonged to the comune by reason and for centuries belonged to all those who lived there. From their forest in the nineteenth century came burnable twigs and branches, trunks of felled trees for timber and furniture, leaves and bright greenish-yellow straw for feeding livestock, all together the material basis not only of an ancient and robust subsistence but of a local economy just as old and that reached farther afield, to a twenty-mile radius[33] around the woods and farther still if one were to count the things,[34] domestic and for the osterie, the boscaioli made and sold throughout the Venetian provinces: in 1825 soldiers found Montello wood as far afield as Conegliano and everywhere along the road between there and the capital city of Treviso south of the forest. In 1848 after widespread famine and the fighting for independence, the contadini held their own revolution by entering the woods en masse and "resuming the [now unlawful] habit of gathering firewood for bundling and cutting grass to feed to their livestock."[35] In what was cleared for agriculture just outside, the incessant need for firewood slowed the implementation of pali secchi, Gasparini writes,[36] so the conversation turned instead to which kinds of trees to plant to uphold vines: for the zone collinare, it was decided, the walnut, the oak, the cherry. During the agricultural crisis of 1880 to 1890, almost a third of the area's 30,000 inhabitants "lived off the products of the forest, and were referred to as boscaioli [roughly, *of the forest*] or bisnenti, a term that was meant to signify *twice nothing*."[37] Throughout, the boscaioli insisted on their rights to Montello's wood and what made up its forest floor—in 1888, 4,453 contadini signed a petition of protest, each identified therein as a "Montello worker."[38] In at least one case some of the controversially illegal wood was found in the possession of a local oste.

Back across the Piave, in Serravalle now a part of Vittorio Veneto above Conegliano and in the easternmost Valdobbiadenese, entire families were arrested and charged with removing material from their own woods,[39] the Cansiglio forest of the high plateau to the DOCG's northeast in Treviso's north and border with the provinces of Belluno and Pordenone. In nearby Premaor and Valdobbiadene to the west, contadini and heads of family referred back to ancient documents issued over centuries by the Republic, which stated their rights and which were held, tightly guarded, in every village, and where the oldest inhabitants and the power of habit passed down—blessed is he—the memory of those rights. Written pleas to current powers likely supplied by the local priest or pharmacist[40] referred to precise decrees of past governments and were passed around to collect signatures or equally inked marks of those demanding protection of their "legal rights threatened by greedy innovators," rights they held as a people older than who else had come, and gone, before. "No authority," one such petition fumed in 1845, "not under the ex Venetian Republic, nor under the ceased Kingdom of Italy, and not even under the present had even dreamed of interfering in that which does not concern it."[41] Of a generation born into these social and economic shifts, who had lived through two famines and three waves of vine-devastating diseases (the few surviving vines those among well-tended hilly sites)[42] in the area by the time he was twenty years old, Carpenè must have believed that his words and interactions, his kind of intervention would help. While Mazzini had left out the contadini, after months of reading Carpenè's writings I have a sense of something else. Is there a sympathy? A place prepared? "There are also bad habits, the ignorance of the contadini and more often of the landowners," he wrote in 1880. "I say that before the contadino, instruction should be given to the landowner."[43] Over and over he attributes their agricultural problems to mistaken understandings and technique, and at worse a hardheaded refusal to move from current ways of doing things,[44] but he never speaks from a disrespect of the growers.[45] "Teach the agricoltore una buona pratica agraria, convince him also of the bontà of this," he wrote in 1875; but for proper application of such knowledge, the grower must underneath it all learn chemistry: "Chemical fertilizers, for example, aren't as trusted as they deserve to be and their use isn't very widespread. This is because they are often applied to terrains and crops that don't demand them. I would say also that [they] have fallen into disrepute from the lack of honesty by the fabricators; there will always be frequent deceptions for the agricoltore who lacks the foundation of science."[46] The chemist-enologist has cast the contadini as not to be disregarded[47] but as part of the force for change, participants, at least their children and grandchildren, in a reciprocal conversation, partners in the grand and promised project that is Italy and finally here. For sure, his articles have none of the disparaging remarks that so many of his contemporaries, including at least once close associate Cerletti's,[48] do. "In the world there is room for everyone," Carpenè writes while exalting *scientists, progress, improvements, fealtà,*[49] and *not denigrating one thing to elevate another*—and, only, upon such a base build competition, which then "eleva, nobilita l'industriale che lavora colla propria intelligenza e col braccio."[50] Along with the suggestions of science, he makes recommendations for which wines all winemakers should send to the exhibitions, how best to represent their lands that they should remember to be proud of, how to package such bottles. There would not be problems in materials and wines, he wrote, "if owners had known how to direct well and if the vignaiolo were a greater expert in carrying out the multiple viticultural operations."[51] Today's many great successful raw-materials- and growers-conscious producers tending to specialized vineyards have to be an outcome he was hoping for. "If in those valleys and those gracious slopes, cultivation was improved, if to vines in closed vineyards were dedicated all the cares and attentions that we in the settentrione[52] are forced to use, then the wine would acquire 'a reputation like no other in the world,'"[53] he encouraged.

With similar encouragement, a mandate from the Società Enologica Trevigiana to get new vineyards planted was carried by the school, shaping the institution from a center of formation to one of research and experimentation."[54] And with the dissolving of the Società and its subsequent reforming into Carpenè Malvolti, finally the chemist could make the wines of the hills, from which both raw materials and the ones needed for transforming them. The "cultori[55] of the wine industry know the excellent lumber used for staves from the woods of Montello,"[56] Carpenè wrote in 1878. "Why do we go abroad to supply ourselves with what we can find in our country? Has the sad habit of appreciating the stuff[57] of others at the expense of our own not yet ended? Let's end it for once and all and try to study our own stuff a bit more patiently and a bit more deeply, at least to have plausible reasons to counter the complaints we might receive due to the preference given to non-national stuff. For our own dignity, if self-interest isn't enough to move us, let's try to avoid the mockery that rains down upon us from foreigners," and not use barrels made "with the staves that the French and we ourselves buy in Trieste as staves of Slavonia, while the oaks that they come from grew majestically under the dome of the sky of Italy and precisely in the Bosco del Montello."[58] Rumors abound of proud Italian goods being sold under more highly regarded foreign labels, as Carpenè himself admitted he was not above doing. "You," he writes in formal *Lei* addressing praise in the thirtieth issue of Ottavi's *Giornale Vinicolo* in Piemonte's Casale Monferrato for the botti built by a certain Fenzi based on his supposed choice of oak from Istria, Bosnia, and Croatia, "have called the time, perhaps without meaning to, to strike down a false assertion . . . born in Italy from unkindness or from the ugly prejudice that is unfortunately rooted in much of the Italian public, and especially among the rich who constitute—with exceptions—the *créme* of ignorance. They pay double, tenfold in preference of foreign stuff, while often it is Italian, and is foreign only by being under another label and because of that it is paid for at a higher price. How much spumante from the Ditta Carpenè Malvolti and of others with foreign labels is consumed by those paying ten for what is worth two! I'll never approve of these games, but how to avoid them?"[59] But, about Fenzi's workmanship and the quality of that specific foreign oak, Carpenè is in agreement, with a caveat. We would like to bring attention to the "excellent quality and high merits of the oak from the public treasury[60] Woods, called of Montello, in the province of Treviso," he wrote. "We assert that the oak of Montello is for its merits equal to that of the woods of Croatia—without any defects that might render it inferior. Whoever would like a specimen to study, please ask us as we are very happy to take the trouble to satisfy you."[61] In the 1890s, *L'Enotecnico, periodico di viticoltura e di enologia*,[62] "dedicato agli interessi degli enotecnici," runs repeated advertisements from Antonio Wallusschnig, cooper in "Conegliano (Veneto)" for "botti e tini of any capacity, of Slavonia, rational and regola d'arte, sent throughout Italy, America, Africa, and Asia."

Upon being selected, wood was to be worried about differently. From forest, treasured and ours, it was now also material for winemaking: "if the chemical means of investigation were greater and were sufficient to clearly discover the different compounds that are generated during fermentation," Carpenè wrote in his *Sunto*, "depending on the temperature under which it is carried out, we would be in possession of valuable knowledge for enological technology."[63] But the kind of compounds that *could* be determined might help with another crucial feature of winemaking technology: to deal with the widely varying quality of wood sold for making botti—how many of the wine-growers made their own?—as with wine's primary materials, one should think about where the trees were grown. Having determined that chestnut and oak were the best trees for winemakers due to their "mechanical resistance, their elasticity, their porousness, their hygroscopicity,"[64] he proposes they continue to research what growing conditions are best for these raw materials. "The oak and the chestnut that come from damp soils, according to [German tree-centered botanist Heinrich] Mayr, contain extractive substances that are soluble in water and less so in wine, much less so the more rich it is in alcohol, while if they were grown in dry siliceous soils they contain more quercina, one of the principal extracts of resinous nature and therefore not very soluble in water, but more so in wine, and all the more so in proportion to its richness in alcohol."[65] The best thing to do was to "rigorously purify all new botti, whether of oak or of chestnut, whatever their provenance." A prudent maker of vats for crushing and fermenting grapes and casks to raise the results was advised to let the staves sit submerged under running water before building the new winemaking equipment. Come autumn's harvest, a winemaker armed with a vat made with wood grown nearby and purified was ready to make the wine that these hills could provide, with one more piece of advice. "Practice," Carpenè wrote in his *Sunto*, has shown that the temperature at which a must ferments is another crucial step to measured glory. Based on "well conducted," experiments, it had been shown "among other important things that a fermentation conducted very slowly, but with ample judgment, yields more delicate wines, richer in glycerin and in perfuming principles."[66] In wood, this winemaking is an immersion in a project that brings him into the world and time that surrounds him, is not the part that carries him away to times others cannot yet see.

In 1893, Carpenè squabbled publicly with the younger Martinotti, born as the new nation was forming, who had also studied pharmacy and chemistry, and who in an "elegant libretto"[67] of fifty-one pages and titled *I vini spumanti* had recently described spumante made by the Carpenè method as "nothing other than a vinous solution of anidride carbonica." Two years later, Martinotti would patent his own apparatus and metodo for making spumante of the deeply perfumed Moscato bianco grapes that grew in his hometown. While over the next decades his "installation for the continuous production of sparkling wines" would be updated by other enologists as they reached new materials, needs, and understandings, his process is considered the basis for the metodo Italiano, today just as easily called the Martinotti method, Italy's own way to make spumante which with adjustments by Trevisian winemakers to fit their grapes and wine would allow them to finally reach Carpenè's Prosecco insight.

By the mid-twentieth century, the wines Carpenè envisioned would be the two denominations protected and defined, the most formidable expression not just of the Marca Trevigiana's hills and grapes but of the metodo Italiano itself.

Mere carbonation? *Your words carry the weight and authority of chemistry, you must make clear that this is your opinion,* Carpenè's published reaction read that year. *There is no chemistry, whose role is like all science for the truth no matter what one's preferences, behind your words.* The metodo Carpenè was nonnatural? Look at the French spumante wines Martinotti calls *natural*, Carpenè continues in rational table form[68] of French "vino fermentato in bottiglia chiusa"[69] finished with an addition of Port, Cognac, Framboise among eight possibilities to add color, lastingness, sweetness, and flavor versus Italy's "vino fermentato in recipiente aperto"[70] and to which is added only anidride carbonica, and only if desired Cognac[71] or sugar: "The genius of spumante is in the grapes and its diligent and rational production and not in the way it is gasified."[72] His feelings are hurt and maybe even his pride, but the true wound is to the elder chemist's sense of rationality, the rules of those who would do science: "Mr. Martinotti knows that scholars of the true must have a great, a noble ideal, to search for the true truth wherever it is found, without aiming at secondary ends, without allowing passion to act as a veil."[73]

As summed up in the news section of the July 1893 *Bollettino del comizio agrario di Mantova e dei distretti riuniti di Asola*,[74] the two enologists agreed that one could obtain perfectly limpid wines with either fermentation in a closed vessel or artificial dissolving of anidride carbonica, but they remained apart on two points: faced with a sparkling example of each method, could one identify how each sample had been made? And, would two spumanti made of the same base wine with either method bubble in the same way as they were being drunk? Carpenè insisted no to the former and yes to the second. A brutally mechanical experiment followed, as Martinotti submerged two such bottles into ice water until crystals formed of sugars and tartaric acid, then forcefully shook them to show that in the first case the crystals dissolved and in the second they did not—evidence, he insisted, that those made by the Carpenè method poured with large, rough bubbles and too-abundant spuma[75] which then, exhausted, quickly dissipated. "I made hundreds of analyses, from which resulted that tartaric acid, determined with the Berthelot and Fleurieu system, was found in the same quantity both before and after the wine was made spumante,"[76] Carpenè wrote, doubled attack of system and scholarship—"it doesn't lose tartaric acid, it doesn't lose one atom . . . if the operation to render the wine spumante is well executed."[77] It was true, he demurred, that his system did have some defects, such as in some circumstances not being able to render some Moscato wines clear and limpid as quality demanded. "Whoever has the chance to uncork a few bottles of vino spumante sistema Carpenè and at the same time wine turned spumante by fermentation (made with the same quality of grapes and the bottles of equal shape and quality of glass)," advised the anonymous newswriter, "could conduct this test and decide the second question for themselves." In 1923,[78] Dalmasso would bridge the two. "In fact,"[79] he wrote, "artificial spumanti should be more correctly defined as (as required by the current provisions of the law) 'spumanti artificially gasified.' This is because they are natural wines, of which only the spuma is artificial."

The solution to natural Italian spumante would be a more generous understanding of the two vessels that stood at the base of the Italian, open, way and at the waiting of the French, closed, method. "You have to think of an autoclave as a large bottle," Asolo DOCG producer Enrico Bedin said when I visited that zone in 2018. An autoclave is strong and sealed and has been used in various forms since the early nineteenth century to contain chemical reactions, including of foods prepared by heating under pressure as was its original purpose.[80] It is a vessel that can withstand pressure like a sparkling wine's heavy bottle but offers greater containment than that glass one's 750 milliliter capacity, and makes possible a system of rapid spumantizzazione . . . with final passage into bottles only when the desired pressure is reached.[81] "In 1851, the very first attempts of Rousseau," wrote twentieth-century enologist De Rosa of the journey there and whose 1964 *Tecnica dei vini spumanti* and 1987 edition renamed *Tecnologia dei vini spumanti*[82] remain references for sparkling winemaking, were "followed in the distance of some years (1858) from those more exacting ones of Maumené, to whom therefore are attributed the first results of an almost-industrial character."[83] Those results were French chemist Edme Jules Maumené's thirty-two-hectoliter (about 3,500 bottles' worth plus a chamber of fifty to sixty liters into which, once wine and gluten plus a little tannins from wine grapes' skins, seeds, rasps to neutralize the fats which the gluten will produce are added, the newly gasified wine will move), hermetically sealed, portable for moving to areas with proper temperature,[84] and cylindrical Afroforo vessel (made of copper plated with silver according to the Christofle method so that yeasts did not die poisoned by the copper, nor the wine "stink like a cadaver, ashen and horrid"[85] and named, in Greek, the bringer of bubbles.[86]) within which the amount of pressure was controlled by way of an afrometro.[87] That very first Frenchman had proposed his own enoforo: a large wooden barrel reinforced with large iron hoops and in which wine was refermented. "It is difficult to imagine how a wooden barrel, obviously made of staves, could hold even medium pressure,"[88] blogs enologist and journalist Lorenzo Tablino of this early attempt's fate, in his brief reconstruction of spumantizzazione vessels built by moving "between personal memories, testimonies of elderly winemakers and old cellarmen, between public and private archives and cemeteries of wine-making machines." It is not clear to me that Rousseau's was actually built; perhaps it was instead a mock-up in wood as its creator knowingly awaited some right stuff to build it from. "Toward the end of the century," De Rosa wrote, "König became interested, too, last came Martinotti, who in 1895 brought sensible and interesting improvements to [this system]. Brillé also in those years applied himself to this quest, devising his unique modular 'tini.' Finally, the Carpenès (Etile and Antonio[89]) in these past decades applied themselves thoroughly, adding a series of ingenious and radical improvements, thereby at last rendering the system of rapid spumantizzazione in autoclave modern and rational."[90] Martinotti's 1895 patent was founded on three containers made of iron, with internal wooden protection.[91] By the time De Rosa wrote, the industry of spumante-making autoclaves—based on his continuous system in which as much of your base wine as you wanted to transform could be refermented, drawn off, and replaced as needed—was working toward today's single versions some built to hold upwards of 600 hectoliters and whose murmuring gas-releasing valves allow spumante-makers a final step in which to determine precise pressure. If Prosecco's CO_2 was to be produced by re-fermenting the base wine, the wine's ultimate aromatic freshness and local grapes' characters that unlike with Champagne's wines tying liquid and gas together was meant to showcase had by then already been long imagined. That Prosecco awaited only the right material. "The first autoclave wine experiments came here in the 1930s and '40s, born with the story of Carpenè Malvolti," reiterated Bedin. "Then the actual autoclave arrived in the 1960s, and then the boom."

Autoclaves make Prosecco Superiore near steel base-wine fermentation tanks in a winery in the hills.
Susan H. Gordon

Beautiful, capable, and enduring. There isn't another other way into this: the reason Prosecco is known so well today is not only Carpenè's insight that it should be spumante and be sent abroad. The vehicle for what he saw is stainless steel, come decades after him, practically rustproof and unreactive to acids, offering various grades of strength, and what's shaped from it generally lasting—without any of itself entering into the composition of a wine. At the end of the eighteenth century, a French chemist had isolated the element of chromium for the first time, thereby rendering it addable to that growing list of elements—"a hard, silvery metal with a blue tinge"[92] that gives rubies their red and emeralds their green and that is odorless as well as tasteless. Quite addable in a physical sense, too, Cr appears naturally in compound form, mostly as chromite (the chemists' shorthand for chromium oxide, both of which terms let you know that it is, combination of an element with also elemental oxygen, a rust), the ore that is mined for its production. In 1888, the school's *Rassegna* reported its chemistry laboratory's research into metals and wine in contact, in the hopes that a nonreactive transport tank, to be drawn by rail, might be created: "The first material to come to mind for the construction of [such] vagoni serbatoi is sheet iron in suitable thickness, or sheets of tin-plated iron or copper. But metals . . . are more or less all affected by the large quantities of carbon dioxide contained especially in the current year's wines and by the free acids constantly present in wine, as they are a natural part of its constitution. After this chemical action, the composition of the wine is changed, its characteristics, its hygienic properties. The metallic radical, having entered into combination with the wine's components, forms metallic compounds that in turn react provoking precipitation of metallic oxide and coloring materials. . . . If the vessel is iron, the ferrous salts that form, even if they remain only as traces in the wine, as soon as they have contact with the air, due to oxidization, they convert into ferric salts which then decompose to form ferric tannate which browns the wine and, if it is white, renders it unsellable."[93] In the first years of the twentieth century, a century knowing about Cr, an English electrochemist began to figure out, literally, materials now far more commonly also proportions on paper, a certain family of alloys that would assuage each of these worries.

An alloy is defined as a mixture of two chemical elements at least one of which is a metal. It is a necessarily impure substance of a special kind: it must retain the properties of a metal[94] and it must be possible to control in order to produce specific outcomes, a new metal desirable and consistent. The criteria for what came to be called in English *stainless steel* (alloy, it would turn out, of two metals) was that it be neither reactive nor corrodible, characteristics determined by the applications that had been awaiting it, and which guided its perfecting after a somewhat long and messy start. The new material's origin story includes miscalculated attempts since the seventeeth century, but the tale that's been passed around with the most general consensus is that the first mix of iron with enough Cr (when at least 10.5 percent of an iron alloy's mass is Cr, "there is a dramatic increase in the corrosion resistance."[95]) was created in 1912 in Sheffield, England, by Harry Brearly, self-taught in chemistry and metallurgy, so that the barrels of England's rifles might remain stronger and last longer than their enemies'. In Italy as in France, the material's name communicates its chemical identity and—with a dash of optimism since under certain conditions such as very high temperatures or prolonged contact with water, the term becomes untrue—is *acciaio inossidabile*, inoxidizable steel. Whereas steels are alloys of iron and carbon, stainless steels are instead compounds of iron and enough Cr, the molecules of the latter joining with elements in mixes such as air and water before the iron can to form a film that is protective and noncorrosive, a layer invisible to us but a barrier to acids including those that are components of every fruit. A rerouted commercial production of Brearly's alloy seems to have begun soon after he crafted it: "The price of this steel is about 20 cents a pound for ordinary sizes, which is about double the price of the usual steel"[96] reads a brief article about an "invention especially good for table cutlery," in the Sunday *New York Times* of January 31, 1915. "It also costs more to work up, so that the initial costs of the articles made from this new discovery . . . will be about double the present cost but it is considered that the saving of labor to the customer will more than cover the total cost of cutlery in the first twelve months."

It would also save the winemakers a lot of trouble. In 1948, the first iron, enameled on the inside, autoclave was made in Canelli, traditional top Moscato vineyard area of Martinotti's Asti, corrodible otherwise and weighty. Today, there is a phrase that is mostly a joke and repeated every time a group of wine journalists visits a winery and that is that nobody wants to see the steel tanks. But it is untrue, at least for me. If you want to truly listen and taste, twice as true for spumante, these vessels, themselves varied and progressions, are an important component of understanding the material and form of the wines made in them. They are also decades of technology embedded in landscape and a steady trickling in, concepts and attempted shapes, since long before, and so an historical and traditional aspect of those wines. This is especially true not only of spumante but even more so of spumante-making zones in which wineries today hold thick protectively curved base autoclaves cast in a steel shape strong enough to withstand the power of metamorphosing wine, form because of which the frizzante Prosecchi often called *col fondo*[97] today, with the wine- and CO_2-making sediment left in bottle to be gently swirled through upon opening, or poured off of according to one's preference, unfiltered cloudiness made of the same grapes, are known of at all outside of these hills. Their claims to *preindustrial*, *ancestral*, made by the people, are sold around the world by tracing the paths made by the spumantizzatori, following spumante's mission to leave the homes in which it was made and drunk and search for fame outside. The pendulum between the two styles has been swinging for a while, a disagreement from Prosecco spumante's start: "As we keep repeating it would be enough to read those old books into the ears of those modern winemakers who declare their desire to go the old ways," Carpenè wrote in 1878. To "make wine as our grandfathers made it, to convince them that they find themselves in contradiction with themselves, showing ignorance of the past and of the present, or with a stubbornness as colossal as the pyramids of Egypt."[98] Most important, it is a conversation that has been ongoing in these hills, and that is now a part of Prosecco itself, also to be exported everywhere.

Thanks to the metodo Italiano and the agreeability of stainless steel, greater quantities of Prosecco can be made more easily. This way to spumante is faster and less expensive than the bottle-fermented Champagne method, and that is where an understanding of these wines can take that one wrong step into, well, complete misunderstanding: the resultant wines are something else that is crucial to Treviso, Italy. Unlike the fundamental Champagne grapes—neutral to lemony Chardonnay, faraway raspberries of Pinot Noir spilled instantly from its grape skins—threaded along the Piave-feeding rivers, the sparkling-wine grapes of these hills, Prosecco, Bianchetta, Perera, Boschera, and Verdiso are subtle honeysuckle, jasmine, citrus blossoms presences with elegant and lifting telltale bitterness that have no need for the weighty bready flavors that Champagne's bottle-fermenting method brings. When Carpenè believed in the grapes grown for centuries at home, he set the course for these hills to eye Martinotti's metodo when it came, for that to be translated 390 km east. It took a while to be stainless steel. That iron Italian autoclave would be used into the 1980s as the number of spumantizzatori of the Venetian hills slowly, steadily grew. Soon after World War II, in Canelli, too, was founded the stainless-steel autoclave producer whose name you see throughout wineries everywhere in the DOCG areas today: *Sirio Aliberti* in letters like red sticks on the fronts of thousands of round-bottomed pressure tanks that transform Prosecco in hundreds of hectoliters.

Since you're reading this now, know that there is an important shape for lines to take so that they might illustrate these beginnings without assumptions brought by the twentieth and twenty-first centuries: as the first commercial believer in Prosecco spumante of these hills, Carpenè is also a believer in industry, but he is not an industrialist. His chemistry is a science not global in outlook the way that chemical wine analysis and market language are today which is a flattening of the earth into the kinds of parallel lines it does not have. Wine, even his most chemistry-rich writings show, was unapologetically a reliance on what was grown and where, winemakers and growers to be taught in accordance with what chemistry (what else was this science boldly and repeatedly Italian, that allowed Italy to talk and exchange with other nations, if not a reading of the world we're in?) showed was at work in grapes and must and fermentation. His project was to paint Italy in relief, hopeful stories for a hoped-for community of scientists, growers, drinkers. Technology, these grapes, Treviso, Conegliano, Italy, this was spumante. He worked tirelessly for how. Read and talked for how to plant the grapevines, read and moved years in laboratory and cellar for how to add reliability to the local wines, push their prized fizziness to sparkling power, preserving them as representative of the hills' most proud traditions, crafting and searching for any technology that might ally itself to him. This is work that is meant to be shared, taught to the new Italy (a second thing he has learned from Mazzini's final book is that there must be educators) which is if not this moment then soon, of and through the sons (*and daughters*, might remind Mazzini) of the growers, who are not among the urban operai the first teacher's book addresses but rural with an agriculture that is, observe, a proto-industry until the metals come and who are the majority of the Italians. He is the bridge between Mazzini and the growers the older revolutionary never mentions. This first true class of chemists, scientists, and academics are prophets, too; will guide what is now geographically and legally one into true Italian land which should be a place for the contadini, too. Carpenè is one of the few to worry about this in his writings, his machines, the school, the Prosecco that will come promising a role and a living earned in this future which will be built of what has been here and what the late nineteenth century is making: "they will have science at their fingertips."[99] The trees made into barrels or straightened into vineyard posts; the grapes sorted by variety and where; the machines for great quantities of the hills' wines made from organization of the thousands of growers with tiny plots planted to their perfect match in cultivated vines. These were the new and ancient rights to be given to the next generations from now on. There is no talk yet of delimitation, as there will be in Dalmasso's 1930s report.[100]

"As Angelo talks about his winter agricultural work it begins to rain as heralded by the leaden sky since the morning," report interviewers Francesca Poggetti and Bianca Pastori of the testimony by Angelo Nardi, grape-grower in the Rolle frazione in the Valdobbiadenese hills and graduate of the school, gathered for the Italian Association of Oral History (AISO)'s first "Scuola di storia orale nel paesaggio del prosecco superiore" in 2020.[101] "Angelo and Roberta invite us in and lead us to the large kitchen table. A bottle of Prosecco is opened, and obviously we ask its origin; and discover that, while his production for the outside market ends when he confers the grapes [to Andreola founded in 1984 in Col San Martino], for domestic use Angelo follows the viti-vinicultural tradition of his grandfather, born in the 1930s, using his same machinery:

> Angelo: Yes, it's a little bit of wine that I make at home.
>
> Francesca: Ah! You make wine at home?
>
> Angelo: Yes, two or three hectoliters, like that . . .
>
> Bianca: For home use?
>
> Angelo: Yes.

Bianca: How do you make it at home, scusami, botti?

Angelo: I have machines made of steel, those that belonged to my grandfather. I crush the grapes, the berries are separated from the stem and also from the skins, yes, in the vat ends up only the juice and from that I do various travasi, and that's it."[102]

Driving the search into how any technology Carpenè could lay hands on might make Prosecco sparkling by way of its material components was a concept still so new it could be named only by tying the existing, but foreign, word tightly onto it. But in 1924, words on that now historic label announced for the first time the new wine's form. And in the 1930s, Carpenè's grandson Antonio began to make his family's wines based on Italy's new way, working to shape the method to its Prosecco, tuning base wines and second-fermentation times in great pressurized tanks which in, so close now, just a few years more would be made of responsive and widespread stainless steel. "Why is Tullio De Rosa never cited in the development of Prosecco?" Canel says to me; before him, if you came here in the 1960s the wine you drank, first was always still and then was scuro, tannico; "tristo," says Gasparini. "They put white wine in botte. And he was one of the very first to fight against this, and push for acciaio." After De Rosa found success with his technical books, Gasparini says, he turned to more of a promotional role—at the cost of his statuto tecnico, and became instead an official taster and communicator, no longer spoken of for his technical contributions. A small note on the website of the Conegliano-Valdobbiadene Consorzio introduces the Prosecco method by noting that "spumantizzazione in the Conegliano Valdobbiadene territory originated from the experiments of Antonio Carpenè, and was then perfected at the beginning of the nineteenth century from the eponymous winery, one of the first grand Italian spumante houses. In the following decades the method was perfected in the Scuola Enologica di Conegliano, with fundamental contributions from the research of prof. Tullio De Rosa, as evolution of the Martinotti method, turned toward exalting the aromas that naturally characterize Prosecco Superiore, which in this territory takes on tipicity[103] and complexity that are unique to it." Chemically equivalent, equal wines or not, artificial carbonation would serve until an upcoming generation of enologists could make Prosecco a naturally carbonated wine that was a rational interpretation and a reasoned expression of the hills' grapes; autoclaves and this new metal necessities for these promised wines. Like their complexities of bitter and sour, a seemingly understated range of flower, spice, and fruit flavors that converge with those of numerous apple varieties, with the jasmine that drifts in later-spring air, of ferrous-sugary taste like a cold pineapple slice served after a summer meal, these Trevisian grapes were an Italian patrimony[104] in need of winemakers to lead them to be themselves, senses and desire which France did not have.

Made close by, in the Veneto, too, are machines to manipulate the air around the bottled sparkle, to create the isobaric environments in which producers of these wines must bottle what they make, every movement from autoclave to bottling rendered akin to one pressure-worried bottle. Isobaric machines are necessary for bottling with the Italian method and, Pussini explained to me by email, they have been here in force therefore since at least the early 1980s. In 1880 in the province of Verona, 93 km straight southwest across the Piave over the Montello woods, engineer Bortolo Bertolaso opened his business that would supply enotechnical machines to the area as vines and producers grew, and so when the time came solved the problem of bottling the hills' spumante. "The right direction is clear," the Bertolaso website describes those earliest days of what is now a company in its fourth generation of creating bottling equipment: "The strong propensity for engineering development leads the company to focus on building macchine tappatrici."105 Like the bottles themselves and the making of the wines the machine will fill them with, it is a question of delimiting, a process reliant upon borders. A bottling line is a machine noisy, long, and for the spumantizzatori, enclosed like an underground subway station completely demarcated by glass. It is built to contain a fast linear procession, with tracks along which empty bottles set off swaying on their way to a filling station, take off again to where the corks and the securing gabbiette cages in threadlike iron are installed, on to the terminal where the labels are affixed, all under appropriate atmospheric conditions. Isobaric means characterized by constant or equal pressure, a system that offers equilibrium, turns the air within its borders equal to the pressure of spumante so that in that transparent tunnel of the line nothing about the wine in the bottles changes, nothing foams and bubbles as it would if the air were instead the one we breathe. "Today they have been perfected," Pussini explains, so that they directly create the atmospheric conditions inside the bottle before it is filled, which in earlier days was not done. Under the same pressure and temperature all volumes of gas are made of the same number of molecules of the substance that makes them *them*.

The trees, the machines, to make wine and a nation, to make them last.

That steel is neutral, as the shortcut saying in wine production and writing goes, is not strictly true either, though it does count as a chemical fact. Steel shapes a wine, in part removing it from its surroundings, in part allowing it an inwardness achievable only through isolation.

"La Prosecca, la Biànchetta, la Marzemina, la Verdisa," wrote Conte Balbi Valier in 1880 for the Italy-wide Inchiesta Agraria Jacini,[106] "are all cultivated, some places more some places less, in the hills."[107] Just what kind of material Prosecco is, is still a matter of reshaping. Until 2009, *Prosecco* was a three-part definition, cultivar, wine, and place substantiated, and for all its life before that a more indeterminate[108] mix still.

From the Treviso province, an illustrated contadina, untranslated as *peasant woman* and *working class costume*.

The Miriam and Ira D. Wallach Division of Art, Prints and Photographs: Picture Collection, The New York Public Library. "Peasant woman, Treviso, Italy, 16th cen" New York Public Library Digital Collections. Accessed February 15, 2022. https://digitalcollections.nypl.org/items/510d47e1-3e25-a3d9-e040-e00a18064a99) / The Miriam and Ira D. Wallach Division of Art, Prints and Photographs: Picture Collection, The New York Public Library.

It makes sense to think of the word, grape and wine both at their best in these hills and whose historical footprints extend eastward following the Adriatic coastline of Venice's Stato da Mar into the Friuli-Venezia Giulia region and into Slovenia and Croatia, as a rationalization. As a composite of grape and wine and hills. Of, too, chemists, agronomists, and contadini who nourished the cultivar sometimes called Glera, likely grown first in those vineyards to the hills' east where, while well respected,[109] it was tied into terrains that could not provide its ideal conditions. But along these upper Marca hills, whose many faces like diamonds unsteady and erratically cut were bathed in sun and enough rain, protected from the coldest winds, and allowed cooling breezes key to these grapes' aromatics and acidity, stretched Prosecco's prime territory and a story of *that* land instead (a time line compiled during World War II of the days of phylloxera advancing through the provinces of the Tre Venezie[110] lists Friuli's Gorizia and Trieste as first detection points in 1888 then the first recorded cases in the Veneto's Treviso twelve years later, a hint that by the late nineteenth century these territories were not very viticulturally connected[111]). The wine that given the right work grew at its utmost here and which, only granted its correct expression, would take its rightful place among the world's great spumanti, was waiting for its name to be made. In the second half of the nineteenth century, time of systematic measuring, sorting into compileable understandings, and making new kinds of sense, *Prosecco* was and would remain uneasy movement from a name for several and/or plural varieties and wine(s) among the Terraferma to a term that meant this zone's best grape and wine, inextricable, to no longer a vine at all but simply the wine of a territory, grape and wine separate entities and substances and the pertinent composite instead an official denomination backed up by quality, historical presence, and law.

A grape's name is an indicator of many things. *Glera* applied more widely from its Friulian town is simply a latest layer. Just as *Italy* has had different meanings at different points in time, for most of their lives, neither Prosecco grape nor Prosecco wine offered a straight line. "Either from Prosecco a variety arrived to Gorizia and was baptized with the name of the locality it hailed from; or to a local variety, or one of unknown provenance, was given the name of Prosecco because of the fame of the wines that boasted that denomination," wrote Enos Costantini in 2009 as he traced one of the word's many paths crisscrossing Friuli along the northernmost curves of the Adriatic Sea; "both hypotheses are plausible and the true truth we will never know."[112] Throughout issues of the region's official agriculture-centered *Tiere Furlane* magazine[113] in which Costantini's words appear there are even mentions of Prosecco, if rarely, as a sparkling wine, albeit in the Champagne method, in the Friuli area. But even taken as grape names, both *Glera* and *Prosecco* historically indicate not a variety, but a collection of biotypes and synonyms which the contadini growing them called *prosecche* and *glere*, not only plural but also by the feminine form in which almost all Italian fruit appears. Focus microscope to *wine name* and as with Friuli's prized Ribolla, Prosecco was there and in the Veneto primarily the name for a prized white wine. "Trieste then preferred to ensure this freedom by giving itself to the House of Austria, with which it then stayed almost continuously; paying Venice only with the gift of oil and Ribolla, that is sparkling white wine which today is called Prosecco," writes Colombo in a 2019 issue of the magazine, quoting historian Cesare Cantù from the second volume—*Storia di Venezia*—of his six-part *Grande illustrazione del Lombardo-Veneto.*[114]

The search for Prosecco was a deliberate one, so much so that it has the whiff of invention, like any good cultivar or wine does. Quality comes from somewhere. Some grapes worked better than others here and so with efforts from the growing field of ampelography to know a grape by what it looked like, what wines it gave, and what it was called by those who grew it, the search for a variety that showed the wine's majestic side, and could be grown reliably even taking into account the difficulties that all of the world's best grapes seem to pose in one form or another, was on. "The problem of Prosecco's varieties has always been complex and difficult to carry out," notes a regional website,[115] "due to the presence of different versions and because the varietal selection work is incomplete." To start with the most important, wrote Sannino in 1913, in table 24 of the *Ampelografia italiana* is "Prosecco bianco[116] . . . the most widely coltivated from Valdobbiadene to Pieve di Soligo and at Maser, across the Piave . . . being the most productive, the most rustic and also excellent for the quality of its grapes, *lightly aromatic* and rich in glycerin. The aroma is so light, that wanting to follow the directions of the Conte di Rovasenda, who considered aromatic only the grapes endowed with the intense and characteristic aroma of Moscato, Prosecco must be considered as a grape of simple aroma. This does not take from that it is endowed with that delicate and pleasing aroma that allows us to include it in the same group as Sauvignon, of Riesling renano,[117] of Asprino di Basilicata e del Privitaro di Cosenza."[118] Prized white-wine grapes of this northeast extremity of the Italian peninsula for decades in continual reformation between Italy, Austria, and Slovenia, varieties like Ribolla, Friulano, and a local variation of one of the many grapes named Malvasia might have found refuge in the name, which in addition covered several types of Prosecco proper: Prosecco gentil,[119] the growers' word for the top-quality type; the analogous round-berried Prosecco Tondo; spicy co-planted and elongated Prosecco Lungo which we now know is a distinct but related[120] variety; the especially notable Prosecco Balbi biotypes[121] of Tondo which Conte Balbi Valier (in the 1892 *Annuario d'Italia*,[122] the Contessa Marta Gradenigo who married him in 1858[123] is listed as both his widow and a wine producer) selected and propagated in the 1860s[124] in Pieve di Soligo—northwest of Conegliano and crossed continuously by the Soligo river, and once twice a day by the Follina-Farra di Soligo coach and which Malvolti cultivated, too, inclusion in the winery's wines. "The Prosecco of the *Ampelografia italiana* commonly called Prosecco Balbi, from the name of the Conte Balbi, who planted it widely in his Pieve di Soligo properties," continued Sannino, "is not cultivated in Conegliano, because at harvest the sparse bunches often held a good number of green berries, unripe. And at Conegliano, instead, for ages Prosecco nostrano has been cultivated, which [variety] in 1895 I found identical to Malvasia lunga of Toscana pictured in table 4 of the Ampelografia Italiana [sic]. And that Malvasia is a grape rich in glucose, but lacking all of the delicate aromatic flavor characteristics of Prosecco or of Prosecco Balbi."[125]

To be the right kinds of material for making wine, wine grapes have specific needs, and Carpenè could now talk about those by way of numbers: "The characteristics that explain the vine and the grape when the latter is mature are known by those who frequent the countryside,"[126] he wrote in an acknowledgment of the language of history and landscape, accompanied by translation into a table of how much heat was absorbed by the grapes of the Marca Trevigiana from soon after flowering—likely in late May unless delayed by common rain or unusually chilly temperatures—to the day of harvest 1838 through 1844, vintages that stood between the area's famines of 1829–1830 and 1846–1847. "All fruit requires a certain number of heat degrees to fully ripen, and winegrapes need at least 2,000," he continued. Wine of the year 1842, when the grapes were harvested after days that added up to more than 2,300, were "excellent," while 1839, harvested too soon, or with too few heat degrees gathered, was "a vintage not very good for the wines' success." A grape's ripeness was the series of changes that could be tracked by looking: when the "berries become semi-transparent, if the grape is white, and in many varieties allow one to see the seeds at their center," he described—"they also take on a yellow color, sometimes golden."[127] The notion of maturity and ripeness was backed by numbers tied to precise substances: "the degree of acidity of a must, added to the portion of sugars of the same, is an indication of the state of ripeness of the grape, the quality of wine that can be obtained in comparison with previous years and relative to the quality of the wine, the land, and the exposition."[128] Wine instruments built on chemistry supplied ways to expose falsified wines, those made with materials other than grapes, including a tool for, "how," Carpenè wrote, "to uncover non-grape glycerine, a use especially prevalent for increasing the value of fine wines."[129] There is Verdiso's acidity measured by an areometer and confirmed that it is indeed the tart substance of grapes; its sugars by the Babo tool,[130] instruments built upon an ethical question, too: wine must be only an enological process, that is having to do with no materials other than what came from the vine.[131] That the crushers should not add their sweat to the must was part of a greater philosophy: nothing should be added to the must at all, wine was a material made "purely of grapes"[132] and this would come to be more important and easier to understand and carry out as Italians relearned what a quality wine tastes like. "What is natural wine, which is made *exclusively* with grapes?"[133] The answer is an amalgam of substances and numbers. "It is a liquid," writes Carpenè, "generally formed over 100 parts of:

Water	88.00
Alcohol or spirit of wine	9.00
Numerous different substances, which give flavor, color, and each wine's characteristic fragrance to the aforementioned alcoholized water	3.00
Totale	100.00."

The growers on each hillside must choose the few varieties that grow there with best flavor and as they work for the Prosecco to come, the wines they make now must be represented with minute care as well: "Get the producers to choose the best bottles in which to send wine samples [for tasting]," he wrote in an article[134] published in the *Rassegna* on New Year's Day 1889. "These should be of a half-liter capacity, not much less, sufficient quantity for a taste and also for an analysis if necessary. The bottle must be a Bordeaux[135] shape for red wines and a Rhine-ian[136] form for white wines. . . . Affix to the bottle an elegant label with the name of the producer or of the seller, that of the wine, the number of hectoliters available, the price, etc. Wrap the bottle, in a sheet of fine white, or colored, paper then inside a cone of straw and, finally, lay it in a small elegant box.

"These descriptions are all things that to some can seem to be meticulousness with secondary importance, but in reality they have a lot. Elegance, graciousness, good taste, also in these preparations, denote perfection and give the measurement of the civility and of the customs of a paese."[137]

In 1880, "peronospera of the vine made its sad appearance in the vineyards of the province,"[138] wrote Dalmasso and Dell'Olio half a century later as phylloxera invaded all the Trevisian comuni. The number of vines of Prosecco, thin-skinned and demanding, would be further reduced from already lowered numbers, the vintage "one of the worst for quantity and quality caused principally by peronospera."[139] Carpenè's hands are shaking not uncontrollably but a little disruptively, and his irreversible misstep on the hill is coming closer. "Now the year 1880 is finished, too," he wrote in that year's *Rassegna*. "The viticoltore especially of the Alta Italia does not feel he can send it off with a grateful spirit, because it was miserly in compensating sweat and labor, in answering his needs with a harvest poor in quantity and quality and among the unhappiest that history has offered in many years.[140] Although in the spring [the vintage] promised an abundance from the vines that had surmounted the exceptionally rigorous winter and bloomed with clusters to make us forget the previous spring," they resulted in barely one-fifth of normal production. "Italy needed to rissanguarsi[141] after the damages suffered from 1879 and filled with hope at the dawn of 1880 gave itself over to work, from which to recuperate what had been lost! but . . . more than anything hail and peronospera of the vine led to so much damage, leaving the viticoltore disappointed by his happiest hopes.[142] 1880 leaves us in the comforting hope that phylloxera, thanks to the care of the Government, the Provinces, the Ampelographic Commissions and the delegates [of the phylloxera commissions], will find it too difficult to propagate in this terrain and destroy entire viticultural regions, as unfortunately it did in France." This particular difficulty is an occasion for hope, a "grand venture" that should not lead the viticoltore to "abandon himself to discouragement but rouses him to improve la viticoltura and increase the production of wine, not only for domestic consumption, which we must hope will always increase, but also for exporting. This in 1880 has increased well, and the wine exportation statistics are very flattering. Let's improve l'enologia, not disregard the perfecting methods suggested by science, work and sell honestly, and the good road which we have happily stepped onto, though not without effort, will [bring us] to the height of wine production and trade that awaits us and that is the destination longed for by all those who have studied and worked hard for years to bring Italy to take full advantage of its conditions so favorable to the cultivation of the vine.

"Coraggio e avanti."[143]

Through the 1880s as the hillside planes of whitish friable limestone clay and in stony calcareous red became too much for him to hesitate along, he continued to write, journals and letters and thought experiments and guidance for Italy's wine and character, places where his sure-footedness stayed key. "Oggidí si cammina colla luce della scienza,"[144] he reminded. Perhaps the sureness needed to guide a pen along the underlying lines that made letters, to move by the boundaries of words through the paesaggio of each page, only slowly blurred into sign of the ataxia that was taking hold; have difficulty walking, slur your speech, have difficulty swallowing. Perhaps this neurological disease that progresses steadily—it can take months or years, and for him it would cross almost two decades—toward paralysis left him the muscle coordination necessary to write and in exchange moved on to the movement of his legs; characteristic difficulty walking, gait abnormalities, stepping slow and rigid. Perhaps he dictated. More possible now anyway to travel by pen than on foot through where the grapes were, and though habits in the laboratory filled with glass and machines and substances always needed new navigating he was often there anyway; difficulty with fine motor tasks, such as eating, writing or buttoning a shirt. Many symptoms of ataxia mimic those of being drunk, introduces one research foundation today, and deepen into changes in speech and unmeant eye movements back and forth; he would have had difficulty swallowing.[145] "The men of science,"[146] he wrote in 1893, "are precisely intended for the highest place in the moral struggle and in that for existence, because scholarship is the first leva[147] of human prosperity." Less than nine years later, he disappeared, at home and surrounded by the family that would use the coming materials and technology to complete what he had seen and left to them. Outside, sign of strength in red-tipped green, the leaf buds of the Prosecco grapevines reached out in slow opening toward summer, sensitive to the first warmth of spring, a variety that buds early.[148]

In their data-filled book "dedicated to the glory and honor of work, and the sacrifices of the Italians," rich with land statistics pertinent to wine-growing in the Tre Venezie and written during World War II to "bring people back to the true life . . . with the fervent hope among the many disasters to bring back love and peace, that is the fruitful activity of the fields, the only one that can ensure serenity and well-being to the people," and due to money and paper shortages not published until 1950, seven years after its completion and after some of the Venetian territories it includes had been drawn out of Italy in an "unjust treaty of peace," Viscardo Montanari and Giuseppe Ceccarelli echo and update Vianello and Carpenè's *La Vite ed il vino nella Provincia di Treviso*. Titled *La viticoltura e la enologia nelle Tre Venezie*, their work is written to defend and improve the area's viticultural patrimony championed by the first duo seventy-five years earlier. It notes some of the same production problems in a landscape that seems to have been little transformed during the years between the two books. Wines made badly by grape growers in unclean and ill-equipped small wineries could be fixed the later authors recommended, perhaps taking the Società Enologica Trevigiana to heart, by "special Stabilimenti enologici,[149] well equipped and well directed." No matter the problems in the land, winemaking as a rational activity, the realm of chemistry- and machines-trained enologists, was making its first and slowly steadying steps, on solid ground and with cleaned feet.

6

The First Prosecco Bottle

Perhaps the only difference is the direction in which they point: memory to the past and imagination to the future.

—Christopher Bardt, *Material and Mind*[1]

"Remember / Ask for / Demand / This label." Advertisement with 1924 label proclaiming *Prosecco* and *Conegliano* for the first time.
Carpenè Malvolti, Conegliano (TV) According to Domenico Scimone, GM, it is believed to be from the 1970s, created in-house and produced by an outside tipgrafia.

Important to write *Conegliano* on the bottles; crucial that they also read *colli*; vital was the third word which appeared largest and drop-cap lettered: here is the label that in 1924 first presented these now household words.

The *Oxford English Dictionary*, the British market then longing for French wines, has found the first English-language instance of *Prosecco* in the November 24, 1881 edition of *The Times*: "Two varieties of white wine, Verdiso and Prosecco, and one red, the Roboso [sic], not entirely unlike the white and red wines of the more northerly vineyards of France."[2]

An entry in an 1887 *Baedeker Handbook for Travellers* lists Prosecco under Trieste's offerings, still "a half-effervescing wine like Asti,"[3] from the land of Martinotti.

7

Promised Lands

In reality there is nothing less static than places.

—Nicola Perullo, *Epistenology: Wine as Experience*[1]

When a quantity is the greatest or the least that it can be, at that moment it neither flows backwards nor forwards.

—Isaac Newton, *The Method of Fluxions and Infinite Series: With Its Application to the Geometry of Curve-lines*

Until the mid-twentieth century, talking about the wines of the hills and particularly the Conegliano-Valdobbiadene zone would have by habit drawn the word *Vittorio*, too.[2] In the colline offset by the Piave on its "calm and placid"[3] run, Prosecco had once been the "principal vitigno[4] of Conegliano but also of Valdobbiadene, Asola [sic], and Vittorio," notes the *Ampelografia italiana* in 1879. "On the plain it has very little importance having been almost completely substituted with Verdiso and in the Coneglianese it is losing ground, disappearing under the scourge of oidium, peronospera and inclemenze climateriche.[5] Now the greater part of its cultivation is only in the district[6] of Valdobbiadene and in the comuni of Pieve di Soligo, Soligo, Solighetto, Farra, Follina, Colsammartino which constitute a series of parallel hills departing from the Piave's left bank to extend from Valdobbiadene to Vittorio."[7] Six years before, "at the end of the past century, and in almost all of the first half of the current one, agriculture was very much neglected among us," wrote Carpenè. *Fu* he slanted that sentence's only verb, a *being* in the remote past for days that were just a very few decades behind him. Now that there was Italy, time was moving faster than it ever had before.

Today's Prosecco Superiore production numbers: 120 million bottles about 82 percent of which come from the Conegliano-Valdobbiadene DOCG (after an 11 percent increase over 2020, the area is now around its full production capacity;[8] about two million are labeled Rive); 18 percent are of the Asolo DOCG. The golden future Carpenè saw as so certain it made the recent post-Venetian disrepair all around him fall faster and further away must still have fallen short of what did happen. "Ma molto cammino ci rimane,"[9] he'd written ten years later in his *Sunto*, within a century of the coming boom. A land particularly well matched to a grape and the best wines it can make is a territory *vocato*—*suited* but with a ringing of *vocation* that carries also the people who make and understand, who take clear Prosecco to full-pressure bottled. Like its vigorous and thin-skinned grapevines and as subtle and as nuanced and as precious, this territory gives just enough difficulty, too. Hills, hard to access, edged inaccessible but not in the way that mountains are, instead shape and allow a specific usage, a giving paesaggio for who has learned to. Forest in the hills is not the forest of the mountains: a forest within and nearby which you live is a very different thing from a forest that does not live among people. As they offered at least one wine-scientist inspiration, are where the aspiration for the finer crops was better motivated, hills also complicate terrain.[10] In this place, grass-terraced, similar in appearance to the mountainous plots not so far away, observes Marino Berengo, a hill is "substantially different in what it yields: the principal crop is the vine, and here we see the few planted to palo secco that the Venetian countryside offers, nor are there missing specialized vineyards."[11] And as these soils and the climate dictate these topographies of hillscape and embankment[12] plotting, so the inhabitants' material interpretations shape the architecture of their place, shelved casere[13]—small overnight shelters built on high and made of hillside, too, for shepherds' runs to greater heights in summertime—or farmhouses called masserie and chiusure by size and here rendered indistinguishable.[14] These hills in which there is always a little wind[15] and made of three geological periods, the extinguishing secondary followed by modernizing tertiary upheaval then morainic quaternary, stones and giant icy movements, broken soils of coarsely crumbling white-gray Biancone, chipped brick-red, greenish chalk veined brown or given over to flinty green pallor, waves settling of layered epochs, are particularly particular. "The same product isn't made in Valdobbiadene as is made in Conegliano, or in Vittorio Veneto,"[16] says Innocente Nardi. "This is not a rationale based on quality, it is a precise logic of territorialità, of entering the territory and explaining the differences." Zoom out, between those first two zones' touchstones of rusted volcanic ferretti and glacial conglomerate and "let's say, the Vittoriese area, Ogliano and Vittorio, that is basically a morainic territory of loose soils and where Glera has found one of its fundamental identities."

In the summer of 2019, much of the Conegliano-Valdobbiadene DOCG was added to the UNESCO World Heritage List, convention created in 1972—

"Noting that the cultural heritage and the natural heritage are increasingly threatened with destruction not only by the traditional causes of decay, but also by changing social and economic conditions which aggravate the situation. . . . Considering that deterioration or disappearance of any item of the cultural or natural heritage constitutes a harmful impoverishment of the heritage of all the nations of the world."—and to be run by an intergovernmental committee charged with identifying sites of cultural or natural heritage in need of protection. Having met "the requirements of integrity and authenticity,"[17] much of the hills is now inscribed on the World Heritage List as a cultural site (one of two awarded that year, the other to the United States for "The 20th-century Architecture of Frank Lloyd Wright") of Outstanding Universal Value based on the following criteria. It "is a viticulture landscape resulting from the interaction of nature and people over several centuries. The adaptation and transformation of the challenging terrain of the hogback geomorphology has required the development of specific land use practices, including: vineyard management by hand on steep slopes; the grassy terraces known as ciglioni, which follow the contours of the land, stabilizing the soils and vineyards; and the Bellussera training system which was developed in the area around 1880. As a result, the vineyards contribute to a distinctive 'chequerboard' appearance with perpendicular rows of high vines, interspersed with rural settlements, forests and small woods; in some of the high parts, mostly chestnut.[18] Despite many changes, the history of sharecropping in this area is also reflected in the landscape patterns." Like a snapshot, the UNESCO designation is an act of preservation. It is also a pronouncement of protection and is a mandate therefore based also on remembering: in Italy *landscape* is another word, it is *paesaggio*, place that I read also into story. It is memory. Lone among its English synonyms—*panorama, view, scenery, visual view*, that *landscape*—paesaggio includes the presence of person. The paesaggio of the chemists I have followed is harder today to imagine than is the work they did. It is helper and hindrance walked and traveled and though source of pride and future for the most part utilitarian, a hilly living with forest that is a managed and used entity. It is an essential part of the medieval borghi that are still foundations here, surroundings reliant on and buffer to what is growing around it, the food and supplies become countryside riches that a nearby city might want and pay for. "Wanting also to examine this region only from the side of the cultivation of the vine and the production of wine," wrote natural history professor Torquato Taramelli in his poetic introduction to *La Vite*, "we find it necessary to begin with some outlines of the nature of the soil and of the climatic conditions, factors which are fundamental to all agrarian production."[19] In the first half of the nineteenth century, writes Gasparini, the Trevigiano was 56 percent "terre alberate, della vigna promiscua,"[20] tree-based promiscuous vineyards against 76 percent of that in the flatlands below. A central concern to Italy later that century was the reasons its wines were being overlooked by foreign markets.

Paesaggio of vineyards, pathways, shelters, and governance: landscape as defined in the Prosecco hills.
Susan H. Gordon

What world has the city been the center of? The school was a fortune Conegliano had earned, Carpenè had stated upon its realization: "Favored beyond measure by nature for its serene[21] position, for its climate and soils suited to the vine, [Conegliano] must be considered destined to become the Veneto region's center of rational wine production,"[22] a bridge upon which to cross the interruptions of Republic fall, misfortune, and wars to reach 1772 when *prosecco* had appeared before the urban agriculture academics for the first time, Conegliano then "still a ferment of research and initiatives" by which agricultural production might be improved.[23] "The area here in Valdobbiadene, with respect to that of Conegliano, can be considered agricultural, says Adami of the Prosecco house his family founded in Colbertaldo, frazione of the comune Vidor where the Valdobbiadenese hills descend to "bath in the sacred waves of the Piave, facing the historic Montello,"[24] based on the viticulture his grandfather Abele Adami, born in the 1880s, did. "Imagine it as two extremes," he says. "Conegliano is the science, the School is there, there has always been the institute of agrarian studies, there was Carpenè who invented a new system of making Prosecco." That *frazione* is an important term, closer to being a number than it is to existing as a word.

One of the first things a united Italy needed was to account for its population by way of a census, notes Italy's census body Istituto Nazionale di Statistica (ISTAT) of the nation's first such count on 8 September 1861: "With the obtained results, in 1863, the legal population of comuni, provinces, and districts was determined." Seventy years later, the census conducted by the fascist government demanded that foreigners without statehood report 'the last citizenship held or, in default, their nationality"[25] at the same time that each comune was made responsible for drawing "its own topographical plan for the Military Geographical Institute's maps at 1:25,000, and tracing on these the comune's boundary. In 1936 came the additional requirement of showing where the populations lived within each comune. A frazione is a minor territorial entity, reads the Treccani encyclopedia, that is individuata[26] "on the basis of its anthropogeographical conditions and legally recognized by an act of the comune," with which it is associated; a frazione is always *of*. The 1951 census defined that *anthropogeographical* as any one of three conditions calculated by structure, people, and gathering place: a small portion of comunal[27] territory that was an inhabited center, a collection of houses with at least fifteen families without the "luogo di raccolta"[28] that characterizes the first, or case sparse—houses scattered along a countryside too far apart to be considered the second.[29] For each it is also necessary that its inhabitants move regularly from the comune center and back home, familiarity and habit that define the space of a comune and presuppose some kind of path between. How to map communal lands, fractured and for generations understood, that exist outside the boundary drawn on a map to define a comune? Does it make sense to read *frazione* as superimposed pattern of urbanization, a tracing of lines of city and administration onto paesaggi complex, older, and unwritten, writing comunal lands that presuppose people into what belongs to the legal comune, which is and whose inhabitants are countable, and locating them *right here*? The census is administrative, too: today, while a frazione is governed by the comune which is also the location of its civil records, it retains a certain amount of its own administration. Obscuring lands rich with unwritten codes and tied knowings of the people who belonged, frazione is also a forgetting that moves its inhabitants into rational nation. "Sometimes, a frazione represents a former comune that was believed to be no longer viable," notes, poetically, site of popular knowledge Wikipedia.

"The boundary of the property is of adequate size, and contains the attributes of Outstanding Universal Value within a topographically distinct and intact landform," describes the UNESCO website. That Value must be an "exceptional example" of a traditional human settlement whose use of the land represents a culture of human interaction with an environment now under threat of irreversible changes. Though it is not made clear which time frame is considered in terms of those changes, the mentions of vineyard practices place what is to be preserved as originating during the nineteenth century, with the hills' viticultural aspect present from at least since the Romans arrived.[30] "Despite many changes and challenges posed by pests, wars, poverty, and the industrialisation of viticulture," reads UNESCO's assessment of the area's application, "many of the attributes such as the vineyards, ciglioni and architectural elements demonstrate a good state of conservation, and the patches of forest have been maintained." Though the people who did so are not mentioned in the UNESCO description, the contadini are almost present in the final determination: "The ingenuity of the *farmers* [emphasis mine], the way people have adapted viticulture to harsh terrain, the mosaic of different land parcels interlinked with woods and the distinct chequerboard pattern of the ciglioni have been clearly outlined." Today, more than sixty million bottles[31] of Prosecco DOCG are exported each year, a global outlook. In a collection of essays published by Conegliano-Valdobbiadene's Consorzio di Tutela[32] in celebration of the denomination's fiftieth anniversary, historical-geographer Massimo Rossi writes that "the contadini's creation of a paesaggio [creates] local stories that resist generalization,"[33] against today's maps that "too often confuse the paesaggio with economic space and not with the luogo of the community." The entwining of landforms and people is also specifically a territory in the sense of an evocation of boundaries: giving hints of the work that might lead to having "riterritorializzato un luogo," a sort of replenishing a place with the aspects that had made it *territorio*,[34] the final page of his essay, fourth of the book's fifty, is an attempt to rescue, a call to the UNESCO inscription as occasion to "activate a process of profound knowing of the paesaggi shaped by the hard work of the human communities that like a palimpsest have succeeded each other and have many times reshaped and reterritorialized these luoghi, in synthesis with contexts natural, historical, social, institutional and economic that have always tied the local to the global."[35] Looking out from a hill onto another, the dark parts are closer up many, many firs. Tucked within hilltops and high-placed landings on the way to them are the small stone casere that were shelter for contadini, shepherd, and wine-grower alike, that held a place for cooking over andiron, glass damigiani protectively wrapped in woven or banded wood and that throughout Italy most commonly held fifty-four liters of wine,[36] set on broad wooden planks cured outside for floors that are otherwise fast drops into the hillside already rushing down beneath them, small trees used for ceiling beams, iron window grates, and outside the curling witches' nails of now that it is autumn a vine that will soon be trimmed back to three, five, eight buds to break into the blue-thin yellow shine of warming late-March days. The stones for these casèi and casère, writes Miro Graziotin, poet, historian, and photographer of the images I am reading from steepest Valdobbiadene just north across the Piave from Asolo, "that draw [their] geometries, in the windowsills, in the corners, in the sills, in the lintels and in the thresholds all come from the valleys in the south between Santo Stefano and Guia ribbed by gentle [*dolce*,[37] he uses] and soft sandstone that here goes by the name of molèra."[38] It is written that popular folklore persists tenaciously in Treviso, wrote journalist Guido Piovene in *Viaggio in Italia*, his accounting of Italy just after World War II. "A doctor . . . told me at dinner about some customs that survive around Conegliano, Vittorio, in the Cansiglio woods. Here people believe in witches no less than in Abruzzo. . . . A primitive moralism persists in the practice of the 'canata,' sinister method of communicating the paese's[39] condemnation [of any of its inhabitants found guilty of trespasses such as adultery]. The young people of the village hide in the trees surrounding the [trespasser's] house. They shout insults all night, moving after each one so that the furious accused cannot discover their hiding places. The revenge of morality thus arrives public, anonymous, as the voice of nature."[40]

"While Valdobbiadene was an area of agricoltori, where there are still the better vineyards in my opinion but where there was not this science that there was in Conegliano, the fusion of these two localities[41] brought an extraordinary equilibrium," says Adami of what by the 1930s was written of in agrarian journals as two subzones of the winemaking hills. He and his brother earned enology diplomas from the school in 1980 and 1977, respectively. Their father, who had never gone to school, "never asked me and my brother if we wanted to go." Nevertheless by doing so, "we simply continued the family's activity," while "clearly we brought everything that a new generation brings." When their father gave them control of the small winery in the early 1980s, they "decided immediately to proceed with spumantizzazione and not to make only the col fondo wine that my father made."

In the twentieth century's first decades, across Italy's once-Venetian lands a dense population filled the little more than 100,000 hectares of land, about one-third of which was Treviso's hills, confirm Montanari and Ceccarelli in their ode to the old republic[42] based on land registries of 1929 and 1942, and not published until the post-war shortage of paper, press, and personnel began to recede, by which time Montanari was an agrarian inspector for the Venetian area. In the 1942 registry, 94 percent of those hectares was "agrarian and forestal" land, and while it varied from comune to comune only one half of one percent of the previous Venetian lands were "nonproductive,"[43] put to neither agricultural nor woodsy use. "Statistical, technical, history, descriptive memory," reads the book's title page—"published under the auspices of the Unione delle Camere di Commerco, Industria, Agricoltura delle Venezie." A work the authors hoped would be appreciated by the "tecnici and the viticoltori," its list of those who contributed financially to its publication includes the "Società vinicola Carpenè Malvolti." In its pages are number tales of the populations of the hills and what they were doing there Prosecco-wise, at least two decades before most but where more now dared dream of the coming millions of bottles: throughout the Colline there were 202,711 present[44] of 206,840 residents on 102,616 hectares, 92 percent of which (94,172 hectares) of agricultural and forested within which

1 percent were specialized vineyards, 43 percent promiscuous. Eight percent, 8,444 hectares, was *unproductive.* In that 4,077.48 hectares of specialized vineyards were grown 5,701,712 vines (1,398.34 vines/ha) and these are noted with a decimal even when, often, to its right a *0*; in the 40,707 hectares of promiscuous, 8,121,400 vines (199 vines/ha), and these, wilder and more fluid, are never noted with decimals. There are 18,352 proprietors (2.44 ha per owner) overall. In the district of Valdobbiadene, 10,279 of 10,425 residents present on 6,070 hectares, 5,672 of which were planted—a 93 percent of the territory that was agrarian and forested, and of which 11.55 percent was specialized vineyards and 9 percent promiscuous ones; the remaining 7 percent, or 398 hectares, was unproductive. On Valdobbiadene's 655 hectares of specialized vineyard were grown 917,000 vines (1,400 vines/ha); on the 520 hectares of promiscuous cultivation were 104,000 vines (200 vines/ha); altogether there are 1,017 vineyard owners (1.16 ha/per owner). "The urban centres of Conegliano, Vittorio Veneto, and Pieve di Soligo,"[45] notes the UNESCO determination. And so, Conegliano: 15,434 present of 16,537 on 3,633 hectares of territory (3,413 agrarie e forestale, 220 unproductive); 18 ha specialized with 25,200 vines, 2,690 hectares promiscuous with 538,000 vines; 638 proprietors.[46] In the Vittoriese hills around the small city to the north, the first specialized vineyards seem to have appeared at the end of the nineteenth century, notes Rossi[47]—in Vittorio Veneto: 23,167 present of 23,475 inhabitants, on 8,261 hectares' territory; 7,313 prod; 948 unproductive); 349 hectares specialized with 487,200 vines, 1,660 promiscuous with 332,000 vines; 918 proprietors. On their summary of vineyard surface across that land, neither year's registry seems to have recorded mountain vines. Specialized vineyards increase from 3,435.00 hectares in 1929 to 1942's 7,388.51—presumably mostly in the hills which counted 4,077.48 hectares in the later year, but the 1929 number is a unanimous one made of hills and plains. The 127,407 promiscuous hectares in 1929 turn to 153,935 thirteen years later, 40,707 of those in the hills.

Examining the play of statistics in nineteenth-century Italy in her *Numbers and Nationhood*, Patriarca notes the outsize role—"almost 200 pages out of 700"—that scientific delineations and descriptions of localities play in the *Filosofia della statistica* of Melchiorre Gioia, who in the late eighteenth and early nineteenth century helped form what is considered the Italian school of statistics. Of the kinds of numbers that helped make a nation, he "insisted that statistics actually comprehended some quite 'stable [invariabili] elements,' namely all the information about the topography of a place," she writes. "As land and climate exercised an influence on economic, intellectual, and moral phenomena, the description of these elements could not and should not be avoided."[48] "I want to introduce a word into this," Nardi tells me of the hills' steep and subtled differences to which the Rive labeling terms are dedicated. "Resilience." The many territori identifiable across the fifteen comuni that make up the Conegliano-Valdobbiadene DOCG "express different organoleptic characteristics through the concept of the Rive." That old word today allows the left-bank hills' more than four hundred producers to convey the specifics found in each comune "across differences in soils but also average temperature, rainfall," and the choices made by the people who have been connected to the more than three thousand vineyards for generations. "I believe that when confronted by difficulty, people have known how to respond to the difficulties of the environment in which they live," he says, "and that the steepness of the hills which differs in Valdobbiadene and Conegliano has contributed to create territorio-level differences which could seem slight for those who see them from a distance, but to anyone who lives with them they are important distinctions." Weekends spent "in riva,"[49] writes Gasparini of those many owners of small vineyards who have for more than a century maintained them after working weekdays in local factories. A vineyard called Perlina, says sixty-three-year-old restauranteur Alberto Resera in a July 2020 testimony for AISO of land his grandfather owned, was famous for its wines. Driving to it, "on the border of Arfanta," Tarzo frazione in the hills just east of Cison and where the Valdobbiadene hills once again rest this time at the Prealps' feet, from the road one could see the farmhouse painted a proprietary red called "rosso Brandolini"[50] for the noble Cison family who owned and rented such properties to sharecropping contadini. "It was a very steep vineyard, everything planted by hand," says Resera. "Every year without fail came the landslides."

How the land moved mountainously in turn determined how the vines themselves were grown and trained. "Undoubtedly in the less steep zones, let's say Conegliano," says Nardi, "it was easier to adopt training systems that I call theoretical, influenced by the proximity of the Scuola Enologica." Of the tendency in the Coneglianese to plant vineyards with plants trained in "perfect examples" of Guyot and of Sylvoz—the latter vines raised on trellises drawn between pali secchi of materials from wood to concrete to steel and named after nineteenth-century Conegliano wine-grower Carlo Sylvoz who drew up this high-trunk, one-armed, up to four annual fruiting canes along it, system widely used in the Veneto. "I call it a more geometric cultivation, more precise, marked by more iron-clad rules," Nardi says. In the Valdobbiadenese instead, "there developed this viticoltura that among other things rarely sees the total uprooting of a vineyard followed by a [full] replanting" that is a practice common to many wine zones. Instead, "producers substitute a single dead vine from time to time, so there is a more historic vineyard, an ordered disorder." This is the "patrimony of humanity" that UNESCO has declared in need of protection, one zone "where the rules of viticulture are interpreted, [and another in which they are] applied in a strict manner. And I think this is a richness of this territorio; we have a way to compare these two realities."

As the nineteenth century gave over into the twentieth, the land was measured. In the small campi, each campo 0.54644808743 of an hectare, of the family farms in the Marca Trevigiana, gelso[51] or more locally, morèr trees in the middle of each vineyard, Primo Franco of the historic Prosecco producer Nino Franco in Valdobbiadene told me at lunch in New York City in 2020 in late February—one of the wine industry's busiest visiting and tasting seasons and as the pandemic was confusedly taking hold both here and in the Veneto—their leaves food for another growing industry that until the 1970s would remain among the hills' most important, sharing space in Valdobbiadene with the grape sellers and brokers it was also known for. From mostly the Alba[52]—*dawn* for the pale color of its buds—variety imported like the industry from China and with smaller leaves with greater sinuses than the black variety native to the Mediterranean came materials that fed the silkworms. Silk was the work of women,[53] who gathered the trees' leaves, tendered them, were hired to sort and grade the cocoons (boiled first to destroy the tiny worker inside), reel as one long thread from cocoon onto spool, or weave into textile in the silk factories built of nearby wood.[54] When it came to wine, women's roles had already been determined. They were expected to weed, to cut a grapevine's unneeded suckers, to pull any leaves that stopped enough sun from reaching grape bunches, writes cultural anthropologist Elisa Bellato,[55] but the yearly journey of the contadine of traditional Veneto ended in the autumn with the harvest, after which they "were practically excluded from the phase of wine transformation." To carry her through the long work of harvest, a contadina might eat some of her gathered grapes, sugary treat that by the end of harvest meant that "the button of her skirt would be moved" to the next thread-hemmed buttonhole—Bellato shares based on the testimony of Giuseppina, born in 1906—"a gesture that reveals other ways of thinking and living a woman's body within essential categories such as strength and health. Feminine worlds are always further away, alive only in fragments of memory."[56] The voices of women in wine in the writing population then must as well be assumed to be contained elsewhere, and especially outside of books. Inside, in the wine cellar the "almost alchemic processes were entrusted" instead to the men in the family. To them "the pride of determining the uniqueness of the [family's] vino di casa," Bellato writes. "The emancipation of women must be continually coupled with the emancipation of the worker, reads, remember, Mazzini's final sentence in his last book, thereby giving our work the consecration of universal truth. Measured and truthful, in every town in the osterie, in the books as men's world, the wine was determined quantity, too: anyone caught using measures either false or different was to be fined for each irregularity. Offenders pouring incorrectly with the right measure, "will be condemned to a fine"[57] for each such action, too. But outside in the vineyards and inside where wine is made and poured there is another order that moves adjacent and it is not the rational one gleaned from observing what the world does or built on rules from new and ignorant rulers but an older one in obedience to what has long been done here when, before, we were Venice, the ombrette long poured into and emptied of the wines made here and whose presence is not diminished by their being also not yet fully realized. In the late nineteenth century as the world here is crossing its own paths again and again, for a moment its scientists and the contadini step together as, zoom out, one walks onto a future road that is being laid scarcely before them and the other along the varied and extant way that is more obviously made also of the past. "An Italian counterbalances at all times," claimed Piovene, "and lives between weights and counterweights, is a revolutionary and conservative at the same time, and this is his law."[58]

This UNESCO landscape is three things. It is the area officially qualified for protection as the *core zone*. It is equally *that* area's *buffer* zone, which in order for the former to go on must be protected, and a boundary drawn up for it, as well. It is also, equally again, a larger *commitment* zone drawn widely about the first two and in which activities, particularly viticultural ones, are undertaken with the scope of keeping Prosecco's rural areas rural. Of the middle thing, explains the UNESCO website, "buffer zones are an important tool for conservation of properties inscribed on the World Heritage List." In the "implementation of the World Heritage Convention, the protection of the 'surroundings' of the inscribed properties was considered an essential component of the conservation strategy, for cultural and natural sites alike. The [committee] valued the concept and included it into the Operational Guidelines as one of the elements to be considered in the preparation of a nomination proposal." The Valdobbiadenese is that core, stepped series of hills from east—near Tarzo just shy of Vittorio Veneto—to west—around the comune of Valdobbiadene—that are in formation stacking and swirls of catastrophic crumbling, glacial push, of waters rushing. Buffering them, the larger zone contained north on the left bank of the Piave river includes the Prealps over the valley and the generally softer hills instantly south, arranged by the same forces with a more scattered sense and making up the Coneglianese. Within the small part of the buffer inscribed in the DOCG, "the plots dedicated to vineyards, established on ciglioni, coexist with forest patches, small woodlands, hedges, and rows of trees that serve as corridors connecting different habitats," notes UNESCO. "In the hogbacks, small villages are scattered along the narrow valleys or perched on the crests."[59]

Out of this. To know someone, it is necessary to know his home, intones Mazzotti in that 1972 Rai[60] documentary in which the hills and their Prosecco and their food. Around and including Conegliano, at the time Carpenè and Vianello were compiling[61] there wasn't much Prosecco being grown, or there wasn't much identified as such: in the Valdobbiadene district the comuni of Valdobbiadene and San Pietro di Barbozza were counting on it alongside Verdiso, a principal grape throughout that district and in practically all of Conegliano and Vittorio. In *La Vite*, the Trevisian land is traversed through grapes and numbers, the latter in the form of averages taken across 1867 through 1871, numerical forms which "are never the true expression of singular terms, but of their ensemble."[62] The future Carpenè saw might have been made also of these kinds of numbers, nation as a tale of rising to the occasion, natural state, and statistics. "The Province," he and Vianello set forth[63] for the tables they made of grape variety and planted land, "is divided administratively into eight districts and 96 comuni. We will follow this division because it has already passed into everyday usage, and the district is by now considered a unit." In the district of Conegliano in the area's eastern hills, the comuni of Pieve di Soligo (3,252 inhabitants; on 1,554 hectares; 1,054 of which planted to vines), San Pietro di Feletto (2,423; 1,863; 1,063), San Vendemiano (2,623; 1,761; 1,174), and Conegliano wherein the city, too (7,614; 3,474; 2,904), predominant white varieties are Bianchetto [sic] and Verdiso, around today in small (the former) and a little larger (the latter) numbers. North, the district of Vittorio grew the same, as did the district of Asolo (steep Monfumo, 1,260; 1,104; 924) and Valdobbiadene back across the river, and in the comuni of Valdobbiadene (4,885; 3,079; 909) and San Pietro di Barbozza (2,761; 2,532; 1,932), grapes called Prosecco dominated alongside them. Verdiso from San Pietro di Feletto was—says its price of 60 lire per hectoliter to prices of the others in the 30s and 20s lire—the most prized variety and provenance. With less than three hectoliters of wine produced per hectare, the districts of Valdobbiadene and Vittorio were far above the province's average yield of 1.28 hl/ha—and further still from today's specialized-vineyards expectations of fifteen and twenty-five hectoliters per hectare for a long-lasting still wine, and up to fifty for everyday or other styles. Of the province's ninety-six comuni only four grew enough of the Prosecco grapes for the vines to figure as primary cultivar—to the two in Valdobbiadene's district add one in Conegliano's and one in Asolo's—while fifty grew Verdiso (23,445 hectoliters a year[64]), and thirty-nine cultivated that Bianchetto (14,466 hectoliters a year[65]).

Venice and its riches were gone. But its lines of aristocracy, Treviso's ancient nobility who were "not disdainful of innovation, who followed the gusts of rinnovazione"[66] remained, links from past glories to future realizations. And were a protective layer against those conservative and unknowledgeable owners. In both written research and in the vineyards they owned, the questions the noblemen had been asking cast the first steps of a golden path a visionary might follow to what would soon surely exist. In commercial terms—counts of, made to be sold—during the four years counted by Carpenè and Vianello, an average of 3,709 hectoliters were produced each year.[67] One obstacle that looms large in "our paese," they wrote, is that many fear statistics are a front for new exorbitant and arbitrary taxes and therefore do not understand how they work. "Some consider averages to be poems, probably they don't realize that averages are the basis for insurance, which brings many benefits,"[68] they added, more practical poetry in mind. Following the line to the future, the "old aristocratic and noble class begins to avail itself of the new technicians: the enologists"[69] writes Gasparini, but in the nineteenth-century vineyards there were at times further partnerships—between some of the hills' contadini, aristocrats, and scientists who overlapped on land, knowledge, observations, though in written records it is only the latter two we hear about and from. To the scientists of chapter 4 add the Collalto family who, Gasparini writes, planted "34 small hillside vineyards on 10 hectares with a high density,"[70] scenery possible only when planted without trees. The Papadopoli[71] who planted "160,000 vines in 30 hectares of surface setting up 25 new vineyards."[72] In Pieve di Soligo that plot of the Balbi Valier family, palo-secco-held vines of Prosecco which would become "an out-and-out cru,"[73] in conglomerate soils that, writes Taramelli, "contain gigantic specimens of Ostrea longirostris [and] correspond to the reliefs in [this] group of hills." Add the red-farmhoused Brandolini who owned most of Cison del Valmarino, the academics and the scientists from Conegliano, and amid the obviousness of practicality, I hear murmurs from those whose names are missing here. "They asked the contadini to try this, try this, and then talked together, scientist and contadino," explains Cristina Munno, then Cison's culture councilor, as we walk with a forest guide[74] along the Strada del Lat, walking trail that today connects a series of older forest roads, and finally I hear their voices. The contadini who had planted and trained and pruned and harvested and watched spoke and worked with those who wanted and provided for the experiments, whose hands and practice did that work of science, too, and together they moved those lines. In the written reports of daily observations in the vineyard, even though the latter are not credited, eno-science must have necessarily been a collaboration[75] between the scientists and noble owners and the contadini, all of whom were there, in plots, attention trained among the vines. "Considered as a class, the contadini were truly another world with respect to, not the nobility and the alta borghesia, but the common people of the city and of the large towns,"[76] oral historian Coltro writes, appearing to gesture toward a similar scenario, those very same entangled lines. "The altimetric measurements, except those taken from the military topographic map, were calculated by the writer following barometric observations made in his excursions," Taramelli offered another glimpse of mixed scientific work, in a footnote.[77]

In 1903, science-fiction writer H. G. Wells would famously note that to be an "efficient citizen of one of the new great complex worldwide States that are now developing, it is as necessary to be able to compute, to think in averages . . . as it is now to be able to read and write."[78] But leading to and through the first days of Italy's statehood, see in such numbers a compilation of public life, statistics "a science . . . very far from being a technical knowledge in the hands of a group of specialists,"[79] a democratic, "useful knowledge" serving statesmen and "members of society in general"[80] and which in this coming state's preunification nineteenth century allowed criticism of local Austrian authorities throughout Lombardia-Venetia then, thanks to Lombardian printhouses, a center of Italian publishing. Then, "all publications were of course subject to censorship,"[81] writes Patriarca: and so statistics could act as secret code, cloaked by numbers' respectability, adherence, and seemingly rational one-meaning. "This statistical weapon was less worn out and blunt than the lamentations of historians and the denunciations of poets,"[82] she quotes Cesare Correnti, patriot, Milanese of noble family, upon unification to be minister of the Kingdom of Italy's Public Instruction department under Ricasoli. "The discussion of averages, tables and numbers . . . allowed us to speak in jargon and in code, and to avoid being mutilated by censorship, which through long practice was capable of sniffing only at phrases and epithets,"[83] he wrote later to a leading Italian statistician. Rationality was no less capable of passing information among exclusive groups than what had been before it, and could be just as silent to the outside. "The thing was done. Numbers told their secret only to those who knew how to read their code; they were the true language of mutes."[84] Here was a way to order the world in accordance with the patriots' vision, a nation belonging to the land and people it was made[85] of. One such ordering model published in 1830 opened with the "*stato naturale* (topography, climate, natural products)," describes Patriarca, "followed by the *stato politico* which included population (its state, increase or decrease, mortality), 'public subsistence and preservation' (food, dwellings, public health), 'public industry' (economic activities from agriculture to manufacturing industry and commerce), and finally 'public spirit,' namely a description of popular mores and the forces able to modify and direct them."[86] Here was an orderly building up of nation from always already existing earthbound matters, and the reliance on numbers carried easily through old Venice's generations, helped by, Patriarca notes, the fact that publishing was in the mid-nineteenth century mostly come from this territory thanks to intensely commercial Milano. When in 1866 the Venetian territories separated from the Lombardian and each joined the new nation as discrete regions, the stories and hopes the codes of statistics, chemistry, and patriotism had been telling could finally be said out loud, and hopefully in conversation with the contadini. And at least among those who would grow grapes, old, old viticulture might be another bridge.

The district of Conegliano, head city twisting into those hills with at 125 newly Italian vineyards by far the Trevisian district with the highest number of them, was also about one-third hilly. Asolo, twenty-two vineyards, was almost three-quarters so; Valdobbiadene with twenty-four including ancient and expensive Cartizze, the golden pentagon from which in the distance appears the small chain of Asolo hills, writes Prealps mountaineer Giovanni Carraro[87] of walking these 107 or 108 hectares entirely of slopes.[88] In Conegliano, the Società Enologica that would at the hands of Carpenè and Malvolti become their winery owned four vineyards composed of 3,510 vines on fifty ari—one hundred square meters. The noble Balbi family held ten vineyards across six comuni east and west—Abele and his fellow growers, says his grandson, "bet on a vine, on a grape that was prized and that had a certain bunch size and a certain length and was called Prosecco Balbi."[89] Other noble vines: Collalto held ten thousand vines in twenty-six vineyards in Refrontolo, same in eight of them in Susegana; Papadopoli with thirty-two vineyards in three comuni. The Brandolini—*Brandolin* is their entry in the table, the zone's habit of dropping an end vowel—with a twenty-five-hectare holding in Cison, and two other vineyards, one hectare each, elsewhere in the Vittorio district. And Malvolti's three in San Floriano near Valdobbiadene, 14,166 vines on one hectare and fifty ari,[90] and the determinative three vineyards including Prosecco (is it Balbi?) in San Pietro di Feletto, 11,250 vines on one hectare, twelve ari, fifty centiari.

Aspiration is part and parcel here.

"My grandfather was a viticoltore," says Adami. "Or rather, I should say he was an agricoltore, because he had vigneti, he had livestock, he had campi of mais, of grain, because the agricultural economy of this territory wasn't made, in the early 1900s, of only viticulture. Viticulture was already, though, important for the territory. Because of the very steep hills you can't [extensively] plant wheat, you can't plant corn, you *must* do viticulture or you can plant chestnut trees, for the chestnuts, but you can't do anything else. And so the viticulture that developed in this territory was already an esteemed one. I think it was called *intensiva*; that is, an area filled with grapevines." Carpenè was lungimirante, *long-sighted*, says Gasparini. But, he added, you still needed a lot of time to pass that science to the contadini and that wasn't a simple task: "the contadino was always very conservative and before accepting a novità he needed to be sure." And later, once Prosecco has become grape, it can enter into a new discussion of the territory, one fully materialized by 2019 when Diego Tomasi, since late 2021 director of the Conegliano-Valdobbiadene consorzio and before then director of the Conegliano branch of CREA, writes that a "vocazione viticola"[91] points to land whose "particular predisposition is identified with a cultivar or a specific method for obtaining a wine."[92] In these hills, he continues, both elements come together so that Glera (we are well past 2009) finds in the climate and soils it is offered the perfect conditions to grow into a grape whose best wine is made via the Italian method.

Day by day, this wine and these hills commune, saving Glera from the "banal and anonymous character" it develops in places that are, flat, dry, hot, not suited for them. Here is what the variety needs and how, well-suited, it tells you about these paesi's natural states: early budding, Glera needs springtimes that bring mild temperatures by late March—"only the exposition, the steepness of the slopes and the barrier of the Prealps that protect from the cold northeast winds guarantee these excellent thermal conditions," writes Tomasi. The vine's high need for water, "due to its considerable growth in June"; its leaves' highly dense network of stomata (tiny passages connecting inside and out by which the plants exchange their oxygen and—dependent on weather and run by water pressure from roots to air—water vapor for CO_2 in the surrounding air[93]); its yield fought back to generous; its varietal scant resistance to drought and dry conditions are all met by the time and amount of rainfall here: which "must guarantee at least 650/700 mm of rain from April to September, which despite climate change are still supplied."[94] A large part of what determines the kind of wine a variety will make is its ability under some available growing condition to produce grapes with enough (not too much[95]) sugars for the fermentation process and hold onto enough (a good amount of[96]) acidity while doing so, two processes in some degree of seesaw relation as the fruit ripens to full flavor. Grapes destined for spumante, wines whose definition is that they be elegant, focused, and taut, need higher acidity levels. The Glera plant, Tomasi notes, does not easily accumulate sugars in its grapes, but too easily drops its levels of green-apple-like malic acid. "Only summers that are not too hot and Septembers already heading to cool nighttime temperatures and mild daytime ones, can guarantee grapes suited to a spumante base wine in which also fruity and floral aromas are the fruit of suitable thermal values." Now that this basic quality expression is possible, so are more precise tastes of the differences within the province's hills, which wrote Taramelli 150 years back, hold 2,325,322 pertiche censuarie,[97] agricultural land of which 1/23 lay uncultivated and through all of which ran 240 "useful" rivers, torrents, and streams. About 173,000 heads of livestock pastured there. A population of "348,473 inhabitants took delight in the products of the soil and of their own industries."[98] Like Venice but in different configurations, water is everywhere in the Alta Marca, and hence so are fish, brought from Venice's sea *and* supplied, like trout, by the fresh waters here (around Vittorio Veneto there is the small bottom-feeding large-headed and unscaled marsòn,[99] fried).

There are two ways to communicate over river: by a bridge or by going in. "One of those harmonies from which nature draws so much vagueness and which are translated in language and poetry, because universally felt," Taramelli splashes off into his topographical outline, "links the vine to the hill and to those stretches of plain which, like it, present variety and moderate amounts of alluvial elements and a steep slope; so as to be markedly distinct from the monotonous, boundless, marshy plains, where, although the ox gets fat and the harvest full, the desire for a more pleasurable horizon stings continuously."[100] The boundary of the core zone, "focuses on the main areas with the hogback landforms and vineyards, which is the area that demonstrates the greatest authenticity in relation to the justification of Outstanding Universal Value," declared UNESCO. "It is more clearly delineated by natural topographic characteristics, and has a more consistent representation of attributes of the proposed Outstanding Universal Value,"[101] traits that have been "documented through sources such as inventories and cadasters, historical and religious paintings, and historical documents."[102] Cima, from the gold of San Marco, the golden-yellow hay mist of dreamed Conegliano where the vineyarded faces of Zuel near Cison welcomed him precisely home, waved the olive groves on the hill of Conegliano's castle, the cliff by mountain path to the church of Santa Maria dei Caminesi di Soligo, emblem of an old noble family of the Marca Trevigiana perhaps relatives of the Collaltos and long destroyed, where at his feet waited viola del pensiero of crinkling bottle-green leaves and famously of San Girolamo in the desert whose dryness translated here could only be rocky and lush thin soils, frazzled lemon sun of the meadow aquilegia comune flower. "With even softer undulation," writes Taramelli, "but with less orographic regularity, the hills of Collalto, Feletto, Refrentolo, Formeniga, Conegliano . . . rising to a maximum of 250 meters above sea level and furrowed by the Crevada, Cervano, Monticano," on their way southeast. The ridge of Montello rises to 282 meters above a base" already at 120 meters above the sea. "It is a beautiful hill," he sighs, "covered with a dense layer of vegetal soil created by centuries-old woodland vegetation. Reduced to cultivation, on its southern side it would be a real garden. As it is, without much advantage of the treasury, it is a perennial occasion of theft from the inhabitants of the villages that surround it."

Here the forest is proto-industrial. The movement of waters for long meant mills powered by the constructed wheels they rushed through and another, rural, Ruio trickles and falls on its way to Cison del Valmarino, crossing that border as a stream along which stood the nation's first cooperative latteria founded by a country doctor named Luigi Alpago Novello before it ends in the Soligo river without ever leaving the comune. On its course that countryside Ruio once turned a dozen mills so that along its banks its inhabitants "have handed down over the centuries one of the most fascinating and ancient professions, that of the miller," notes the twenty-first-century guide to what is now a trail named La Via dell'Acqua e la Via dei Mulini (all of the latter inoperative now), that follows the small valley the river grinds there. "Only imaginable [today] is the frantic coming and going of donkeys and mules loaded with sacks of grits and flour, cloths, coal panniers, the busy shouting of the artisans, the inhabitants of the villages and the merchants who frequented them." It had flooded in 1882. Abundant public fountains[103] linked to the modern aqueducts of the late 1800s and early new century, underground canals encased in cast iron pipes strong enough—the metal's thickness and the tube's circumference a function of the internal pressure water exerted, as told by the 1686 formula[104] of seventeenth-century physicist and church elder Edme Marriote[105]—to carry three hundred liters of water per second from drinkable springs and lakes. Born in 1943 just south along the Soligo river, from age ten Ada Mazzero has lived in Cison's Rolle (wherein one of the steepest and most picturesque vineyards among the hills' many) and when she marries years later, her husband buys seven hectares here, forest and vineyards of Glera, Malvasia, and Bianchetta. Women were essential for managing a vineyard, she says in a July 2020 testimony for AISO, with precise duties that included "resteàr su par le rive tra i zopài"[106] and "far i mar."[107] Turning willow leaves to ties that would fasten grapevine to training wire, she helped her husband "far fòra e sache," and "tor su i stèch," bundling what was removed during grapevine pruning, to be used as firewood. In the same interview, her sister-in-law, Ambrosina Gallon, born around 1930, remembers being sent "a netàr el bosch," to clean the forest underbrush and gather branches and leaves for their animals and for the fire, and that their vineyard held pear, apple, and fig trees alternating with the vines while, hidden from the landowner—the Brandolini family owned most of the land in Rolle, notes another resident interviewed by AISO—beans and tomatoes were planted at the base of grapevines so that the family would have more to eat. Close by, a torrentello runs a small mill wheel, wooden and stone steps behind it carry you by forest in and out of Zuel, and since the 1990s there are olive trees on a face of the hill below. Before the Prosecco boom, Gallon notes, the families here mostly lived on what they grew and made; what money they earned came from selling wine, milk, figs, and silkworms, and so, reminder, there were also many, many mulberry trees. The advance of woods is the loss of conoscenze.[108]

The hills Carpenè walked bustled with another movement aquatic in nature: daily, women went to the fountains where important roads met in the towns left of the medieval borghi to get to water that they could carry back home, until starting in the 1950s a new infrastructure built of iron and wood began to bring it to them, from the paesaggio's potable springs and riverheads, from reservoir lakes, instead. In the spring of 1882 comes an alternate latteria for the hottest months: at more than seven hundred meters above sea level it will be a summer rendition from May to October, "when the one in Cison is closed partly for the

In Cison di Valmarino, preserved remnants remind us of the many mills powered by waters of the Prosecco hills.
Susan H. Gordon

higher temperature and partly because almost all the cows are brought to the mountains' summer pastures"[109]—extensive grazing land and the customary presence of the cows, the use of nearby cold, cold water for cooling. I have considered rivers only for their flooding, their fish and their connective tissue, stoma-like exchange of cool and warm air, entering into conversation with air and land and people. But they are industry, too, given to power and transport. Bisol, those noted ironworkers[110] with a wheel in Vidor (where "in the 1800s there was a presence of Verdiso, Bianchetta, and Perera," says Innocente Nardi) turned by the torrente Teva on its way into the Piave at the meeting of the UNESCO core and buffer zones—Berto in Piave, clouds over the river that mean rain.[111] "In the limestone or mountainous region, just as landslides are frequent and vast, so there were in various eras frightening cascades of rocky masses protruding from mountain slopes,"[112] writes Taramelli. Like those masses, rivers are dividers, too, and hence place and reason for gradation and divergence, advantageous when it comes to great wine's reliance on particularity. "With the phonetics of our dialects as well, of Valdobbiadene, of Miane, of Col San Martino,[113] of Pieve di Soligo," says Nardi, there "are differences in ways of how to express oneself."

In 1920, Abele Adami bought a small plot from Conte Balbi Valier,[114] a land with which to begin again," notes the Adami website today, and whose fruit he began to ferment and vinify on its own, against what was the larger practice of blending sites, early tastes of what is today those Rive-labeled wines. "My father continued to keep them separate," Adami says, "and this became the philosophy of the azienda. We *continued* [emphasis his] to do this because every hill is different from an other, and then there is each hill's east, south, west, there are different altitudes, and you need to gather each one's grapes in the right moment." In 1933, Abele carried two bottles of Prosecco—"what to him were the two best wines," narrates Adami—to the first Mostra[115] of the vini tipici d'Italia, held in Siena, Chianti's centuries-old tuscan rival: one from the Cartizze and one from Balbi's plot, *wines that represent, per mo di dire, the climate and the terrain upon which they are made.*[116] That plot is in Colbertaldo, trained along lines on stakes made of the chestnut wood that has been here for so long, and rooting into limestone soil, thin layer over the rock it is weathered from and which at times can be seen through that earth. The vines cling to the chestnut poles, winding in parallel rows, all facing south, "with the nervous gait typical of these slopes."[117] Descendants of the mostra, local exhibits still take place in the paesi nearby, in Vidor, in Santo Stefano, San Pietro, Adami explains: we producers bring our base wines, "that in the springtime are not yet spumante," to be tasted. "I didn't learn to taste at school," he says, "I learned to taste from my father," who carried the conversations between *his* father and the other producers of the area. Ninety-five percent of their vineyards are in Valdobbiadene, and when there is heavy rain, "you have to pick up a hoe and go into the vineyards to direct where the rain goes," to head it off from somewhere that could damage the vineyard, particularities known on foot of a luogo where the vines are being grown. "It's not the rigor of quality," he says of Valdobbiadene's especially structured and lifted Prosecco, "which would be offensive to Conegliano, but it's the rigor of provenance, of having learned and been here for generations. I am very, very tied to this territory." Blessed is he, of knowledge passed down.

Behind the houses in tiny Tóvena, frazione of Cison, rise sharp hills rounded at the top in suddenly tightened diameters, covered in what looks like pristine green, the numerosissime trees that make up forests. But once, a lot of that was vines. "In the middle of the 20th century, depopulation was a determining factor in the decline of viticulture, leading to the abandonment of inaccessible vines in the wider area and increased forest cover. There has been a revival in the central region in recent decades due to the commercial success of Prosecco,"[118] reads the UNESCO determination. That we have always lived in relation to forests is a seeable thing here. In testimony collected by AISO in July 2020, Augusto Vidor, who lives on six hectares bought by his grandfather in 1893 upon return from emigrazione to America in Rolle, remembers the many apple orchards ("Presso Dolle verdissima di meli," *in Rolle the deep green of apple trees*, wrote Zanzotto[119]) that stood where now the forest does, until the 1950 brought changes to the market: "There, there, over there, there was no woods! It was all meadow," and fewer vineyards. "Twenty years ago, maybe thirty, they began to make vineyard, more vineyards, there were some before, but they were organized in the old way. They say it's ruining the environment, that it's de-foresting, but that isn't true, before the 1960s these didn't exist. You'd have to pull out at least sixty percent of the woods here to return to how it was then." In World War I, he explains, as a year of battles raged[120] along the murmuring river, the trees that were cut down in order to dig trenches left the residents with large tracts of earth ready to be cultivated. Of Rolle, too, and born in 1940, Oreste Gallon told AISO historians in July 2020 that his father had owned vineyards, too, and was one of the few small owners in that land largely held by the Brandolini family. In its steepest reaches they grew Verdiso and Bianchetto [sic]; stables held cows for milking in the earliest morning before the daily vineyard work. Which came first, the viticulture or the vine? "The area in question," writes Tomasi, "or at least its western sector (the Valdobbiadene area), was thus formed and shaped over the course of time from a difficult morphological and historical starting point; in fact only the vine has enabled a rural economy to evolve in this area, which was followed by human settlements built with respect for local natural resources and the characteristic aspects of the surrounding landscape."[121] In the hills of Treviso, what had gone wrong were components of the political, but newly the natural state was suffering, too. From trade with the outside had arrived those two new and devastating additions to the pressures on Trevisian vineyards, with the third one still silently gathering, all ending the great vintages of the times before the crittogama, lamented and general-termed vine mold that fills Italian wine literature of the nineteenth century, came, and that would be mostly dealt with later by the element of copper.

Agricultural, mercantile Valdobbiadene—where "along the slopes of the limestone chain there are also some marls, red or greenish, alternating with limestone,"[122] wrote Taramelli of these not very compact rocks easily erodible and in places changing mix percentages from step to step—is steep and still more fragmented, lands stitched together by undulations thrown into leafy or lignified relief by the traditional girapoggio planting patterns that direct the rows of vines around a hill into each stumble and lift to reach back around in joining, or dead-ending into forest where a hillside's slant gives it instead an exposure to the cold and shadowed north. This western side of the Conegliano-Valdobbiadene DOCG is in a sense two-centered, in the sense also that geologically it spills east of the Soligo, roughly thought of as the dividing line of the two subzones, if one drawn guided by conventions and proximities. Students from the school in the technical heart of Prosecco Superiore cultivate vines and make wines here with what knowledge they've gained, as they do in Asolo where in addition a centuries-long line of writers has flourished since the Rinascimento,[123] but as a luogo of nearly inaccessible places Valdobbiadene maps onto that center its own old vines, thick trunks and unevenly bent limbs replaced only upon death and never by divorce of row or plot for the higher yields that younger vines supply. Of the nineteenth-century vineyards here, writes Gasparini, 44 percent were deeply measurable, specialized onto palo secco, 56 percent a wilder if just as counted series of plots of patient tree trunks and branches and leaves and fruit imposed upon by another woodiness as grapevines infamously festooned through them. The latter are, remember, plants with an instinct to either lay low or keep on climbing. And in Austrian-ruled 1846, a few of these hills' wines, joined in the region only by ancient Valpolicella, Bardolino, and Soave, could survive the bottles and time of export.[124]

According to 1938's *Tradizioni popolari della Marca Trevisana*,[125] on Good Friday upon night's fall all the hills sparkle with lanterns that are the shells of snails, filled with oil and supplied with wick. This little book is published during the fascist years, ones that "reignite interests in an oral world," folklorist and oral historian Daniela Perco said at AISO's Cison project in July 2020, and the writing inside is "like tourism," with no attention to social structures, to families, to thinking, just "showing the most obvious." Until recently this ahistorical view in folklorism persisted, she explained: "good in facts, but what does it give back to this community?" What does it restore? When I read through the book, the scenario is even worse, contadini presented as if on display; the ugly word *zoo* comes to mind and I move over each page quickly, to glimpse those facts in the hope of somehow hearing out of them all these people now gone. The contadini also cultivate, Carpenè and Vianello wrote, granoturco, frumento, bozzoli,[126] soil and industry companions of the vine. "The training system and corresponding planting patterns, both variables related to the soil types," writes a century and a half later technical Tomasi who also points out "Sylvoz in the [softer] area of Conegliano [where rows—each with vines 1.2 meters from each other—3 meters apart are a planting pattern that allows about 2,800 plants per hectare[127]], Cappuccina stretta [two arms tied back down to the trunk for the older, heart- or person-shaped vines with which growers fight the crittogama peronospera by pruning tangles of leaves into passageways for breezy air] in the high hills of Valdobbiadene and a form that almost incorporates the two in Valdobbiadene's lower areas." Zooming out again, the Bellussi brothers' eponymous vineyard-training method conceived of "to fight peronospera[128] at the end of the 1800s, provides a geometric arrangement of planted vines,"[129] trained on wooden stakes of three or four meters crossed with threads of iron and shaped into a larger pattern that "seen from above appears as a giant beehive, geometric needlework of nature, which draws the territorio in an almost abstract way."[130] Training techniques, the number of fruiting buds left to each vine, the management of the leaf canopies, whether in hilly vineyards rows run rare rittochino,[131] up and down and easier to access, or following that girapoggio jiggedly perpendicular to a hillside's shifting vertical axis and against erosion in the steepest and therefore soil-poor sites, "are some examples of viticultural choices selected and adopted explicitly for Glera in its elected environment which in this way becomes still more exclusive for this winegrape variety," writes Tomasi of the contadini's mutual fine-tuning of grape and sites throughout these hills.

Shaped by their use, the hills are nevertheless for looking at as well, for taking pleasure in the aesthetic made from the centuries-long work by one's own arms: "Che bel panorama! che ti prende sempre il cuore,"[132] says one grower in especially steep Rolle, industrialist whose grandfather and father had grown vines here and who slowly found his way back, metal sharecropper, industry from publishing to kitchen appliances during the week then on the weekends tending a small family vineyard whose grapes he sold to local spumantizzatori. "Io è vent'anni e più che son qua: ho ancora da stancarmi, ho ancora da stancarmi. Perché la collina, qui, a 360 gradi . . . guardo di qua, guardo di là . . . è come il mare, io sono appassionato di mare, se andate al mare: ti siedi su uno scoglio, guardi le onde . . . guardi il caminetto, alla sera, seduto: non ti stanchi mai, stai lì ore. Così è questo panorama."[133] Turning to look at this grape, in the vigneti of his grandfather there were "vines of every type, uva dolce, Malvasia, varietà scomparse (but now almost all recuperated)," various clones of Prosecco to which Rolle is promised, "due to the clay-ey subsoil, which induces mudslides but at the same time confers a very much appreciated minerality[134] to the wine."[135] In some accounts, Perera is the grape of Valdobbiadene; Verdiso, grape of Conegliano and first to flower and last to harvest; Bianchetta Trevigiana, grape of Asolo, whose population, Dalmasso and Dell'Olio note of the 1930s, is 10,003 present of 10,042 official and as such is the most stable of the hills'; here, 1,129 vineyard proprietors guide the vines of a 4,543-hectare territory (of which 4,335 hectares are productive, 77.38 hectares hold specialized vineyards with 108,332 vines and 3,200 hectares are planted promiscuously with 640,000 vines[136]) within the province's 232,532.2 hectares held by more than 40,000 possidenti. In 1873,[137] the 21,806-hectare Asolo district[138] accounts for seven thousand possidenti, "therefore the average possidenza in the Distretto is of 3 hectares." Two hundred and eight hectares of Asolo are nonagrarian and unforested.

In Carpenè's century and tapering into now, movement has been not just of the land, but *in* the land, grapes to tini to botte to osterie, to the bottles that take the place of the last damigiane and move the wines beyond the rivers. The slow-chomping and then turning spinning silkworms whose resulting bozzoli's importance was on par with what was taken from the woods—in 1915 the school's director announced that a forty-five-day "Corso teorico pratico di bachicoltura e di gelsicoltura" was to be offered there; for eight students, "young people chosen from the agricoltori-bachicoltori of the Province of Treviso," who would also be offered a small stipend and travel expenses "between home and Conegliano."[139] The cows, in this land since at least the fourteenth century of sheep for giving milk and wool,[140] swaying as contadini shepherded their livestock high in the hills to eat while they themselves rested, drank a cloudy wine, moved back down to where on the hills the flatter lands were countable steps away. Lines are the railways and the rows and the peronospora-fighting vineyards and are also the slippery contours of the hills and training wires and vine trunks drawing choppy but accurately cleaving and protective girapoggio, and vinous arms reaching through the trees. "Supposing the Province divided into eleven parts," wrote Carpenè and Vianello in their book's "statistical summary": "there are seven planted to vines and four without vines. Among the non-vined there are the fields, the pastures, the Montello woods, the hills' other sides [what given hill placement and/or surface angle faces north, too cold and sunless for crops] almost all forested . . . the foundations of the houses, the vegetable gardens, etc."[141] White and red grape cultivation was five to four, white grapes the ones of the hills; the general system viti a filare, grapevines guided high up into trees and rooted where they are, too, the paesaggio use and moving and for now still beyond the reaches of someone else's industry. The *coltivatori* are fond of their patria, Carpenè had written, but the word is a substitution, a translation of Mazzini's workers who were not to be found here. "There are no real *proletariats* [word as word mine]," the scientist-winemaker writes without adding an unnecessary *here*, "by which word we mean to indicate individuals who live almost as nomads and day to day by the work of their own arms. The tenants of the chiusure are also called braccianti, because they work with their own braccie.[142] . . . But they have their home, their land, and their share of products. . . . Our contadini carry out agriculture according to the traditional understandings to which they are tenaciously bound."[143] In the hills, trees for the vigne a filare were the ash-like orno,[144] the acero[145] of the city's plots, and sometimes the cherry. As in the flatlands, poplar was what is planted in substitution of a dead tree.[146] When the denomination for Conegliano and Valdobbiadene's Prosecco was created in 1969, "defined and uniform criteria" came with it and "there can no longer be fruit trees"[147] those small vineyards, in these plots of differences now to be written (and *filare* co-opted) with trellises and wooded stakes.

The podesteria of Treviso that is now the province (the boundaries might have shifted, but the concept as entity and territorio has not), across 365 maps is "measured and made visible," writes Rossi of a late-seventeenth-century work by around a dozen surveyors[148] of Treviso's 251 villages, "operating also in Asolo,"[149] at the decree of the Venetian senate—and who laden with their instruments and recording paper traveled from town to town, moved intently and daily along hill to hill, for nine, twenty-seven, forty days worked to access views obscured by groves of cultivated trees, interrupted by canals or streams, homes flanked densely by "broli, orti, edifici artiginali,"[150] writes Gasparini. "It is not by chance," notes Rossi, that the hill territories, including small, steep Colbertaldo, were among the last to be cartografati;[151] "the harshness of the luoghi, the steep drops of the ciglioni and the depressions" required the surveyors' extra technique of marking point by point onto a horizontal paper surface where a lead-weighted line fell vertically along an fresh-air x-y axis—"with an extraordinary precision that remains verifiable today." Colbertaldo, an early eighteenth-century map shows,[152] saw a difference between how the hills were planted and how the flatlands. Running downhill from a central villa, "the immediate hilly zones" soared up and down from a masseria, from a few instances of casere in sizes from pointed-roof square to rectangular compound fitted with windows and stall and occasional tree, from at least one Venetian country villa, main body between two agricultural wings, and where the land stayed steep it is drawn into small wedged and numbered plots. It is a place where "generations of contadini have given form to a paesaggio above all useful and expressing a material culture and a patrimony of knowledges and management practices that are a resistance to the generalization of maps," writes Rossi of this *paesaggio*-reflecting image that tells the "importance of understanding local stories that make places noninterchangeable." Where it widens temporarily, Colbertaldo's survey breaks into "pascolo vitato," spreads into more spacious pastures still to the north, and where it is flatter and more uniform it is recorded as one large—drawn by one line and a single number—parcel. In the measurers' hands Colbertaldo is a mosaic of plots in straight lines and craggy contours both, each marked with a number between four hundred and six hundred, land already more counted than it seemed.

"There are a number of elements which currently suffer from a poor state of conservation (particularly buildings and vernacular architecture in the property and buffer zone)," reads the UNESCO determination. "A detailed and systematic assessment and monitoring of the state of conservation of the attributes of the proposed Outstanding Universal Value is recommended as a priority."[153] "With extremely rare exceptions," wrote Carpenè and Vianello, revisiting the Marca's building types, "all of the terrain of the Province is divided into fractions of two types; one which extends one to three hectares, to which is added a small house with stable and hayloft for keeping two animals, and these are called *chiusure*; the others extend ten or twenty and at times twenty-five hectares, with a larger house and a stable with hayloft for eight or twelve bovines, and these are called *masserie*. These fractions of the provincial land form many farms one independent from the other."[154] In a chiusura, one or two bovines, cow and calf, and the small size of the land under care means the work is all done by hand; in a masseria one can have a bull, for working the land, too, which is large enough for that. But these distinctions vanish in the hills. "In the colli there are instead the smallest masserie, only a few of which surpass 10 hectares and run smaller and smaller until they reach the chiusura and are confused with it."[155] While the contadini account for 5 percent of the owners of such structures and accompanying land in the plains, 40 percent of the hills' farms are contadini-owned—in both cases they hold the smaller farms which in the flatlands measure ten hectares, in the hills less than one.[156] "The workers of the upper flatlands and the hills are more active than those of the lower flatlands,"[157] Carpenè had offered in unabashed favoritism, in support of which I think he would have pointed to their vineyards. In the plains there is one chiusura to every three or four masserie; "while in the hills [fitted with those casere, places of rest and restoration for the contadini[158]] it is impossible to set a proportion as the two structures are interchangeable" structures for living, equal exigencies of living here at all.

As I look for writings on the contadini, what there is are mostly echoes, grasping sounds of places that in the end had mostly not included them. I keep coming back to walk in the moment where Carpenè was, too, in which there was a place in this gap between worlds and into which more fit. Traces easier to find lingered for a long while. When in 1959, Maffioli, author in 1988 of *La Cucina Trevigiana* of recipes unmeasured and written in local words, organized that first festival of Trevisian cuisine, there was still pellagra, Gasparini tells me: the event was created "to elevate the state of restaurants here, through all the dishes of the feste of the contadini." Ambrosina Gallon remembers eating polenta and milk, pestarèi, turkeys that her family raised, chicken, soup, and mushrooms, still pride of the hills. I return here to walk a paesaggio made just as much of time, at a moment along which writers freshly Italian were moving in the world that had been and in the one coming, worlds in a moment that is same place and time, cultural plates moving against each other, one heavier and far older pushed by one younger and lighter but fast, dense, nervous into catastrophe. The mountains made would remain presiding but unreachable, gone but for echoes here and there and then the new generations that would, today, base their return to the land on what little they could still know of the contadini. What are he and she to whom knowledge is not passed down?

It is not enough that this be beautiful. You must know that the woods are reclaiming land, making into another landscape what was for so long a place where entire populations lived. Streets ran through forests, which the inhabitants steadily maintained, which housed hundred-year-old chestnut trees. In Conegliano, Valdobbiadene, and Vittorio, Verdiso was the grape most grown, and in Asolo the vineyards were more noticeably mixed.[159] Shrines that are vertices of crossroads, lines of pleas and dedications by objects to tell supplication stories that are based on vineyards. A strada bianca takes you to Premaor, frazione of Miane, where in 1940 one of the great small producers will begin making wine commercially from grapes his family had first grown and passed down in 1600. Here living is acknowledged as a daily pilgrimage, a shrine more usual is a shelter in stone and white plaster, housing for paper prayers and flowers at the foot of a little painted likeness, in respect, of saint, sometimes Mother Mary, directed to this near-vineyard moment where paths and people crossed. Passing back and forth from vineyard to home or stable was movement recognizable to contemporary carpenters, ironsmiths, shepherds, and clothes-washing women alike, habits of walking that the zone's new scientists followed, too. For at least as long as there have been photographs, these hills have been filled with vines. Or had been. Again, often where you see forest once was cultivated. On walks I have seen patches of sloped forest betraying familiar embankments covered in grass that once someone had planted and where that day surely there were vines. At least one new producer, in Tarzo,[160] is making Prosecco by replanting abandoned land like this. Searching for vineyards in books and forest, they can be found, specialized on palo secco or fragmented by corn and wheat and trees and either way worked and observed and sometimes in partnership recorded by the contadini. On the hills' north faces, the woods were always there.

In the hills a similarity of language tied local chemists to contadini to their shared history. "The language that most resembles Trevisian is Venetian," wrote Alessandro Pericle Ninni in his 1890 *Materiali per un vocabulario della lingua rusticana del contado di Treviso*. "As soon as they open their mouth, a Trevisian is recognized in Venice, a Venetian in Treviso." Under the empire, Venetian dialect was the common language of trade, used by some speakers of the dialects throughout the Venetian maritime and Terraferma provinces. There is a poem by Cecchinel, born in Lago where the Soligo river forms, that is written not in Italian but more quietly in the words he hears around him which are Belluno-inflected Trevigiano, too, poem of strame[161] and startled oak trees. Against the numbers, let the poets try: *de là de i càrpen e i rore spauridi / del sas rucià su la lasta slisada / te na onbrìa umida de fià sfinidi / fa qualcheduni che 'l dròn e che 'l sfiada / de là de l'ultimo ciaro del tràgol / del sas rucià su la lasta slisada / fa sanjut de stran patòc fin tel médol / l'è 'n vèrs inrauchì che 'l va e 'l vien e 'l sfiada / na ànema che la ol restar viva / drio 'l sas rucià su la lasta slisada / andé che gnesuni pi romài riva / se no l'é da tant che 'l dròn e che 'l sfiada / par sanjut de stran patòc fi n tel médol / de là de i càrpen e i rore spauridi / de là de l'ultimo ciaro del tràgol / te na onbrìa umida de fià sfinidi.*[162] Where the hills share space between castles and casère, strame is layered, foraged dried grass and hay bedding for livestock. By the sea, *strame di valle* is what growers call the mollusks and crustaceans of the lagoon beds. Between the pianura and the hills there is a wide invisible line, and it is an immediate place one can drive and walk upon. It took the contadini centuries to make the luoghi of this protected place.

In 1958, the father of Rolle restauranteur Pancrazio Brun bought a farmhouse, the latter told an AISO interviewer in July 2020—a "casa bianca,"[163] a color exception but regardless one of the Brandolini's many farmhouses—and over decades turned it into the restaurant it is today. By 1960, "comincia a mòvarse pae strade la prima gente che caminéa che poteva 'ndar in passeggiata . . . aven comincià darghe un bicer de vin, un panin," he explains of the people who began to be able to stroll here and to whom they would sell a glass of wine, a sandwich, as the restaurant came to be then grew, from ten seats to more than two hundred today. As children he and his sister had walked, carrying heavy two-handled metal jugs filled with water back to home. "The transformation of the stable is a turning point," notes his interviewer. "The definitive liberation from a peasant life, which sends him off, at the cost of hard work, to become a small entrepreneur." Mazzero's father—Rolle once rich in stables—had worked to gather milk in the outlying frazioni and bring it to the latteria.

Where the first latteria is, the Strada Caldarment (*armenti* in Italian, *herds* in English) is the first road we take up on the Strada del Lat.[164] It follows an east-west-long ridge in the Valdobbiadenese core zone and drops off on its south side to vineyards and fields and houses, and to forest a mix of unhindered and remnants of once maintained trees on its north and Prealps-facing side as we walk along the crest. Here is a chestnut on my right. Walnut leaves are indeed strong and odorous. This section of wooded pathways is devoted to Sant'Eurosia and in within it is the Brandolini's Oratorio di Sante Eurosia e Irene church, tucked into where as "more compact formations unfold, such as marly limestones and conglomerates, there correspond [in the landscape] as many rows of mounds," wrote Taramelli. "Where marly or micaceous-sandstone soils emerge, parallel combe [the ground depressions typical, for these calcareous reasons, of northeasternmost Italy] or crests are formed instead. . . . The region admirably translates its geological structure into its hilly appearance."[165] Walking along the forest roads grown over, leaves and stumbling stones, a curled and rusted segment of what unguided I would not have known had been a sickle, you begin to see this not so much as *nature*, but as (as in the towns where forest-filled hills rise suddenly behind homes whose occupants, if they're still there, once tended family plots with climbing vines in the steep parts while flatter ones were planted with mais and grass for grazing until, abandoned, this agriculture was reclaimed by forest) the ruins that they now are. Growing here are pungitopi, which families once placed like tiny green barbed wire in the cantine near the salumi to keep the mice away. Families once ate as well the sparasi da rust, the pointed plant's purple-brown-edged green flower buds, preserved under oil and vinegar. In the combe, quick sunken ground of collapsed calcareous marl, chestnut trees grow naturally. Around the trail, for that is what this road is turned to today, the equisetto fern that grows throughout is rich in silica (SiO_2, the dioxide of silicon[166]) and so has long been used to sand wood or, more often now, made into tea for strengthening hair and nails.

Cal Zetera is the name of the road we descend to return to where the Soligo is running past wide and ragged hornbeam along that invisible line. It is a steep path in the forest, stones set as staircase to carry you back down. In Cecchinel's most substantial collection of poems, *Al tràgol jért*,[167] the title itself holds the word for this kind of steep path which the contadini took to drag down to the valley the trunks of the trees they felled. The poems within, written between 1972 and 1984, are (in the poet's words) "*scaje de tràgol,*" splinters on this steep and heavy path. Like Cecchinel's use of this particular dialect, borrowed into writing and otherwise local oral language of the upper reaches of the Soligo river difficult also to Trevisian speakers (closer to Bellunese noted

Once, a vineyard was here at the edge of trees; now there remains only a grass terrace ciglione made for holding it.
Susan H. Gordon

one speaker[168] of it when I visited in 2021; "hard, dry"), this is about sounds heard, leaving traces and not an invitation. In a July 2020 testimony for AISO, grower Angelo Nardi explained that the forests were once managed and cleaned: "Angelo showed us vine shoots, he talked to us of the techniques of viticulture and of vine diseases, observing that these are tightly connected to the ecosystem," records his interviewer. Some diseases were there during his grandfather's time, others like today's ongoing challenge with the lymph-blocking flavescenza dorata which turns attacked vines an malevolent yellow had not yet come, says Angelo Nardi—the high numbers of the insect that carries this newest arrival to the vines a result of inhabitants abandoning the forest. Once, he explains, when everyone had their own livestock and their stall in need of hay, the woods "were cleaned up, tidied up, always cared for; now no one enters anymore, it's not really worth it." In those past autumns, his grandfather had gathered the trees' falling leaves and brought them to the barn, "to keep everything clean."[169] Map of now, map as now, at the tri-intersection between the Cal Zetera, another road up, and a flatter path facing toward where the Soligo is running, there is a structure at the hills' crossroads that is both usual and unfamiliar. A large and plain cross made of wood, unadorned but for a large scallop shell that might at this crossing be a symbol of the world's best-known pilgrimage which ends at the Galician city of Santiago de Compostela 29 miles from the Atlantic ocean, shrine to the apostle and patron saint of Spain Saint James and a likely tie to Sant'Eurosia who was a Spanish saint picked up in northern Italy for the protection she is said to offer from the thunderstorms and hailstorms that strike a particular kind of fear in wine-growers. There is talk that San Giacomo passed by here, which would mark it as an Italian section of the famous walk. A once-a-year crop delicate from bud break to cantina, the Prosecco grapes make each protective and gathering walk through the vineyard and every crossing of the distance from vine to winemaking another devotional journey to a sacred place, practical acts that are also homage, respect. In early July 2021 less than three months before the harvest began in early September, I walked with the brother-sister team of Prosecco producer Col del Lupo high along the Rive of Vidor and Colbertaldo to that 360-degree view that is one characteristic of this land, as a thunderstorm wove and curled around us. When at last the sky grayed and cracked, the opening rain and thunder were met not only by Sant'Eurosia but also by the steadily shouted warnings of hail cannons, shots now usually of lighted gas fired from stretched cones made of steel the height of three adults open-end into the heavy air to hopefully—we don't know if they work, every producer I've met who uses one has told me—disarm the clouds about to bring the hail. They were developed in the late 1800s, writes producer Giovanni Gregoletto in his collection of wildly pensive essays, *Vite ambulante. Nuove cattedre di enologie e viticultura*,[170] and championed then by the school. Gregoletto's homage includes a postcard of an exhibition of the latest top models of hail-proofing instruments[171] held in Italy in 1900, by which point four scientific congresses had gathered to study two thousand stations, twelve thousand installed cannons[172] across the north and center of the nation; between legend and reality, the line is very thin.[173]

"In the 1900s, there was a wine that was called *asciutto*, a wine that was sweeter. It was bottled and in the spring naturally it fermented and in the bottle sometimes the wines were spumeggianti," says Adami, echoing the hills' long hinting toward a greater sparkling. "The marriage of bubbles with the grapes of this territory was understood immediately. In the meantime Carpenè had been studying it." As Carpenè mixed rationalized wine-growing and the contadini, someone else was talking to them, too: Checchini of character-name Massarioto and writing in dialect works that were meant to be read out loud, scenes of un-idyllic countryside. "Ma rassassassi de siori, quando verziu i oci par vedar ste sacrosante verità?" the latter wrote in the diocese's *Vita del Popolo*,[174] outlet of Catholic social justice. "Don't you see that the contadini serve as the foundation for manufacturing poor people while you are tasked with making the walls and roofs? Whether they want to or not, if these signori don't work to improve the contadino's conditions, by helping him, encouraging him, respecting him, and watching over him and letting him know that the art of the contadino is the most nobile because everything comes from his hands, in a few years they too will be the materials from which are manufactured poor people!"[175] The declaration he had made to the stalls "is already a definition of poetics, a selection of language, and a pointing to his public,[176]" writes historian Mario Isnenghi. Both writer and audience, he points out, "ascertain and indeed privilege" the Comune as a natural gathering place, while seeing its officials as occupiers at odds with their older paternalistic rules. In the work of Checchini, "this non-writer"[177] self-taught after completing the second grade (public Italian school an invader, too), "dialect was one with everyday life, not a populist affectation, nor a verbal mimicry and not even a pedological tactic, an academic "farsi popolo," *popolo* understood in a rural sense, even though it is a result of "decisions about communication and therefore of education,"[178] reading not just out loud but in episodic publications, narration placed as well in slowly unfolding time. In those moments the "osmosis" between author Checchini, character Massarioto—"Paron Stefano, witty, kind, a good speaker (and an equally good drinker),"[179] described a local paper in 2014—and the contadini of the "subaltern public" is complete, because Checchini is already inside the world he is enacting. He is not a contadino and neither of course is he the only one to write in dialect (for his part, Isnenghi points out the pharmacist Pittarini in nearby Venetian Vicenza), but here "life and enactment become one, because writer and character are what they say they are. . . . For once, the written word, the culture, do not remain outside, do not manifest as a thing of others."[180] In those moments oral culture has made itself a written culture.[181] Checchini's work is rare today in public libraries (places "without the books of the non-readers"[182]) and found mostly in the Seminario di Treviso, Isnenghi points out. I had a hard time finding, then keeping for very long, a library copy of the 1982 *Paron Stefano Massarioto: La crisi della società contadina nel Veneto di fine Ottocento,*[183] from which Isnenghi's words, and there were no copies that I could find to buy.

"Some objected to us that they do not know where to plant the vineyards, having occupied all the space possible with vines a filari," wrote Carpenè and Vianello of their conversations with the contadini. "We ask these gentlemen to calculate how many hectoliters of wine they make per hectare with vines [grown this way, trained on trees, sharing space with plants to eat and sell]. We have seen that the maximum production obtained with this system occurs in the Valdobbiadene district, where they produce on average 2.33 hectoliters per hectare; and let us admit that those diligent agricoltori obtain a special production more than double the average, and reach five hectoliters per hectare. But one hectare of vineyard[184] produces at least twenty-nine on average; it therefore seems to us that they too have an interest in researching whether a few hectares of vines in filare are worthy of replacement with a vineyard, and so to go on, little by little, covering with low[-trained] vines all the space they find most appropriate."[185] Measurable means something else to the contadini. Balance, enough, abundance. Combai, frazione of the comune Miane and in a series of combe and easy peaks, through which one passes from one valley to another along those mentioned streams, depressions and outcroppings within, the color of skies and heralds, azure marls more or less compact and forming powerful banks, more or less wrote Taramelli, where some vines are still today planted the ancient way, trained high on trees[186] among which the giant chestnuts that yield the famous Combai marroni each November, today a local festival that joins the yearly mostra of Verdiso wines. The small precipitous plots of vines on the narrow ciglioni, on each one up to two rows in this farmed forest, each defined by thick covers of tenaciously rooted grass and herbs and Prealpine flowers into small ridges to make these hillsides as steppable as they can be, growers caring for and harvesting the grapes whose very name is golden. From solid *Balbi, parseco, gentil, tond*[187] into liquid by the same soft yellow light landing on vines planted in that other kind of filare that ran along with trees used for food and fire and building. Everywhere the churches' bells rush in. Through the Fontanassi wetlands, forest, and tree-lined campi chiusi[188] in Treviso's southern plains when I walk there in the last days of 2019 as the pandemic's first traces are appearing elsewhere, across blue Soligo digging south from the Revine lake just above the core zone and striking across the road that divides DOCG from DOC,[189] the river's ancient words measured in those compact bluish clays that alternate with marly sandstones and calcareous layers in Combai and reappear across the Piave in the Castelcucco hills of the Asolano.[190] The layers are a language, too, already said into poetry, their own varied depths translated into the economy and rebelliousness of numbers. The moment is a litany of colors and the collapses of landslides that it seems have always happened here, vines, agriculture, pasturing, and roads. You must resee the forest we are in, for it is a language like that of the half-molecule, *natural*, observed and organized into new meaning, rationality that was never meant to lose what it referred to, should still be bound to what it began as beholden to. "This is the very interesting area of . . . chlorite sandstones, more compact than the previous rocks and therefore emerging in the hills of . . . Arfanta, Rolle, Farrò and beyond the Piave,"[191] wrote Taramelli. "The grain of the rock is varied and the color goes from intense green to yellow and brown. Often white flashes of lighter and harder rock are observed," the bare limestones, to paraphrase him, on which sections of the shells of echinoids and mollusks are drawn. The Gallon family is the first to plant olive groves in Rolle, notes a AISO interviewer: "today they have 250 plants with six different qualities," planted in 1997.

Beginning in the later 1990s, notes the industrial think tank Fondazione Nord-Est concerned with economic and industrial development of this part of Italy, small aziende agricole[192] with limited technology and instruction and strong ties to the land gave way by the 2000s to technical skills, deep attention to top quality wines and then to a focus on selling and marketing. Who held the vineyards down until then were families who tended their plots, long farmed and usually small, on the weekends, made their own wines and often sold their grapes to among others the historical Prosecco houses. At the turns of both centuries, these small vineyards—"maintained by owners more for passion and love of the land than as true instrument of sustaining the family which, vice versa, often found a professional outlet in the world of industry."[193]—and the barest winemaking corners began to be transformed by the hands of a new generation taught at the school to invest ever more in technology, to renovate the cantine, introduce new technical expertise, and to focus on their product: only wine, only a certain type of wine, aiming ever higher for quality understood as the need to work always to improve Conegliano-Valdobbiadene Prosecco, of the importance of the market's interest[194] in a 150-year-old dream of deserved stardom waiting in the quality of grapes established by those who over centuries ripped these inaccessible hills from the brambles.[195] "We could increase production by five times," Carpenè had pleaded for replacing vines a filare with vines a vigna in an 1880 article, that pushed as well for better drinking. "In the Veneto the contadino and the worker don't drink wine regularly;[196] it is still a luxury; from this are derived misery, crimes, immorality, while statistics show that where wine is drunk the population is stronger, happier, more moderate and measured.[197] Dedicated to Conegliano's new specialized elite who are the coming generations who will plant and train the vineyards and teach and guide their neighbors to do the same, in Carpène and Vianello's reintroduction to the vines of the Marca Trevigiana a modest note in little typeface appeared on their first page: *Studia*, it reads, *di conoscere te stesso*.[198]

"My grandfather was lucky because he had studied a little bit," Adami tells me. "He knew how to read and write. He was an extraordinary lover of wine and he was on very good terms with Carpenè.[199] He followed these things but he never wanted to produce a lot of bottles because he was an *agricoltore* [emphasis, spoken, his]. He didn't dream of a large winery. His story was also that he bought wines from this territory on account of Carpenè who loved [Valdobbiadene] as a source of quality wines. My father would say that in their conversations, Carpenè would tell him, 'Abele, sell four cows, sell a stable, sell two campi of mais and build a winery. Because you have many children, and the youngest, if you want, we'll send him to school in Conegliano, he can come live in my house,' and that was my father. But my father never wanted to go to school. These weren't times when school was considered important, what was important were arms for working the land. It was an era in which there was a very humble approach and very little commerce, it was, very agricolo, all tied to the territory. But they were starting to place their trust in the plantings in the hills between Valdobbiadene and Conegliano. It was very, very hard work, because these very steep hills have to be worked by hand, planted by hand, trained by hand. But it was understood already in those years that [with this wine] there was an extraordinary connection between the climate and the earth, the luogo and the people. But no one knew that Prosecco would become a phenomenon like it is today."

8

On the Prosecco Road

Estesa coltura della vite per noi non s'oppongono ne il clima, ne il terreno, ne le braccia, e la vigna bassa o presto o tardi dovrà divenire la risorsa di questa Provincia.

—Angelo Vianello and Antonio Carpenè,
La vite ed il vino nella provincia di Treviso

Per la nostra zona, per il nostro territorio, è il Prosecco, indipendentemente dalla moda o come va il mercato. È il vino per eccellenza. Per questa zona.

—Luigi Gregoletto

In 1972 at the boom's first passes, Mazzotti had starred in that Rai documentary set in the Marca Trevigiana he had been promoting for decades: "I say you cannot know a paese if you don't eat the foods that are made in that paese, and above all if you don't drink the wines produced in that luogo,"[1] he toned therein. For more than half a century the countryside's cataloger, he was also timeless collector of unrusted whole iron and wood artifacts abandoned mid-use like so many campi and pastures and by this time despite their old-fashioned heaviness mere traces of people and hills. Like his choice of public and unlike Paron Stefano, he stood outside the world from which those objects he held admired and no longer of purpose had come. For Mazzotti is named the province's State Technical Institute for Tourism,[2] and with his extensive photography[3] and many festivals to promote the area's restaurants and wines and rebuild this heavily damaged area after World War II as the contadini grew ever quieter and invisible he had been, unavoidably I found, one of the most heard voices coming to Italy from these hills.

In 1950, Montanari and Ceccarelli had described through grape variety and Prosecco typology the area that in nineteen years would become the legally defined Conegliano-Valdobbiadene denomination: "The white wines of the subzone of Valdobbiadene are made from mixing the grapes of the Prosecco, Verdiso, Bianchetta, Pevarise [an older name for Perera; green-shadowed, spotted and soft-skinned, gently aromatic], Malvasia Trevigiana, Riesling Italico varieties," they wrote. "Of these the first three are fundamental, the other three complementary" in both the 90 percent of white wines produced here as the still, asciutto[4] typology and the 10 percent made amabile and frizzante. While, "the whites of the subzone of Conegliano result from mixing the grapes of the Verdiso, Prosecco, Bianchetta, Malvasia Trevigiana, Riesling Italico varieties. Of these the first two are fundamental, the other three are complementary."[5] Territorio-level differences which could seem slight for those who see them from a distance, but to anyone who lives with them they are important distinctions. During the three decades after, De Rosa ("of the Stazione Sperimentale di Viticoltura e di Enologia di Conegliano"[6]) brought Carpenè's work within one step of the materials it needed, then in the next two decades spoke extensively, too, about the taste of this wine that was extraordinary connection between grapes, wines, technology, and people, beginning with advice on how to open a bottle of it. "To speak for a moment," he wrote in a posthumously published book on tasting wines, "of how to remove the cork enclosure from a bottle of spumante (the only kind of uncorking that is carried out in the presence of the guests), not valid therefore for all the other types of enclosures, we underline that the shooting of the cork is to be considered a vulgar gesture and insulting to the spumante and to who has lovingly and wisely produced it. Spumante is not an object in an amusement park; it is instead the best [expression of] oenological wisdom. Shooting the cork can be allowed 'semel in anno'[7] only in the extremely brief space that signals the birth of the new year. It is a custom, and let's leave it at that. . . . The cork is to be unblocked by hand or with the proper pliers. The bottle is held tilted against the chest with the right hand. The left firmly grasps the unlocked cap and holds it while the right slowly and continuously rotates the bottle itself. When there is a clear sensation that the cork would jump out, at that point instead it is held and allowed to emerge very slowly until a puff[8] (I said 'puff,' not 'shot')[9] signals the successful conclusion of the operation." His *Tecnica* textbook is a recording eighty years after Carpenè's by which time particulars of spumante wines are laden with further details, and there is a clear distinction between carbon dioxide and carbonic acid in the prose to match that of the consistent chemistry codes: "In regard to the constituents of the aromas," De Rosa wrote therein in 1964, lines into sprawling 3D scenario, "it is necessary to underline the exalting effect of carbon dioxide, due to the phenomena of adsorption[10] tied to the superficial tension of the gas bubbles, on the surface of which the mentioned substances concentrate, with a clear increase of the prized organoleptics of the wine itself."[11] Those mixed raw materials thus transformed are another language, the form of which makes worlds: Prosecco asciutto is sold within the year in wood casks or glass damigiane; compared with the wines that come from the Coneglianese, it is "milder,[12] less deeply colored, fresh, properly tannic, and therefore more ready for drinking"[13]—the latter it seems is stored in cask, and likely, hand in hand and balancing act, with longer grape skin contact, bigger and bolder and more tannic than today. At the end of World War II, "there is a strong recovery of the cantine sociali," says Gasparini, "because at that point wine becomes a reliable source of earnings for the contadini. I bring the wine to the cantina, you guarantee me a price. . . . The same thing that happened with the latterie." By 1964,[14] the set of activities on which Italy's now standing autoclave-based spumantistica operated required a new and specific spatial organization: from cantina the base wine must flow to a cold-temperature stabilization station (precipitating troublesome solids like sugars and microorganisms from the liquid wine) to autoclaves (where the wine is refermented and CO_2 dissolved) from which the spumante is passed isobarically to bottling to pasteurization[15] (the solution to microbiological stabilization until later decades of the twentieth century; to be replaced by sterile filtrations before and during bottling[16]) to a reserve room where the bottles could have time to rest, two to three months suggested, as each one gathered itself into discrete good.[17]

It is in this moment at the start of the second half of the twentieth century that the boom begins, gentle tremor announcing the coming full motion, first gathering direction and then outward everywhere. A paesaggio is also a percorso, a configuration of space relative to the implementation or duration of one or more moves, movement in a measured time. From this moment comes the sharp forward flow of the industrialists and the counter of the ruralists. But since then, too, more and more small-scale producers (an example today: fifty thousand bottles filled each year by serious growers equipped with temperature-controlled stainless steel in scaled-down twenty-hectoliter and fifty-hectoliter autoclaves[18] that are a mix of new, some built in the past two decades, some from the past century) have been learning and making and experimenting and observing land and apparatuses to bottle some of the best understandings of Prosecco the world has seen yet. By De Rosa's time, the wine chemists have a more complex tale of both CO_2 and pressure, and Italy is strongly assumed. Carbon dioxide in vinous solution causes an increase in the wine's volume, he fills out: in its standard 750-milliliter-capacity (which he measures in solid cc; one cubic centimeter = one milliliter) bottle, the wine undergoes an increase of volume of about 4.5 cc for pressure between 4 and 5 atm of CO_2, "a colorless gas, with a pungent smell and sourish taste, which in solution has a weak acidic reaction, forming carbonic acid according to the following reaction: $CO_2 + H_2O \rightleftharpoons H_2CO_3$," the right side an "extremely unstable" acid that tends to quickly sunder, "moving the reaction to the left side again."[19] The moment is dissolving back into continuous time, pouring backward and forward. The base wine, "limpido, giallo pagliarino chiaro," endowed with singular and fine aromas, and "preferably still equipped with a little residual sugar, around 1% (interesting organoleptic character because to it is tied the elegance of Prosecco's characteristic aromas)"[20] is now a reliable technical feat that includes routine "travasi tempestivi." In the "campo of spumanti," if one such wine were rid of its bubbles "the various organoleptic constituents, composing a harmonic picture[21] based on the elevated content of CO_2, would betray immediately a noticeable imbalance" if the wine had also been tasted prior to eliminating that CO_2. The production numbers just as precise, but changed: to Carpenè's four hundred bottles per day, De Rosa describes how to create a spumante winery space with equipment for yielding ten hectoliters, 1,300 bottles, per day—working three hundred days per year to send 390,000 bottles into the world. It is a scalable feast, with the numbers and sizes of autoclaves, now sites of re-fermentation or resting or bottling feed, given for how many bottles the spumantizzatore would like to fill each day.[22]

The point being not just transformation but across this particular process, there is a "fundamental criterion for the base wine," De Rosa wrote of the now-existing Italian method's series of equipment and moves. "For this system's cycle are suitable only wines that organoleptically require a minimal rest on the fecce[23] left during refermentation, wines that is whose quality is based on 'youth' and 'fruitiness' and not on being full-bodied, while being also refined and elegant."[24] As basis for his chapter on how to use this system to produce spumante, "for our convenience, and for love of simplicity, we refer to the typical spumantizzazione in autoclave of Prosecco, already a component in a highly notable percentage of the complex of Italian spumanti obtained by refermentation in autoclave."[25] By the word *autoclave*, "or even 'tina chiusa,' we mean a metal vessel of various capacities by volume, whose fundamental characteristic is that of being ideal for containing, without leakage, a liquid (as well as a gas) with a pressure different from atmospheric pressure."[26] As De Rosa wrote that sparkling-wine–based *Tecnica* complete with autoclave, stainless steel was still arriving. Autoclave materials, those which came in direct contact with the wines, he clarified, were two types: acciaio inossidabile and "common iron" enameled on the inside. "Without doubt, today the preference is given to the latter, and not only for its marked economic advantage (an autoclave in enameled iron costs about half of one of the same capacity in stainless steel), but also for clear technological reasons." Half a century later, producer Francesco Drusian would tell me that the enameled iron cracks and contracts, but in 1964, the "excellent material" of stainless steel was best used for winemaking equipment that would have limited contact with the wine, "showing itself instead to not be sufficiently unassailable[27] after the days of rest necessary to spumante during an entire cycle." (Its unassailability was coming soon: today a specific range of stainless steel grades[28] can withstand spumante's acidities.) And with this method, the wine enters the bottle already under its final pressure, and with its full range of flavors.[29] "Sensations, for example of sweetness, are subdued when a wine is enriched with CO_2, and vice versa," De Rosa notes, severing sparkling typologies from Champagne and tying them to chemistry and drinking instead. "For this reason, the terminology used in spumantistica regarding the different sugar contents in the various types of product is noticeably shifted with respect to what is usually used for other types of wine, and precisely 'secco' (or, more commonly, 'sec' in French or 'dry' in English) for a spumante that contains from 1% to 3% [ten to thirty grams/liter][30] of sugars, which for other wines would be enough to classify it as 'amabile.' A spumante with 4% to 7% [forty to seventy grams/liter] is demi-sec, and would be a sweet wine otherwise."[31]

Eight decades earlier, with no mention of typology the wines were becoming officially matched to the grape varieties they were made of. At the 1878 Esposizione Universale in Paris, the Carpenè-led Società Enologica had presented five wines including a "verdiso bianco" and a "prosecco," "made of grapes called verdiso" and "made of grapes called prosecco," respectively, definitions less sure than the description of their red wine, "called raboso and made with grapes from raboso vines." While per best growing practices the latter grapes were grown in the Marca's flatlands and the white grapes in its hills, they are, the exposition catalog confirms, all from vines trained on trees and sharing soil[32] with the usual mixed plantings of grains, beans, vegetables. The *adaptation*[33] of the terrain to vineyards, Carpenè had written in 1880 in hopes of it coming true, though he might not have imagined a widespread reliance on specialized plots of only vines and stakes and one-variety rows would be brought by a fatal mix of devastation by the aphid phylloxera[34] "now a matter of foreign affairs" as Italy imposed a border quarantine in 1875 four years before the nation's first officially reported sighting, in Lombardia's Lecco,[35] with local northern reports in 1851 of the pest hitting Piemonte and Toscana hardest, less of it in the Venetian provinces Verona and Vicenza, still less in the Padovano and Veneziano, and almost nothing beyond that;[36] and by the shattering and innumerable deaths of World War I while that pest is still raging, the fascists' slow tearing ruin after that and the hills shaken in defense by the hopes and shots of the partigiani and then of the allied forces. But keep in mind that on the most abrupt faces of the jumbly hills specialized vineyards had always been true: arduous climb, difficult to bring what you'd grown there to where you would prepare and use it especially daily meant that *mixed plantings* referred to diversity in vitigno alone, that Verdiso, Perera, Bianchetta, Prosecco Tondo and Lungo, and other ancient[37] varieties made up what was densely planted on each ciglione, while places elsewhere were meant for the foods that grew best and most appropriately accessible in *their* places. "Once," said a woman from Tóvena during my visit in 2021, "everything was paid close attention to. There was the time of chestnuts, of mushrooms, of legume." In the decade before Mazzotti's show, Veronelli in his five-hundred-page *I vini d'Italia* published three years before De Rosa's spumante textbook wrote encyclopedically with the kind of personal and wondering of the previous century's written searches into chemistry and agriculture, this time in service of philosophy and farming. His written questions into the nation's wine and the places and timings of its cultivation included these hills' wines. Though as in the osterie the wines he lists from here were mostly still, note the exceptions below for they are part of the root system of what we drink today.

"Bianco dei Colli di Conegliano," he wrote, echoing, was made of: "verdiso 60%, prosecco 25%, malvasia trevigiana, riesling italico, bianchetta, and other grapes 15%. Drink it within the year, because as it gets older it loses freshness and rapidly oxidizes. Color is golden yellow with greenish reflections, very clear; characteristic fresh and pronounced aroma. Dry with clean bitter[38] finish, very pleasant, good body, a little tannic, fresh. Alcohol, 10-11 degrees, total acidity 6% ["In addition," De Rosa explained, "anidride carbonica noticeably influences the impression of acidity, augmenting it, so that normally spumanti tend to not have more than 6–7% acidic content, expressed as acido tartarico."], and particularly noteworthy with fried fish." The nineteenth- and early twentieth-century poet Olindo Guerrini, who had sung *V'adoro, v'amo, / sogliole fritte e vin di Conegliano* to Venice,[39] "had learned to love [this fried sole and wine of Conegliano]," Veronelli hummed, "tasting it in [a legendary Venetian osteria]. . . . Serve it at 10°C." The latter author recommended the following Coneglianese wines: Collalbrigo frazione of Conegliano; Barbisano frazione of Pieve di Soligo; San Pietro di Feletto; Collalto frazione of Susegana. *I'll go to alpine Carnia*, Carducci is, widely sung apocrypha,[40] said to have written to a friend of the summer of 1885, "where there are mountains, and valleys and forests of fir trees and cool waters and excellent meat and wine of Conegliano and trout, all for six lira a day," inspired some say by Guerrini (possibly escaping, too, the descriptions of pellagra in 1885, cholera in 1886 in Brunello's *Ribelli*), wine-heightened conjecture. In Mazzotti's Rai episode, cuisine as collectible as artifacts, a camera's view wobbles over empty plates displayed along a table, wine names in paint on their encircling lips, the words for foods that are made here listed in the plates' purposeful wells. *Proseco de Conejan* reads the swirl bordering the letters that spell some celebratory dish of the *disnar trevisan*, perhaps *risi e luganega* or *poenta e osei*. While both episode and plates preserve and transmit these words and ingredients to me, searching, I wonder what can they do beyond this cataloging for collection, and I am reminded of Perco's words: where is the history and the people, and what does this give back to them?

"One of the characteristics of Treviso is the osteria, the going out for[41] ombre," presented studious Mazzotti. "An ombra is a glass of wine, a glass of white wine." Noting the futility of looking for certain meaning in the long and intoxicating history of wine here, Trevisian writer Giovanni Comisso wrote that it is still worth repeating certain stories: "one example of intoxicated uncertainty," he sets out in Veronelli's *I vini d'Italia*, is how the word *ombra* came to be used in the Veneto as a way to order a glass of wine." Someone, he wrote, "says that as the osterie were once heavily frequented by painters, one of them had the inspiration to substitute the objective notion formed by the glass with the figurative one of the shadow that the wine defined on the table, when the glass was filled."[42] "These places were open almost 24 hours, they opened at 6 am and closed at 2 am, and every day," says Canel. "And therefore there was an assumption of availability on the part of the oste, a total tie, a total attention to the client. Hospitality was sacred." Until mindset and framework to export their wines began to take a stronger hold in the hills in the 1980s, these places played another local role: some of the zone's best and most serious wines were to be found in osteria. The host had a close tie to local authorities, to the police but also therefore to each diocese's priest. But the osterie were tied to alcoholism as well. "As a luogo, they were always a place of sociability, of business, of meeting, but also always a place of perdition," says Gasparini. And of organizing. "I have very vivid memories, in Vidor there was a section of the communist party, and in the 1970s their reunions were in osteria, with 'gran bottiglioni di vino.'" What there wasn't a very lot of was cuisine. For that there were the restaurants.

"A good restaurant serves the wines of the zone," said Mazzotti. The restaurants built from wanting to or needing to but also as grown out of circumstances include Canel's, the Ristorante Da Brun (founder Pancrazio, notes his AISO interviewer, tells his own life history as a story around the history of his restaurant which he refers to as the "ambiente"), and Resera's Andreetta, on what is colloquially called the Strada del Prosecco, founded as the Strada del Vino Bianco in 1966. That year, Pieve di Soligo's pharmacist Giuseppe Schiratti (who with his famous wine cellar appears in Soldati's *Vino al Vino* as does the "oste Condo") along with oste Balliana and Mazzotti set off what will be Italy's first wine trail. The percorso is memorial and celebration and economic sense: "the Trevisian hill is the self-portrait not only of Giovanni Battista Cima, but of the contadino of here,"[43] Schiratti wrote soon after. "The Strada and the Consorzio have been able to save, perhaps for once and for all, this precious viticulture that the pressures of these times were about to make disappear. The vineyarded hills of the past are still here, therefore, and as we can now enjoy them, our children will be able to enjoy them, too!"[44]

Ten years earlier the first Mostra del Vino was held in Col San Martino—it was Mazzotti's idea, says Gasparini—showcasing the wines of the paese from March to June, wine in botti, botticelli, fiaschetti[45]: spring is Prosecco, time of the wines' greatest freshness. In this frazione of Farra di Soligo is the Locanda Da Condo—in Soldati's book it is a stop he calls by person, "the oste Condo," who was Giocondo Canel, father of Enrico now oste and chef of those protected foods. Mazzotti was a good friend of the elder Canel (a theatrical presence, says the restaurant's sommelier Beatrice De Mori married to his son, who became a reference point, drawing people to the locanda), who came here for the "vera cucina veneta," as did Maffioli, husband and wife tell me. "Since Col San Martino is in the middle between Conegliano and Valdobbiadene, it has always been a paese of disputa," says Canel. It's "in mezzo a due fuochi." An in-between land of the two systems of hills and cultures, in the crossfire with a sense of being torn but which made it the soul of the hills' two sides of wines, Conegliano-Valdobbiadene's informal but daily and vital "centro vinicolo." His enterprise's name gives another hint about the frazione's past: *locanda* means that once there were rooms here, which in turn tells on Col San Martino as a "zona di passaggio," Canel explains, crossroads that as a knot gathered the hills' roads and daily lines, place that people traveled along those through but stayed a while, entangled, when they did. It has been a place of many nourishments. The two entrances to the locanda's building were then a butcher and the original osteria, and across the street with shared clientele the Ballianas' Osteria Dei Colli, where celebrities from Marcello Mastroianni to Ernest Hemingway headed regularly as did, most importantly, singer Toti dal Monte and poet Zanzotto and Arrigo Cipriani[46] who came here to fish from the Soligo.[47]

"If you don't drink the wines which are in turn nourished by the sun, by the air, by the earth, by the water of that luogo," said Mazzotti in the Rai episode. "You can't comprehend[48] a paesaggio. You can't know the inhabitants well if you don't adapt to their way of eating and drinking." In 1971, and soon this will change, along the Strada's 42 kilometers were fewer than twenty drinking

Don't change it, said Giocondo Canel: Locanda da Condo's by-the-glass Bottega on the Strada del Vino Bianco, now del Prosecco.
Susan H. Gordon

places authorized to pour by the glass[49]—where you drink "Bianco dei Colli [which "extend from the Castello of Conegliano to Valdobbiadene"], secco, and amabile; you drink Prosecco, the king of the Strada, in amabile, secco, and frizzante; you drink Cartizze ["like a Champagne," observes the voiceover in a higher, excited, note], "rustic and warm-hearted."[50] When the documentary is filmed in 1971, and soon this will go away, the wood barrels raised just above the ground already offer a greater and closer complexity, as do the tables they surround. Visitors are seated there, outsiderness evidenced by among other things a group of two men and a women drinking, languid and relaxed. I am reminded of what might have been written in more encircled days, in one of those four words that were understood and expected, il vino bono, Gasparini had said during my visit, one tristo, and then the bianco, and the rosso. From the surrounding casks are filled the pitchers for pouring into glasses. Each bears a great word in chalk. *Buonissimo*, and *normale*, *fresco*, *buono*. Each is pinned also with a handwritten sign and together the markings tell you stories like this: one cask chalked *normale* offers other more official words: *Prosecco, Uve Miste, 1971, Secco, località Costa*. A *fresco* is signed *Cartizze* [no *Prosecco* written]. *Misto. Pastoso. Località S. Stefa. Buono.* The pitchers are painted clay and are conduits for moving wine into small bowls and little glasses made for quick amounts of wine, diminutives that hold the shadowy ombre. High above casks and tables and visitors there are bottles, secured with spumante-wide corks squeezed by bottle necks against the pressure, you'll see upon opening, into something like a mushroom and lined up on a wooden shelf behind a small bar where three people can stand for a drink. There is another sign. It hangs from that shelf of bottles and reads *Cartizze secco—Dry*[51] the traditional fuller-bodied typology in this subzone where those violets and fistfuls of radichiella are still growing on the same steep hillsides of higher-altitude cooled summer nighttime and heartily ripening sunny daytimes. In chalk, *500 Lire*. Hanging close by and larger, another sign: *These bottles are ours and are not for sale*, and above that yet another, forbidding anyone inside to, *codes* I think or *loyalties*, sing or play an instrument.

Bianco di Vittorio Veneto is an "analogous," wine to that of Conegliano, notes *I vini d'Italia*, made of "verdiso, prosecco, bianchetta, riesling italico and pignola" on the hills of Vittorio Veneto. The entry is a more mysterious and respectful carrying, a visit across sentences so short that some are a phrase, or a single word or place name. *You'll have to find the rest out for yourself, now that I've told you that these categories of wine are here.* "The most important fact to remember about wine labels is that most wine names are place names, and that you are therefore being told, most of the time," guided Alexis Bespaloff in his 1971 *Signet Book of Wine*, "just where the wine comes from." (Under the Bianco dei Colli di Asolo heading: Veronelli does not like them, which I will take as indication that north of the Piave the winemaking was starting to widely enough excel, and in the Asolo hills just below not enough of it had yet.) On July 12, 1963, Italy's denominations will arrive. The nation will make legal

"Standards for the protection of the denomination of origin of musts and wines,"[52] under the official system of the Ministero per l'agricoltura e per le foreste (today, the Ministero delle politiche agricole alimentari e forestali) that today includes the two Prosecco DOCGs—remember, *denomination of controlled and guaranteed origin*. "By *denomination of origin of wines*," announced the *Gazzetta Ufficiale della Repubblica Italiana* of the new system on July 15, 1963, "we mean the geographical names and geographical qualifications of the corresponding production areas—accompanied or not with vine names or other indications—used to designate the wines that originate from them and whose characteristics effectively depend on the grape varieties and the natural conditions of the ambiente."[53] The Prosecco wines of these hills were defined and protected by a DOC[54] six years later; then the great changes of 2009. The legally termed *Prosecco Superiore*, spumante of the two DOCGs[55] only, a widened road toward which this book moves. Definition and a kind of creation, but also of destruction through untangling synonyms; choosing from among the many, many options; straightening branchy multiplicities. The hundreds of grape cultivars that for a century have been the nation's will begin to be defined, first by the points of nineteenth-century ampelography then by the repetitions of twentieth-century DNA testing—and cataloged in Italy's national grape registry, wherein all wine grapes whose existence is acknowledged and therefore allowed to become wine. Some names of wines including the three mentioned by Veronelli above will change or disappear.

As will *Bianco dei Colli di Valdobbiadene* for the wines that Veronelli describes as "prosecco 60%, verdiso 25%, bianchetta 10%, malvasia trevigiana, pevarese, and riesling italico 5%," and which are made into two typologies: "Asciutto is ready to drink which is to be done within the year, amarognolo, low alcohol, light, balanced,[56] a good wine with fish. Amabile, suitable for aging, fresh, frizzante or spumante; goes well with fruit." Alcohol is 11 percent, total acidity is 5.5 percent, and particularly recommended are those from the frazione Soligo of Farra di Soligo, Vidor, and Valdobbiadene's frazioni Guia and San Pietro di Barbozza.

Located in San Pietro di Barbozza, regal Cartizze will retain both its name and its right to appear on labels. For its entry, Veronelli writes that a wine thus named is "made of prosecco, verdiso and other local grapes, is a pale hay color, clear, delicate, persistently perfumed. It is gently sweet, soft, fresh, light, balanced; a natural spumante. Exceptional for fruit or without food, served at 8°C. This wine, truly excellent, has just started to be produced more, also in the dry typology." *Prosecco*, he introduces, "is a wine of such fame that it established the saying: 'amphora of Falerno or of Greek wine—neither is as good as a glass of good Proseco.'"[57] The grapes, he continues, are only Prosecco. "Golden yellow, sometimes deeply hued, of intense bouquet, from amabile to sweet, pretty astringent. It is a wine for drinking fuori pasto,[58] chilled to 8°C. He recommends this Conegliano wine, for "the vinification of the prosecco grape, either by itself or with the verdiso grape, is highly esteemed for the production of the so-called *Prosecco spumante* [emphasis his], which has reached well-deserved and vast fame."[59] Now, De Mori explains, "we have clients who come here, have lunch, and then go to the family they've been buying wine from for, say, 30 years; that is every family has acquired its own clientele." But when Veronelli wrote, fame moved more slowly and had not yet reached beyond the few cognoscenti: "very often" the contadini would come to his father, Enrico remembers, "from Treviso, from Venezia, from Padova, and ask him . . . per favore, Condo . . . because they had a small cask of wine, or a tub, to sell"—and he would make his own blends from the wines the contadini brought and label them *super*, or *extra-super.* "It was an aesthetic, you kept everything clean," one of the oral historians, from outside Venice, told me in 2021 about the complicated places that the vineyards once were. "You kept everything arranged a certain way. This was lost; industry was easier." Remember, in 2021, including Italy, the global wine market demanded more than one hundred million bottles of Conegliano-Valdobbiadene Prosecco Superiore (another fifteen million are held as reserves if more was asked for), one hundred thousand of Cartizze, and four million labeled one of the forty-three Rive, old understandings official since 2019 and the collection of which draws a map of this DOCG's taste by, its greatest claim to fame, its steepest parts. "I understand now," Canel says, "to be careful because times change quickly."

Today, the maps made about wines are stories about geology, rocks shaping outside of people time an almost acultural topography, soil compositions and definers that are hills and valleys and waterways that do not carry or separate anyone, just a few timeless portraits and habits. These geologists' tales perhaps inevitable of where the wines that are grown and made can feel like an exploitation of an seemingly otherwise unpopulated land, though the Italian word *sfruttamento* comes closer in both meaning and bio-etymological basis for wine. If a map can exploit, how can it also or better, instead, work to (akin to UNESCO and the Conegliano-Valdobbiadene and Asolo consorzi if of another form) protect through delineation? What might a map that can protect its source material look like? So near the end of my book I needed another guide and found a jolt in the imaginings of mapmaker Wood whose work is also a "thinking seriously about a poetics of cartography . . . [a] phrase [which] wants to capture something about the way a certain poetic specificity manages to resonate." If a map is a tool it can be a bad tool or a good one, to paraphrase him, brutalizing facticity recast as warm, connective specificity so that I can usefully reimagine and re-place his words: "The map as poem: this is something maps have labored to deny from the beginning." Although the expression is a common one today, *reading* a map is not a literal wish, especially when the material in question is several maps, a composition that portrays a planet, a nation, a neighborhood, or what leads to and from a wine. "Consult your atlas, find the fact (fast), reshelve the atlas,"[60] Wood writes. "Do not read it, certainly do not read it cover to cover like a novel." New understandings of reading that allow us to understand, vice versa, a storytelling like a map, one that can correct misunderstanding a fact or a place as a discrete instance. Cartographers tend "to deny any continuity with painting and printmaking, stress 'scientific' measurement, stress ever greater 'accuracy,' . . . stress systematic order . . . or deny any continuity with narrative," he writes. "The problem is none of this works."

The particulars of a denomination are in a legal document called a disciplinare. Along with delineating grapes; production methods; and wine characteristics by color, aroma, and taste, the disciplinare defines the DOCG as a border. In it, Article 3, "Zone of production of the grapes," is a well-defined series of walkable pathways written in the image of the kind of map you can hold, places meeting and crossing each other in words. As part of my visit in 2021, I tasted along Conegliano-Valdobbiadene's roads and rivers, poured from more than 120 Prosecco Superiore bottles, many made of the specifics of the Rive. I begin at its beginning: "in particular this zone is bounded,"[61] the document's latest update in 2019 reads in the *Gazzetta Ufficiale della Repubblica Italiana*'s 160th edition, "as follows[62]: for the description of its borders we take as the starting point[63] the località[64] Fornace (at 175 altitude), about three kilometers from Valdobbiadene to the west, where the administrative border between the comuni of Valdobbia-

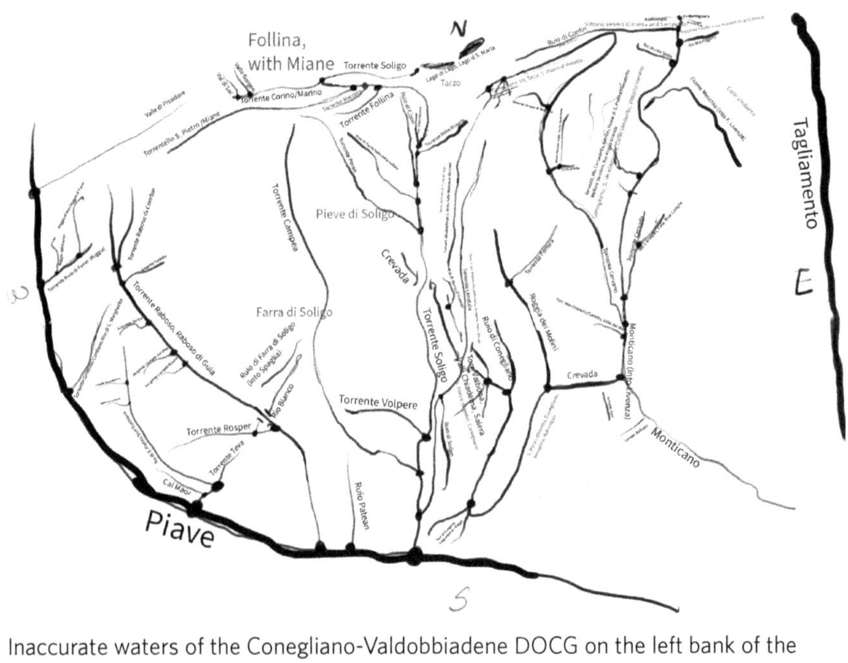

Inaccurate waters of the Conegliano-Valdobbiadene DOCG on the left bank of the Piave river
Susan H. Gordon

dene and Segusino meets the Valdobbiadene-Segusino road. It then follows the administrative border between these comuni up to Col Antich, where it meets the altitude 500 level curve, which it follows up to Ca' Pardolin, near Combai, from here it leaves the altitude-500 line and continues on the path, which leads up to the piazza of the paese first along via Cimavilla and then through via Trieste. Here, following the road that leads to the church, it reaches the Duel casera, then, along the ridge of the hill, crosses the Miane-Campea road, climbs up to Mount Tenade and, still following the ridge of the hill, reaches the località Tre Ponti on the Follina–Pieve di Soligo road. Having crossed the road, the border goes up the Croda di Zuel hill and following the ridge it passes upstream of the church of S. Lucia at 356 altitude, upstream of 'Zuel di là,' and upstream of Resera; the border then follows the Resera-Tarzo road until it joins the Revine-Tarzo. From the aforementioned junction, the border, still following this road, reaches Tarzo and then Corbanese up to the intersection with the Refrontolo-Cozzuolo road, in the località of Ponte Maset, it then follows the border between the comune of Tarzo and of Vittorio Veneto until it reaches the local road called 'dei Piai.'"

We have been moving eastward and have roughly crossed the core zone of the Valdobbiadenese. As wine has always been technology, too, pick up here with the Martinotti method and another definition of the DOCG by instead those forty-three named steepest places carrying their own criteria for demarcation, which is taste. On a map instead of tasting notes, we will step through some of the Rive, names that are either one of twelve of the DOCG's fifteen comuni or one of the many frazioni attached to those fifteen (remember that they are not *in*, but outside, *of*, a comune).

At this east end of the UNESCO's core zone, the Prosecco Superiore of Vittorio Veneto's frazione of Manzana, 2020 (vintage is a grape's recording of the year's weather, taste of time and place) and Dry is perfumed of grapefruit and apples with a lacy foam, sweetness but acidity a tart strong force; from its frazione Formeniga that same year an Extra-Brut is a yellower gold, rich and bold fruit balanced by sharp and grassy, no foam by the time I taste it; from Farrò, a gold Extra-Brut with lasting bubbles, dried white and yellow flowers, lemon and green and yellow apple and a lot of pear (richer than you'd think, fresh and vinous, broad acidity, finish is long and lemon and relatively simple but well done). Walk south, loosely with a little push east, moving first out of the Valdobbiadenese and into the hills of the Coneglianese that skip up from the train station first alongside roads and buildings and then soar up and west in greater and emptier gradients. Northernmost of this complex is San Vendemiano, an Extra-Brut is golden and quiet coiled herbs, a sense of being something that is mysterious, pear and soft acidity and bitterness, apple very much a presence, yellow and red, otherwise a gentle wine.

Continuing south, Ogliano, frazione of Conegliano, 2020 fruit made Extra-Dry, is bare green gold, concentrated fresh hay and wheat, juicy without loss of focus (Brut is green apple, terpenes that quickly turn to grass, green pear, shades of green and white and pale yellow, floral and with an underlying weight not immediately evident); Brut from Conegliano's frazione Scomigo is pretty and apple-vinous, sharp and focused then drying grasses, lime and a crop of sharp and bitter citrus, showcase of the trademark bitterness (2020 and Extra-Dry is very perfumed, of sweet jasmine and honeysuckle, rose and above all orange blossom; great presence of acidity, a gentle complexity and I'd love to see it again with a little age).

In Rua, frazione of San Pietro di Feletto, Extra-Brut that year is a gemmy gold; deep, deep pear nose and a saltiness even in aroma; there is a cilantro soapiness that can come up; beautiful and wild, fine and linear, richness of dry extract and concentrated flavors, bitterness used beautifully as a balancing agent and a favor component; Brut still carries a sense of sugar left. Extra-Dry of frazione San Michele, lighter, herbs and white lemon; palate is all pith, flowers, green crunchy pear; sharp acidity, austere with great balance with residual sugar. Just north in Refrontolo, Extra-Brut, light gold, final bubbles at the rim, suddenness of sawdust with very pretty citrus (ripe lime juice, ripe lemon, blossoms of both) and apple blossoms, gentle lemony acidity; Brut is a persistent foam, rich, creamy, acidity, plenty of blossoms (apple, citrus), bitterness there but thought of as among all the rest including the rs;[65] the Extra-Dry is a wide rim of foam, hay and sun-warmed lemon, rich and twisted with thick green-yellow rind, polite bitterness, sweet lemon and apple, herbs; all carries a sense of sweet fruit, lasting, acidity, very present wine; Dry is expressive and floral, wow on palate, the acidity and texture, the sugar another component, nicely done, bitterness in there like candied pith, blossoms, apple, herbs. South to Susegana frazione Colfosco's Brut, veil of thickly lacy foam like a lagoon; an Extra-Dry is lemon sorbetto, lemon lime and apple and sage, acidity racing through holding a straight line, rs is there buoying and balancing, acidity and bitterness to clean and refresh. An Extra-Brut from frazione Collalto is gold, pear and apple, rich and if simple is fine and firmed with acidity, gentle mint, a bitterness that lingers; in 2020 the grapes for a Brut that is shy then lovely wild, gold with absolutely no green or white, rich while acidity is still of foreground, a gorgeous ripe greenness, and the bitterness jumps up as it warms, long finish with all of that, mouth-cleaning; a 2016 is dissipated foam, at rim, deeper gold, youthful development, nuttiness, tea, crushed herbs, palate is lovely, great acidity, the development is gentle and on the same levels as the citrus and the unripe apple and pear, long long finish.

To Pieve di Soligo's frazione Solighetto; a Brut's foam is steady with a little shake, savory lemon then taste is expressive in rich ripe lemon and lime, sweet lime tea, lime leaves, broad pronounced acidity.

Leave these Coneglianese slopes, northwest across the Soligo river we have crossed back up into the Valdobbiadenese, an Extra-Brut from Farra di Soligo is a soft deep gold, pear and apple on nose, gorgeous pear blossom and bitter herbs, sharp upslanted vertical texture and pale yellow gold, wild lemon nose, concentrated in delicacy, play between generous acidity and extract, just perfectly ripened grapes and masterful vinification, soft lasting foam with some little airway; Brut is dissipating foam, but just barely, herbs and hay, and savory spice, palate is creamy and concentrated and with a little sense of width, fantastic acidity, vinousness still there if you look for it, so much lime, acidity, zest, sweet zest, floral, finish just keeps piling it all on, yes.

A northward slide to Col San Martino frazione of Farra di Soligo, Extra-Brut is steady enthusiastic stream, a sense of white gold, gentle green apple, sage, a much fruitier note, palate matches the nose, and though it's sharp there is broadness, sweet herbs and deeply orchard crunchy fruit with a gentle softness, a perfectly balanced spumante, gold with gleam of yellow, texture and fat sharp acidity, more floral than fruit, spices in the background, stone and mineral and bitter greens; Brut is a very shy nose, pear blossom, it has a glycerol presence, a little sweetness and heft, sweet ripe lemon and apple, a sense of roundness and creaminess, a gentle bitterness on finish. Extra-Brut of the Soligo frazione is golder; quieter pear; palate is more sharp and pronounced with rich pear and bitterness playing out from lemon and lime behind, honey, and an overall wow, concentrated and lines like violins through it upward, beautiful, savory, and serious and concentrated in texture and fruit; Extra-Dry complex but new, fruit and flower just elegant and telling, rs is immediately evident, then lemon and so much apple, the saltiness cuts across after, Adriatic air or trick of the spumantizzazione?

To Colbertaldo frazione of Vidor, 2020 Extra-Brut is white-yellow gold, quiet and robust, touch bready, rich foam and pretty creaminess, a relatively powerful wine, lemon, apple, celery, wild mint, power in spite of reticent parts; Brut is lemon, green apple, and unripe yellow apple, fresh and broad, lush bitterness is part of the ripeness. Down into Vidor, steady ripening, Dry is plush green lemon softened by the sugar, spices, broader fruit, bitterness comes along the finish, savory aspect, too. The wealth of Valdobbiadene frazione San Pietro di Barbozza, in 2020 an Extra-Brut is white-green gold, foam like a veil, sweet herbs, lemon, and citron simple and focused, gently creamy, a sharp acidity, relatively simple, green and bitter herbs and wild apples and lemons, concentrated; from 2018 a Brut savory and rugged, wild lemon, well woven acidity, mid-palate falls off a bit, finish is fine in texture, with plenty of acidity, lacks some concentration; 2020 Brut with a veil of foam, shy apple, rich and broad and fruit driven, tingle of salt of anadride carbonica, finish a little vinous, soft spice and white grapefruit lemon; a 2018 Dry is pretty herbs, intense bergamot, lemon, savory and serious, sweet fruit, lime now, too, and blossoms, the bitterness is tamed, though I wish it weren't, the grapes are gorgeous the winemaking attentive and controlled but never loses sight of the hills, rich and austere, and understanding; 2020 Brut is vinous and apple blossom, lemon leaves, a sense of weight behind the acidity; 2020 Dry, is complex and expressive, lemony and star anise, coriander spices, bitter in the background, rs in perfect balance, savory and much acidity, fruit flavors a backdrop to this intensely textured grape.

A number of wines again in Valdobbiadene's frazione Santo Stefano where a 2020 Extra-Brut is a white-green gold, a pretty foam and small bubbles that are relatively slow, deep pear and some yellow apple, floral honeysuckle, deeply perfumed pear and lemon and flowers up to sepal, white-gray soils sharp and rich; in 2019, foam is a gentle margin, bitter sugar and a creamy bitterness enveloping and strengthening the acidity, savory preponderance. Gold with green tinsel ribbons woven through, hay and allspice and a sense of honeyed richness, texture matches then add rich acidity, just-ripe complex lemon and red apple, add anise, add pear, straw, green almond, plum apples, constructive bitterness, elegant, savory, concentrated delicacy; Brut is floral, pear and bergamot, creamy with very fine bubbles, a bitterness and hay and dust, if you look, creamy but acidity always overcomes it, lemon and lime and sweet bergamot and more complex as you go tasting; Extra-Dry is ripe fruit, softer acidity, apple and lime, more peach than flower; Dry is floral, candied mint, complex and concentrated blocks of flavors, in apple and sweet thyme and after a while pear, juicy acidity, rosemary. Turn right to Valdobbiadene frazione Guia and Extra-Brut is white gold, pear and that pretty soapiness on nose, cilantro, acidity with beguilingly jagged edges, uplifts bitter pear and apple, complex and concentrated but also a pleasing solid simplicity, foam at rim; Brut is lacy dissipating veil, sweet herbs perfumed nose, limeness everywhere, creaminess is at a level just wonderfully showcasing the acidity which is gorgeous, looong finish lime and cream and apple, complex and concentrated. Lovely apple finish; Extra-Dry with shy florals and where the rs gives a texture and a platform you start to see the lemon and the herbs as the bitterness follows, nice finish floral and appley; Dry is a more structured wine, acidity driven, with rs folded in underneath, lemon and bitter citrus so lovely here and showcased beautifully, hill herbs and flowers, ginger and fruit is closer to quince. A Dry from the northern ridges that are Miane frazione Campea is silver and pear, palate shows the rs, but also the acidity and pulls out the saltiness, lemon meringue, apple, with some floral specs and light cinnamon. "Traveling the main Valdobbiadene-Segusino road, reach the Fornace località again thereby closing the perimeter of the bounded zone."[66]

Across the last three decades of the twentieth century arrive the withstanding apparecchi a pressione in stainless steel, embedding themselves into this territory. Today, armed with grades of stainless steel incorrodible enough, wine-contact surface smooth and hence far fewer places for bacteria to rally, Prosecco Superiore is also still an experimental wine: in autoclave base wine changes to spumante generally over around thirty days, but many producers now play by additional weeks and months, an occasional year in transformation, to see, pushing that freshness toward body. "Prosecco spumante, vino spumante tipicamente veneto," wrote De Rosa in 1970's *Andar per vini*, part guide part personal, before the machines' outspread use, is "very similar taste-wise to Prosecco da pasto—still wines, opened at table and meant for meals—except for the amabilità[67] that here is well evident and of course the presence of foam which makes it strongly frizzante."[68] The CO_2, he ponders more technically in context of the Italian method, "(which among other things gives rise to the characteristic perlage,[69] the gaseous crowns on the sparkling wine poured into glass) must be considered in terms of the duration of the froth that develops on the surface of the wine when poured and of the diameter of the single bubbles that develop in the glass, with regard to the total number of these in a given unit of volume of the wine and the overall duration of the phenomenon itself, therefore with regard to froth formation, artificially caused at considerable distances of time from the moment of pouring into the glass (the precious phenomenon, that is, of the wine's inertia ceding the gas contained in supersaturation), and finally with regard to the organoleptic influence on who is tasting." Before going into detail on "these parameters of evaluation," he writes, a necessary observation on that final vessel in the spumante cycle: "you would do well to remember that the most rational glass for spumante is the flûte (flute,[70] in the French tradition, since it is vaguely similar to the vertical flute [musical instrument]) and that instead completely irrational and therefore to rid yourself of, is the coup."[71] Vino tipico, here that *tipicamente veneto*, typical to a place, usual in the sense of usage. What follows is by no means an exhaustive list of the Prosecco spumante of Conegliano-Valdobbiadene. For one thing, it is barely a list, It is a few small tastes of who you have been reading about in this book. For more of the producers—and there are ever more small, obsessed and supremely capable (usually by way of the school) ones, grower and spumantizzatore, opening—you should walk the hills, visit the restaurants, stop by the wineries, get lost, spend the time. For now, along the intercomunal roads, the local paths, the city streets and roads of paese, wines brought to the outside by rails outside the hilled delineation, here are some of the points in those maps, shelter and landpoints casera-like, and with them the smaller stories within this larger one which you have almost finished reading, a story about wine which must include the wines or you will not understand. So, tastes of the ancient wines here and of the now traditional search for technology which created them as spumante.

The Gregoletto family who in 1600 signed a contract with the Abbey of Follina to grow vines on sites that they still work today; their Glera grapes are growing in Miane, Refrontolo, San Pietro di Feletto (where they grow Merlot, Cabernet, Marzemino; red wines for tipicity, too); Verdiso (from which an important varietal bottling) near the ancient and providing woods of Premaor, frazione of Miane, and with at least one olive tree and one old mulberry. Centuries ago, their winery was a farm: in 1970 walls were built around the hayloft to enclose what is now the cantina, work of founder Luigi Gregoletto, who passed away in early

2021. In late 2019, I was fortunate to speak with him with the help of his daughter Antonella, winemaker and whose brother is the book- and museum-maker.

"It is a continuation," Luigi Gregoletto told me of Prosecco Superiore. "It is the natural path of an era but it has roots far away [in place and time]. Today Prosecco is in style, but it wasn't invented today, it existed for centuries back."

Antonella: "Yes, but in his career he was influenced by the scuola enologica being here, by the instituto sperimentale, no? That is, he made the wine and always had these collaborations with the experimental institute, with the school, with research, and even in his career this . . . the wine that he made and the wine that his grandfather made were a little different from each other."

Luigi: "But not much."

Antonella, laughing: "But not much. In the color, in the filtration, in the cleaning, in the method as well after a while, no?"

Luigi: "Some things, because they were relevant, some things we took from the modern art [of making Prosecco], for example filtration, there was a time when it wasn't done. The first filtrations weren't like now, mechanical, they were natural, by sack, a sack with another sack inside it, a type of botte, and [the wine] wasn't pushed by a pump ma cadeva per caduta."[72]

Cultivating their vines on the Rive of rocky Soligo and limestone-clay Ogliano in sites long owned by their family, in 2014 the Moschetta siblings founded BiancaVigna specifically as a spumante house, and their Prosecco includes Perera and Verdiso in small amounts.

The Collalto family still makes wine, and especially Prosecco, and remains experimental in focus, including many monovarietal bottlings of numerous cultivars native to Treviso and several varietal crosses[73] developed here by researchers starting in the nineteenth century; first Bellussi trained then replanted in the early 1980s with Sylvoz then heart-shaped Cappuccina immediately north of the Piave, cool wind in the summer comes from the south across the plain and in winter the Bora wind of Trieste waiting on the Adriatic sea.

Giovanni Marchiori is a self-taught chemist, says Gasparini of the second generation of the Marchiori family spumantizzatore in Farra di Soligo. Here, Treviso's grape varieties are chemistry's costar: Glera Tonda is blended with Glera Lunga, Perera, Verdiso, and Bianchetta Trevigina each fermented according to what its components do best, and each from vines grown, soil and plant maintenance, following the phases of the moon.

In the 1980s, he and his brother "immediately," says Franco Adami of his twenty-year-old self, set about transforming the azienda from a local one into one "with the ambition to become" an international one. While there wasn't yet a market for Prosecco, their vision was to work to expand into north Italy, then all of Europe, then all of the world, a dream they shared with the few nineteenth-century seers and that has made standards of their wines, including two Rive bottlings, one bought from count Balbi Valier.

Nino Franco, now in its fourth generation and founded by Nino in 1919, as a second business of grape-growing to supplement the family's silk business (gelso in the center of vineyards), both silkworms and textiles, but mostly negozianti[74] for the grapes, bought either locally or—WWI having destroyed many vineyards and growers—by train in six-hectoliter barrels. Family ledgers in the 1930s note *white, red, Cartizze* sold.[75] In 1972 their first exports to the United States, in 1979 their first autoclave. Nino's grandson Primo who along with his wife Annalisa I met with in chapter 7, began growing their own grapes, planting old clones; his daughter Silvia now heads the operation, and their Prosecco has for a while been a widely enough available example of top Prosecco Superiore in the United States. "Carpenè was always trying to do a better job," Primo told me that day.

In some of the younger vineyards of this old-vine biodynamic grower in Colbertaldo and Cartizze, broad concrete telephone-poles hold up rows of co-planted Glera and Bianchetta, reminders that in the mid-nineteenth century and into the boom, winemaking materials were what surrounded you, be it woods or new public utilities: Francesco Drusian (whose great-grandfather belonged to Vittorio's cantina sociale in 1890) founded his eponymous winery in 1988, vines and know-how passed down to him from a long line of growers. He grows Verdiso, too, in a plot in Combai where "there's less sun, so it's hard for Prosecco [Glera]," and guards centenarian vines, fat-trunked in Colbertaldo but almost spindly in the Cartizze where, he explains his father told him, they have instead always looked like that. An outspoken voice of the DOCG's growers, he pairs his grapes with a younger tradition and more personal obsession of how steel can make them most themselves during the "metamorphosis"[76] of base wine to spumante, a feat accomplished and available for tasting (for example) in his Rive Santo Stefano bottling.

Riva Granda, whose rows of vines spiraling to the top of a sandy and rocky conglomerate hill in Col San Martino include interplanted Verdiso (10 percent of owner and sommelier Silvia Spadetto's vineyard), one vine of which made its heightened pale-green appearance in chapter 4. Spadetto makes savory and intent Prosecco Superiore.

Col del Lupo, where the hail cannons thundered out that day on a morainic hillside where winds from the Adriatic seas arrive as well, bringing with them sprinklings of salt and passed down with its casera by great-grandfather Abele Adami, the brother-sister team of Marco and Giulia farms sites long held by their family, and makes spumante from Rolle, too, ancient vines and grapes, including an astonishingly floral Extra-Dry. "You must make a wine that is good but also lasting," Marco said to me of Prosecco's staying power. "If a wine becomes old before ten years, it means you did something wrong in the vineyard or in the cellar."

In its fifth generation, today Carpenè Malvolti thrives on the site where founder Antonio Carpenè experimented with Treviso's wine grapes, built winemaking machines, wrote innumerable books, papers, and letters about the new nation and its ancient wines, set up his spumante apparecchio, and dreamed Prosecco. His son Etile, born in Conegliano, continued his father's work building eno-machines and pursuing raw materials and eno-techniques "with the passion of a scholar."[77] and in 1924 brought *Prosecco* and *Conegliano* into commercial existence with that historic label. Since 2019, there has been a "1924" bottling, made, as it was in that time, of 10 percent Verdiso, Bianchetta Trevigiana, and Perera (massale selection[78] vines, in Guia and Santo Stefano in westernmost Valdobbiadenese; cofermented with Glera) but—dream reached—transformed into spumante with an Italian method, and millesimato.[79] They have also made grappa (and brandy) since their early days, and today they offer two labels of "Grappa da Prosecco," made from Glera vinacce. "We need to be jealous custodians of public health," Carpenè had written in the third edition of his *Sunto*. "It is up to the agricoltore possidente to remedy any of wine's deficiencies, it is his duty to humanity to make it difficult for certain industrious speculators to sell off liquids under the name of *wine*."[80]

The path to Asolo is the next and final chapter, *If there were not wine what would become of you, what would become of us?* painted on the walls of the osteria of the Scuola Enologica in the city of Conegliano behind us now, on a map to the east and north.

9

Old Mixed Vineyards of the Twenty-first Century

> Il terreno qui da noi è bruno-argilloso, se vediamo il terreno del Montello è un terreno rosso, s'è differente.
>
> —Luigi Gregoletto

> E non si torna indietro. Mai.
>
> —Danilo Gasparini

"The Rabbiosa grape is a kind of Durella,"[1] says Paola Ferraro about the Prosecco wines her family grows in the Asolo hills thirty kilometers from Conegliano, on the right bank of the Piave, and just west of the forested and compound Montello hill.

On his second visit to Italy, the English poet Robert Browning, "walking in 1844 by chance from Venice, as one did in that walking age,"[2] saw Asolo for his first time. "How many a year, my Asolo / Since—one step just from sea to land—" he would write four decades later in his final work, *Asolando*,[3] published on December 12, 1889 which is also the day he died in Venice. On September 15, 2021, along with its countryside the town of Asolo on the Rocco hill (home to once-queen Caterina Cornaro turned Domina Aceli by the Serenissima in exchange for adding her Cyprus to its Stato da Mar, and Venetian Bembo who wrote of life in her Rinascimento court in the tuscan volgare of Dante; muse to Tiziano who painted portraits of the Queen and of Bembo[4] and to twentieth-century English travel writer Freya Stark who wrote of the heightening orange-yellow haze that so moved Browning and Titian, too: "I have seen this splendor sometimes on a spring day when every blossom seems to be alight like a small candle round the hyacinthine altar of the hills"[5]; host to writers like Carducci and Ernest Hemingway) was included in UNESCO's approval of Pre-alpine and World War I battle-site Monte Grappa as a Biosphere Reserve[6] in the organization's MaB[7] (Man and Biosphere) program. Asolo is a middle land within the Monte Grappa territory, the UNESCO designation says, its hills mediators between the plains and mountains, a bio-ecological bridge and "a point of union-division and, therefore, a balance and exchange between climates, ecosystems, cultures and economies."[8] Asolo's inhabitants, wrote the Austrian officer[9] in charge of cataloging the eponymous Kriegskarte map section XI.12 in an appreciation tinged with the telling complaints on osterie we saw earlier, "are loyal, peaceable and live well, but they are very uncultivated and strongly dedicated to drinking. . . . the fruits of their fields is barely sufficient for six months, while the production of wine is in good shape." "Asolo continued to show sympathy," Stark wrote clarifyingly, this time of the fascist years, "hiding escaping prisoners in spare bedrooms or bathrooms and allowing the Resistance, the partisans, to hold their meetings in its attics and gather its supplies. When I came back to my own house I found their lanterns and boots still scattered about my study, an unobtrusive room, high up under the roof with a far view over the plain."[10] "In general," the Kriegskarte document admitted, "the air here is healthful and pure."[11]

The Ferraro family has been making Prosecco wines here since the 1970s, and matrilineally in the surrounding area for generations before that. Through their azienda Bele Casel, Paola's parents, now with Paola's brother Luca as winemaker, have been thinking about Asolo for decades, searching for what Prosecco wines are when they come from here, taking in hand again the land, the beast that is the vine, the story of the vineyards here; *every one with its own perfume*. Their most recent proposal rests on the Rabbiosa grape, grown here since at least the nineteenth century—"The [white-wine] vines most cultivated are: in the Asolano and in the Maser-Cornuda hills Verdiso and Rabiosa [sic] dominate; then, in lesser numbers, Prosecco and Bianchetta"[12] confirmed Dalmasso of the 1930s—and a small but defining percentage of their newest and most ambitious Asolo Prosecco DOCG, which is their "Vecchie Uve" bottling. There are two vintages available so far: the first one, made of the 2016 harvest, was 3,381 bottles released in April 2019, from a base wine kept more than a year sui lieviti[13] then transformed by an especially long Italian method lasting one year in autoclave, the resulting spumante resolved over four months in bottle before anyone was allowed to buy it. In Manhattan at the end of March 2020 as we moved deeper into the pandemic I opened a bottle brought to me one panicky month earlier by a visiting mutual friend, and tasted it over three days. The wine was a pale daffodil, bold, fresh, toned, supple lemon and lime restrainedly complex, unrepentantly cut through with acidity. On its third day, it remained giving, stayed focused, creamy, and austere, and was throughout this journey above all and in both senses *textual*. Drinking it, the wine asked of me a fourth category of analysis—knowledge about Prosecco in Asolo's district of seven hills lined up just south of the Piave's turn after its southwest run from near the Austrian border to divide the Asolano from, closest, the Valdobbiadenese and the hills of Conegliano. "Vecchie Uve" is from two vineyards in Asolo's Monfumo comune on these hills' north faces, each vineyard a steep hectare of mixed-variety rows of vines around a century old, some of which are Rabbiosa. Today there is not a lot of this wine grape in Asolo and there is, at most practically and likely literally, none in the Prosecco vineyards on the river's left bank. "In the hills of Valdobbiadene, Vittorio and Conegliano for example, Verdiso Bianco dominates, but you find here and there other cultivars disseminated in lower numbers; and here too," Carpenè and Vianello further compared banks, "as in the hills of Asolo and Montebelluna, where varieties mixed together are more abundant,"[14] jagged lines and manyness that today "we are starting to think of . . . as something valuable, not as an obstacle to unification."[15]

Rebellious and difficult in acidity, Rabbiosa took the Ferraro family time to understand. "We made mistakes," Paola had written in a letter that accompanied that bottle to my home. Those mixed vineyards held unclear percentages of old varieties. Glera, which might lose acidity in some of the hotter growing

A view of the Asolo hills from within, by a vineyard in Monfumo
Susan H. Gordon

seasons Asolo can have, stands alongside vines of Bianchetta Trevigiana, Marzemina Bianca (with greater resistance, too, to peronospora[16]), Perera, and that Rabbiosa whose acidity can take up for Glera when it's needed—"All of which with the exception of Prosecco, Bianchetto, and Pignolo, must be considered ordinary varieties, due to their nature and also for the almost savage system in which they are trained,"[17] had misstepped Vianello and Carpenè. The Ferraros never counted how many of each variety they have, Paola says, and anyway "depending on the vintage sometimes one produces more than the other ones." Vineyards that include this grape so old it is at least plausible to think that the Romans knew it here as Duracina, tell the story of a very different vitiscape as much through language: Rabbiosa is an "older synonym no longer in use," writes Italian wine expert Ian D'Agata about, instead, Durella.[18] But where the Asolani vines remain, the name holds strong and the local biotype remains insistently acidic with, it is said, a greater sugar capacity. (It is also being considered on its own by a handful of producers here: Rabbiosa-wise, it would be impossible to not mention, in the nearby comune of Castelcucco also on the Asolani hills' north side and where, notes the Kriegskarte map, two public woods supplied lumber, dedicated third-generation Pat del Colmel who is the local champion of this grape through their all-Rabbiosa traditional-method[19] sparkling wine made in addition to their Asolo Prosecco.) An experimental story about Asolo's stalwart Prosecco vineyards, the bottle I drank was also a declaration by the Ferraro family: *Now we got it right.*

"To describe how we arrived at making wine, we have to go back a few good decades," Danilo Ferraro, father of Luca and Paola and who cofounded Bele Casel with his wife, Antonella, told me when I spoke to his family by phone that first pandemic summer. As a young child he would run away some days to spend time with the contadino who lived across the road, "to look at the vines, the animal stalls, and what not." It would be a few more years before he and Antonella met, but she was learning the same things then, without having to leave home. "My father was part of a family of contadini who also sold racehorses," Antonella shared during that same phone call: mid-century, Ilario Pellizzari and his wife Corinna ("but everybody called her Gina," Paola said of her grandmother), a nurse at the Ospedale di Milano who had fled to Treviso in 1943 during the Allied bombings, moved to his family property in Caerano near Montebelluna in the plains directly south of where their descendants would farm those two vineyards. "My mother began to work in a cotton factory . . ."—"Until she became a contadina," Danilo teases—". . . exactly, and then she settled into the family economy," says Antonella. "Little by little, [Ilario and Gina] began to produce wine, to work small pieces of the terrain near the house, to launch this little azienda agricola," and to grow a family with the addition of Antonella and her two sisters. In these flatlands just outside of the Asolo DOCG, the Pellizzari grew Merlot and Malvasia, raised those wines, a red and a white respectively, in a small cellar beneath the house fitted with some rows of botti, a few salami hung to be cured by time and the cellar's old and fertile air. "In Asolo," notes the Kriegskarte, "of particular note are the subterranean Roman waterways,[20] by which water is still brought into the town."[21]

In their *La vite* continuing inspiration today for being as data-filled as it is personal and poetic, scientists and viti-reformers Carpenè and Vianello had listed 3,364 hectoliters of *Rabbioso* [sic] from the province of Treviso, 3,100 of which in seven comuni of the Asolo district, the rest from a comune apiece in Montebelluna and Castelfranco, also in the plains nearby. From the materials of Prosecco grapes were made 3,709 hectoliters more, from one comune in Asolo, two in Valdobbiadene, from one in Conegliano. It's not clear why the gender of the variety's name changes. The two authors refer also to *Bianchetto* and count it present in more than 14,000 hectoliters across eight districts roughly spanning what now makes up the two Prosecco DOCG areas. That last variety serves as another Rabbiosa anchor: a 2010 study on the synonyms and parentage of 1,005 *Vitis vinifera* grapevines[22] "suggested that Durella is one of the parents of another Veneto native, Bianchetta Trevigiana,"[23] writes D'Agata. And in that Trevisian ampelography,[24] the "big book" in both object and local nickname, published in 1870 by the agrarian commission of Conegliano, "Grape N°35" is, handwritten, "Rabbiosa Bianca gathered in the District of Asolo," is commonly called "Rabbiosa Bianca," and is maybe corresponding to "Cruia del Vicentino" and, very handwritten, to what I can only interpret as "Cagnina del Capo Trevigiano," which does not point to a grape name recognized today. Along with illustrations of grape bunch, leaves, woody parts, and a list of terrains, the big book notes that this variety produced "vino ordinario," a category it lists among the "Collection of the best white grapes," skewing this quality category that runs from *plonk* to *lunchtime*, *casual*, or *informal* to its definition's right side. And, maybe a clue to its uneven gendering lies in the Piave. The many streets (and a coastline fragment by which the river emptied into the Adriatic sea until the sixteenth century) still named Piave Vecchia testify that the river was long thought of in the feminine—as it continues to be, notes the Italian encyclopedia Treccani, in local dialects and the speech of older Venetians. But as the place of Italy's at-last victorious battle in World War I, it was crowned river of the nation and its definite article changed from *la* to the masculine *il*, so that Italians might celebrate heroism and nationhood as "masculine" concepts instead. Perhaps, risorgimentali, Vianello and Carpenè desired that same flavor of fame for Treviso's grapes?

In 1892, the closest train station to Asolo was Cornuda,[25] eleven kilometers away, built eight years before as terminal of the Treviso-Cornuda line and an obvious marker of flatland before the latter comune's hills, which with those of Monfumo and Castelcucco are the DOCG's steepest: up to a 40 percent incline and an average of 20 percent, and where the Fagarè forest, tall trees, hardwood, stand.[26] In 2012, a zonazione study of Asolo led by Italy's preeminent wine grapes geneticist, scholar, writer, and professor Attilio Scienza noted eleven subzones in this DOCG: clay soils is an element for most of them, and there are also morainic aspects and a conglomerate mix, all linked by a Mediterranean climate punctured by pockets of cold Monte Grappa air. Confined to New York, I move over the hills by way of maps that I am trying to turn three-dimensional through imagining. A distinction from hill to plain here is measured first in iconic wines and paintings, winds down next to occasional villas, and then slides into industry sprung from local imagination, practical and squared solidly scrappy and tied to its agriculture—in the nineteenth century literally so as masserie farm compounds on the plain were converted to small factories (reserved and utilitarian, Veneto's industrial buildings are still today in a departure from the northwest's modernist beauty repeatedly signed by enviable logo and typeface designs), while mulberry trees stayed participating in the landscape. Upon walker's descent, these bubble hills drift also into small flatland stretches where another type of building is found instead. For two hundred years before the fall of the Serenissima, wealthy Venetians came here to stay in villas designed by architect Andrea Palladio built on plains, his Venetian interpretation of Roman grandness rooted in an agriculture closely tied to that nearby urban culture and the pleasure and ingredients it took from its Terraferma, to which Asolo had belonged since 1338. At the feet of the Colli Asolani's south side four kilometers from the two vineyards, in Maser (whose hills like those of Asolo and Castelcucco, described Taramelli in 1873, are made of both sedimentary micaceous sandstone given to ground depressions and, studded with gigantic fossils, the calcareous conglomerate materials—granites, granitic gneisses, tourmaline diorites, angitic and feldispathic porphyries, paleozoic schists, and quartz puddingstone, he clarified—that form the rises across the Piave from Soligo up past Vittorio Veneto[27]) and built into space that is neither hill in levelness nor flatlands in unremarkable results, the Villa di Maser built by Palladio in the 1550s on principles from Vitruvius's *de Architectura* retains its agricultural past and boundary position as an azienda agricola producing traditional wines like Prosecco, Verduzzo, Trevisian crossing Manzoni Bianco, and red Merlot and Carmenere. You can visit this home, also called the Villa Barbaro for the two brothers who commissioned it, as the partial museum it has been converted to, to remember some area basics: in the sixteenth century, to protect itself as its maritime power began to give, Venice expanded farther into the Terraferma then used the land to build and think deeply about grand agricultural centers. Inside, commissioned paintings by local Paolo Veronese were, still are, meant to start the Barbaros' guests contemplating the ways in which Palladio's building was a way to organize a farm. One wing—a barchessa, an agricultural place—flanks either side of the central house, columns like a temple in the front, home of bread, wine, and thought for the family and their guests. Granaries formed the east wing, the winery the west as connecting porches made all three part of an enterprising whole. The outside is in the ancient bugnato brick-based facing, smoothed by plaster painted the shade of yellow that was reserved for country houses of Maser's Venetian noble family. Most Palladian villas no longer include ownership of land—in the 1950s and 1960s, agriculture was considered unimportant, our guide explained when I visited in 2018—but this house's two hundred hectares and their agriculture have been maintained. The *Annuario* places the town of Maser seven kilometers east of the town of Asolo and on a fertile plain. In Maser, there is one winemaker listed; products on offer were wine, bozzoli, cereals; two roads led to Cornuda in one-quarter of an hour, presumably by horse. Among its shops no liquorista, no drogheria, but a pharmacist, an engineer, one doctor-surgeon. In 1580, Palladio had died in Maser.

The two vineyards that grow "Vecchie Uve" climb slopes as steep and round as teacups: karsik, sedimentary, like others of Treviso's hills and in Friuli where the grape grew, too, this terrain has been prone to sinkholes—rainwater running regularly through fractures in this otherwise poorly draining and calcareous rock to form rounded breaks within the hills, as where the chestnuts grow on their own, too. Where you walk here, the stones throughout, supposes the Kriegskarte entry, contain numerous traces of a far older volcanic mountain in the area. By now, the sunken spots have hardened into green amphitheaters crowded with forest and some vineyards, planted for southeastern facing toward the sun. A large part of Asolo's hills are forest that have for a very long time been so, woods of black hornbeam and a mix of downy oak, holm oak, beech, chestnut in the hills' highest reaches where people do not go, and differently from the Conegliano-Valdobbiadene hills where the old shepherds' steady shelters appear as brief terraces in higher and more rural steepnesses, few structures pin these slopes rooted into by those other shades of green. The paths here upon which you can walk into these hills are sliding scales of landscape long sharpening the brushes of local painters into instruments that measure and report on the famous yellowing Asolano density of air and light.

Today the town is an official Città Slow—a Slow Food designation for luoghi that have successfully tied that organization's philosophy to their own community by setting "ecogastronomy as daily life practice, seasonality, healthfulness, craft, landscape, religious rites."[28] Until an economic rise in the later twentieth century, this hilly medieval town and containing comune might still pass for their description in that 1892 edition of the *Annuario:* the district that surrounds and includes a town girded by walls, at which rises, atop a height, a vast castle. ("Until now," wrote Stark in 1971, "its hillside setting has been unspoilt, with vineyards coming right up to the walls.") Wine shops belonged to Angelo Bittante, Felice Rostirolla, Luigi Seraglia. One of these men triples as a droghiero; another also sells grains. One ran an hotel named Alla Corona; a few are unsurprisingly also sellers of liquors; one has a drogheria, still today a specialty grocery shop whose spice-shop origins lie in *droghe*, a synonym for the spezie which were once more officially used to treat matters of Italian health. Asolo's fairs were held the third Monday in August across three days, and on the first Saturday of every month. Nearby runs the Muscone river which crossed a territory rich with grains, grapevines, olive trees, and other fruit. Rock hewn from various nearby quarries is duly worked. The ways the Asolani measured weights, pours, lengths of things were one component of criteria which varied within Italy even down to district: here were apportioned a large libbra grossa (0.516 kg), a slight libbra sottile (0.301 kg), a 0.73 hectoliter mastello. Shared with the city of Treviso, Asolo's piede for factory measuring was about five-sixths of a piede for land surveying; an arm's length of cloth (the 0.676-meter braccio di panno) greater than an arm's length of silk (the 0.634-meter braccio da seta).

In Asolo, the *Annuario* lists thirteen producers of silkworms sold for seeding new raising centers, at the heart of "the complex agricultural, industrial, scientific and social world that for so long revolved around this very useful insect,"[29] beginning here in the nineteenth century. A few of those producers are listed also as winemakers. There are silkworms on great farms and in small family operations. By 1936, the province of Treviso was home to forty thousand contadini families whose work included raising silkworms—the money they obtained from the sale of the cocoons (gaéte, echoes a local dialect) was the first money in after the long winter (when "these hills come near us, dazzling in their snow," dare I guess wandering walker Stark noted, "but in spring they take on this faint, far colour and become remote and intimate like dreams."[30]) and one of their few cash entries. The mulberry trees whose leaves fed the larvae until they could begin to spin—from each cocoon come five hundred to one thousand meters of silk thread—fit seamlessly into the landscape: vines trained upon them, grains were grown beneath both, business and domesticity one single focus. Fresh young leaves, which silkworms ate just loudly enough that from the sitting room one might hear crunchy murmuring from the kitchen (they need warmth and humidity) for a month, day and night, came from those mulberry trees both imported and native. Who planted cleverly by following the hills' varying altitudes to ease into macrozone site selections offering cooler and cooler air could stretch first leaves' appearance over that whole month and assure a steady day-by-day supply. Asolo's women who sorted the bozzoli into the three, ascending, qualities of scoatína, ingropína, and mistra (very few made this top grade), spinning into thread by pulling one constant thickness, uniting thinner filaments into that as needed sat for hours at their stations at great flatland factories built of wood. Antonella's grandparents raised silkworms. Her parents did, too, she says, her tone softening further into an easy childhood memory. "We had the stable and above it we had the granary where we kept them, and we would go out to gather the leaves of the mulberry trees to bring them to the silkworms to eat. We had an azienda with everything, pigs, hens, and cows and vitelli." For the children, her father kept a pony, in remembrance of the old family business and his passion that remained of it. The mulberry trees were theirs, planted, as everyone did here, at the azienda's borders instead, family farms attached to campi chiusi. Local merchants came to buy the cocoons. She remembers the grisióe. Later, Paola sends me a translation: "trellises specially made to hold silkworms," from her *Italian dictionary of the Caerano and Montebelluna dialect*, published in 1989. Both of *her* parents were born in Caerano.

"We can't really talk about *a* Veneto dialect because every town has their own words to describe things," Paola says. Asolo and Valdobbiadene, separated by a twenty-five-minute drive but especially the Piave and made up of those hills that make movement between them as difficult as leaving, offer differences: "The willow branches that we use to tie the vines, here we call them *strope*; in Valdobbiadene, they call them *sache*"—in Asolo the tree is called *stroper*, *sacher* in Valdobbiadene. And, "*sput* here means *to spit*; in Valdobbiadene, it's *a little bit*." Words but no less meanings according to local need and humor that are keys here to where the long dominating Venetian empire could easily or desirably reach. In Italy, dialect differences are landscape features drawn on the same lines that rivers run, that steep hills and mountains climb. As Trevisian linguist, wine writer, and promoter Irene Graziotto, who carried Bele Casel's bottle to me in Manhattan, points out in a thesis-in-progress,[31] throughout Italy geographical factors are tied to a dynamic of dialects that feature Latin as a mostly unifying[32] substratum (analogous to post-phylloxera rootstocks, she offers) and local languages as highly diverse superstrata (vinifera varieties grafted, it is obvious). This interplay makes "the differences in dialects by location greater [in Italy] than elsewhere." In ongoing conversations, Graziotto steered me to the shifting Piave as one such site, or agent, of union-division, of balance and exchange. A centuries-large obstacle to travel and cultural transmission between the two banks, today the river offers linguistic differences between and within its two DOCGs: the influences of Venice including by its language of international commerce in the places tied to it by either roaded proximity or simple loyalty (in center of science, business, and agriculture Conegliano, I have visited with researchers, growers, and wine communicators who talk to each other across Venetian, Italian, and one of the Trevisian dialects) versus the Friulian notes that begin to mark its speakers farther north and east. In the centuries on their way to the Italian nation, academic battles raged over whose dialect should become official Italian language, and Cornaro's court in Asolo, Graziotto reminds me, was a center of debates over whether a tuscan dialect (considered appropriately melodious) or Venetian—accused of being too "ugly" to satisfy potential national pride—should fill that necessary role. In Asolo these debates were immortalized, reasoned and in support of tuscan, by Bembo in his *Prose nelle quali si ragiona della volgar lingua*, commonly called the *Prose della volgar lingua*.

Driving north along the SP667 Italian provincial road, you see almost too easily the seven hills drawn out of this fluctuating karsic range's rises, solid and comforting along your left as you gain on the Piave river. Look right, onto from here the steadied and softened mound that is Montello, "a beautiful hill, covered with a dense layer of organic soil elaborated by centuries-old woodland

Inaccurate waters of the Asolo DOCG on the right bank of the Piave river
Susan H. Gordon

vegetation,"[33] home to vitigni of red-skinned grapes and to the oak trees so prized by Venetians and then Austrians that those opposing reigns were linked by a continuous rule that no one, and especially not the people who had long lived here and claimed the trees, else was permitted to remove even fallen branches. To where those boscaioli and bisnenti were jailed and ticketed into various degrees of submission, to where at last no penalties would be as severe as the new Italian rulers': by stricter arrests and fines and via the emigration that followed, after unification the boscaioli disappeared. Here there are still more stories told by trees, longtime partners of vines until stakes and monovarietal rows and vineyards became the twentieth-century assumption. The vite maritate,[34] arms encircling their trunks and intertwisting with their branches, were often a few of one variety, several of some other, in promiscuous vineyards a filare. What trees the vines moved up are stories about a wine-grower's ambition and place. In the plains, flimsy poplars dropped vines once they reached about thirty years old, older and heavier than they had been. Reliably strong enough instead were providing walnut which held for decades more while giving wood for furniture, walnuts for easily stored food, but left an odorous mark; and faithful mulberries flashing leaves for silk-industry livelihood. But even these two were more often a matter for the flatlands. Here in the Asolo, lithe maple (replacing larger-canopied trees), ash, and cherry have held the hills and vines.

Within the two Prosecco DOCGs, wines of Conegliano-Valdobbiadene have been working out their own varying land complexities for years, knowledge long both passed down and researched. Producers of Asolo, until recently suppliers of grapes and makers of more local-sighted wines, are working to find their land's own complications now. The Ferraros think of Monfumo as Conegliano-Valdobbiadene regards its grand site Cartizze, both for the quality of the grapes, and for the intense demands that go into farming it: steep and crumbly sites, here of Asolo's typical more reddish clay, demanding climbing and maintenance and harvesting entirely by hand, for generally very old and related vines. Again and again, Asolani producers point to two characteristics that unite their zone's Prosecchi in indication of these clay soils: a noticeable saltiness; and with 240 days from moment of bud to fall of leaf, higher amounts of dry extract—those solid components that make up flavor, color, and body—for relatively lusher wines. While Asolo's disciplinare mandates at least 16 grams of dry extract per liter, more usually the wines here have 17, 18 g/l. One of the two Monfumo vineyards from whence "Vecchie Uve" is the ninety-year-old Vigna dei Confinanti in which vines are an average of seventy years old, further adding concentration of flavors—older vines produce fewer but more intense grapes. The Ferraros bought the plot in 2013 after it had remained for sale for four years, its owner a patient farmer who sold grapes to a cantina sociale, and then "decided to sell it because he couldn't do it anymore," says Paola. When a family friend who grows wine in the plains turned down an offer to buy it and instead showed it to Danilo, "because he knows that my dad is a little bit crazy at times, we fell in love with it. My dad, who is someone who sees the future, he said, Okay, let's do it; it will be tough but it is also a very good idea."

That Asolo is thick with forest and other wildly occurring plants is not always a check on several kinds of human deeds. At Christmastime four years ago, Paola and Luca visited that vineyard, and along a boundary with a neighbor's plot found at least two dozen of their prized old vines uprooted and thrown against the ground. "The neighbor had decided that it was his land and not ours, so he removed the vines," she explains. "So as a joke we decided to call the vineyard dei Confinanti." In English, the word would mean *adjacent* if you could reshape *that* word into something of a person, into the people who share a border. The strip the neighbor was contesting remains theirs, but without vines. "We decided not to [replant it]," says Paola. Now it simply grows anything, a dot connecting to Asolo's wide diversity in green and animal life; chestnut groves, mentions Kriegskarte map section XI.12 of Monfumo. The second vineyard of "Vecchie Uve" is the slightly older Vigna dei Piei in Monfumo's Castelli to the northeast, owned by family friends. The Ferraros have been farming it since 2014. To get there, imagine a straight line from the first vineyard to the great flats of the Piave: on that line, dei Piei is less than three kilometers from dei Confinanti and is an amphitheater not center-circle as the Greeks saw but like a Roman redo in an encircling curve interrupted by wide stage that here is a descent into flat plots and trees. In both vineyards, at about seventy years old are some of the oldest Bianchetta vines within the two DOCGs. On, impassioned and rational as his brother, Giuseppe Carpenè's enographic map in *La vite,* those varietal blocks sized by volume show Verdiso and Rabbioso as the most countable *yellow* white-wine grapes growing in Monfumo. Both of these vineyards are mixed-row, and the Ferraros patiently harvest both by variety: Rabbiosa late-ripening with no threat to its acidity so powerful it needs the winemakers' thinking through is picked last. Picked first is tricky, thin-skinned Marzemina Bianca which "we could lose overnight" to overripeness, loss of acidity, or worse. Glera and Bianchetta Trevigiana are picked in between the two. Luca and his father blend the results around November, while in both vineyards fig trees explode into sweet yellow and purple fruit and have been doing so for a long time—their thick trunks tales of their ever-deepening roots that have also been keeping the soil of these steep sites in place. In June 2016, a bad landslide happened in the first vineyard anyway. The Ferraro family rebuilt. The sound of the wind that comes from nearby Monte Grappa and runs through the grapevines' leaves is loud like music.

Danilo's youthful escapes would teach him a love for a world that stretched from soil through air, for the stories what grew there could tell him about where and who he was, for how the grapevines had their own certainties about this land. By the time he married Antonella and fell just as deeply under her father Ilario's spell, he had studied in Conegliano at what was now Italy's oldest winemaking school, had "learned, and learned to appreciate, the techniques of winemaking and wine itself." For a while, he worked at the Brotto distillery in Cornuda, so he could "see all the details that machinery required in order to get good results." Each harvest and pruning season he would stay home to work in vineyards and cellar with his father-in-law. There had been an augury: years after Ilario and Corinna had assembled their farmhouse, at the Caffè Bramezza in nearby Montebelluna since 1911, as Ilario drank and talked with friends the city of Milan came into his life again. Orlando, Milanese salesman of personal goods, walked in, says Antonella, and asked for a Prosecco and a Merlot. "The best Merlot in the world you'll find in my winery," Ilario told him. Back at that cellar sotto casa they launched a friendship and partnership that would last for the rest of their lives, years of old-truck rides that included Danilo and Antonella to Milano to deliver that Merlot while Paola and Luca stayed at home. "Unfortunately," says Antonella, as the world took its first steps toward a fixation on Prosecco and after a particularly destructive August storm in the early 1980s, the family replaced both Merlot and Malvasia with Glera, grapes which today they sell to others—the land here is flat, and Bele Casel makes only Prosecco DOCG. *Unfortunately* because it was a wine they were very fond of, a high-quality Merlot from small yields. "Our clients still remember it," Antonella says. Around 1977—personal and thoughtful with "the slowness necessary to do things right," the exact year is hard to pin down—Antonella and her husband began to build Bele Casel. "Curiosity is king and the more involved we became with wine production the more we realized how much pleasure it brought us," says Danilo. With an eye to what Asolo should do, they made Prosecco, fermenting in bottle for the older sui lieviti style. Bele Casel inched. The couple added a small piece of hand-bottling equipment, fixed each one of their labels in place by hand. They bought an automatic bottle washer. Luca had been born a few months after Bele Casel, ongoing witness matching the winery's very first steps. The first son in two generations, he was a natural for Ilario's attention, his arrival announced far and wide by a large blue bow tied to the antenna atop the house. "Anything he wanted, my father gave him," Antonella said, and what he wanted was, like his father had, to do what Ilario did. "When my grandfather worked with the tractor, I was on the tractor," Luca told me during that family call. "When my grandfather went into the vineyard, I went in the vineyard." Half of an obsession began to take hold. Following his father into the cellar filled in the rest: "with the passion with which he worked, with which he still works, it was nothing to say, wow, how beautiful, I want to continue this, to understand, to try, like that!" Ilario passed away later in the 1980s and soon after, his daughter and her new family moved into the house. "We just went forward with my father's work. We kept up with the vineyards; we followed everything my father had been doing." They reworked one side of the house, where Antonella's parents had maintained that stall: while Antonella kept an unrelated day job and Danilo worked as a winemaker for another producer (when your work is doing things for other people, he warns, they get to decide your future), every Saturday and Sunday they rebuilt what they had. A small cellar on the stall side; in what had been Corinna's parlor, a storefront to sell the wine. In the early 1980s they had bought an autoclave and added Prosecco Superiore to their line, becoming one of the zone's earliest spumantisti in order to make sure that the base wine transformed to sparkling was indeed theirs, made entirely from the grapes they were choosing as foundation. Paola was born soon after.

In 2003, they remodeled the winery into its current form. They bought a secondhand 1970s Bertolaso bottling line, now resting regally in one corner of Bele Casel, an all-mechanical ancestor of the brand's spumante bottling lines that are the coda of Prosecco houses on either side of the Piave—"It felt like having a spaceship," Paola wrote to me of *that* eventual upgrade. "We devoted all of our free time to it," Antonella says about Bele Casel. In the background all those years something had been taking shape though need: while Luca went to school and Paola was still young, they made what would turn out to be a crucial concession, that would slowly lead straight to the bottle I drank: time had to be found somewhere and so for a while they had been buying grapes from a good friend in the Santo Stefano frazione of Valdobbiadene. While honing his Prosecco-making skills on these prized grapes, "my father always had a vision," says Luca. "He was always searching for a structure, a [certain kind of] drinkability." Was Danilo missing something from the wines of Valdobbiadene? Yes, answers

Luca. Within Prosecco's grandest wines, Valdobbiadene's beauty is a soaring kind, floral, citrus blossom, skyward. But "Asolo has a lot of structure compared with Valdobbiadene," Luca says. Paying close attention to those grapes of Santo Stefano, Danilo had seen instead what his land of Asolo wanted to say, what he would come to understand as "the saltiness" typical of the wines on this side of the Piave. In the late 1990s, with a Conegliano degree to match his father's, Luca joined Danilo in mission and daily work, as they moved back to Asolo's grapes and Danilo began to think about local vineyards. "The most important jump for our little azienda was to buy sites in places that were not very comfortable to work," Luca says. "Everyone thought we were a little crazy, or strange in the choices we made for viticulture as well as enology." In both

21st century Prosecco Superiore came off an isobaric bottling line of the 1970s in the Asolo hills.
Susan H. Gordon

they were after "salty wines, more structured wines, bigger-bodied wines, not for thinking about but for drinking."

"The vineyard in Monfumo, we can say that it is the flower in our lapel," says Danilo. "The effort it demands it more than repays in a unique, a constant, quality that I have not found elsewhere. We found in this old vineyard a series of grapes that were once part of the territorio. Monfumo was a very particular wine; in *all* the vineyards here it was once normal to find this diversity of grape varieties. The Prosecco we make from them is a Prosecco that is as old as I am, several good decades, and there are vines here that are a hundred years old. We bought this first vineyard in Monfumo, because it was so difficult to work that no one else wanted it." They now own other Monfumo sites: Vigna Longon and Vigna Col degli Ulivi which they have planted according to this old system, planting one with Glera only and the other with all the old varieties—all selection massale with cuttings from the Vigna dei Confinanti. "As it was done traditionally," says Danilo. "We continued in the footsteps of our great-grandparents and grandparents."

"We did it ourselves," says Antonella.

The climate is still warm enough to grow olive trees here, among Italy's very few northernmost spots for that. When I visited Asolo in 2018, it was late October, and fresh olive oil was poured from refillable tableside bottles in the town's restaurants. In 1709 a frost had destroyed vines and olives, and between the varieties of both species that offered higher yields and those promising top quality, many twice-shy growers chose the former for their replantings. Once Conegliano's school opened, wines of quality began to be made on a larger scale here again. But oidium and peronospora arrived here, too, and soon after decimating and implacable phylloxera—to be solved within decades with an entirely new planting method. Immune to the louse's damage, American species of *Vitis* supplied replacement rootstocks for almost everywhere *Vitis vinifera* grew, new roots for cultivars that would now grow grafted onto them, and so many *vinifera* varieties did not make it onto growers' replanting lists: a century later, the process of recuperating and safeguarding almost-lost varieties like Rabbiosa is part of Asolo producers' work. After all those years searching history and hillsides and terrain's soils and winds, how did they know the Vigna dei Confinanti was what they had been looking for? "All you have to do is see it," Antonella says.

If you are standing at the vineyard's road height, the path out is a twenty-degree swerve off from the driving road along a descent brushed with grass. Set off walking. This close up the way distorts into tight curves and through-fields, and wobbled and ridged ribbons of herbs and grasses rich with pebbles that make it easier going on foot then the paved roads you'd share with cars that appear slow and steady on the crowded horizons; a minute off-road, forest patches between the hillside vineyards and the flat patches that hold bushes and smaller crops. On this small northern piece of the Asolo hills which rise from just below four hundred meters to just below five hundred above the Adriatic, the wines are said to be more floral, more lifted than their south-side counterparts: Monfumo and Castelcucco elegance to plusher Cornuda and Maser, an analogy up to body that carries roughly across the Piave to respectively the hills of the Valdobbiadenese and those of Conegliano. Monfumo is also marked by cherries. In dei Confinanti, "there are plenty of fruit trees around the vineyard and I'm pretty sure that they were used as they did in the past, to keep the vines standing," Paola says when I asked how the vineyard we are looking at and drinking now was planted then. There is a pear tree, a loquat, two sour cherries trees one of which is dead, remains upright and memorial, and "at least three willows that they used during pruning, for the ties."

For those "Vecchie Uve" mistakes mentioned months ago, "we were mainly referring to Rabbiosa," Paola says. Tradition is made of both the plants and the knowledge about them, but the latter had not come automatically when the Ferraros bought the vineyard. It especially wasn't clear whether to harvest all these varieties together, or when. "When we bought that vineyard we tried to identify the different varieties but we didn't really know what to do with them," Paola explains. "We knew that Rabbiosa was very acidic, so we decided to almost let it dry on the plant. We picked it, I think it was October twenty-something, then we crushed it in the same way we crush the Glera grapes, not very much, we have a very gentle crushing machine [the Italian Siprem VS80]." They threw that first attempted vinification away. "It was too much for Rabbiosa," Paola says. "The tannins were very strong and it was too acidic." But the stumble taught them to harvest separately, to pick Rabbiosa for its acidity instead of waiting longer in hopes of taming its strongest trait. "In the hands of less-than-talented producers or those who let yields creep too high, these wines can make for some rather tough going with tear-inducing acidity," writes D'Agata.[35] "That's why we care about our small producers," then-president of the Consorzio Vini Asolo Montello Armando Serena told me in 2018 about the role Prosecco makers like Bele Casel play in protecting and growing Asolo's native varieties. Efforts like theirs and Pat del Colmel's[36] have lead the consorzio to apply for Rabbioso to be included as an official variety under their denomination. "We work for Bele Casel, but we also think of ourselves not just in our small part of the world, but as a bigger entity," says Paola. They are members of the Vignaioli Indipendenti Trevigiani, inspired by the nationwide Federazione Italiana Vignaioli Indipendenti (FiVi), along with producers from both Prosecco DOCGs and from the larger Prosecco DOC area. "The greatest thing is that we are not fighting, we're trying to learn from one another, we do a lot of tastings together to find differences and to learn new things and to try to raise the quality of our wines, and I think, and all my family thinks, that that's the most important thing right now, working as a group, and not as individuals because you go a lot further," she says. "Historically," says Luca, "Asolo has always been considered Valdobbiadene's little brother. Because we were smaller . . . for a thousand reasons." He shares a comparison from a friend: Asolo like Champagne's Aube, thought of as minor but that yields extraordinary wines to those who go looking. "In the past few years, Asolo has raised its head and is showing that it has no need to envy the other historical zones. There are many small vignaioli who are doing truly beautiful work. There are new generations, they have studied deeply, they're learning, they're trying different things, and we're managing a high quality across the zone," he says. "This is our culture, there are few of us, the consorzio has maybe fifty producers, but we're all united and we are lifting everyone in the same direction."

In a "brief final reflection" at the end of the 2012 study, the "perfect zonazione," wrote Scienza and his team of scientists, is the one that "is never finished." It is more successful "the more it continues to develop over time, to improve its methodologies and its 'User manual' to take root in the territory and, above all, to increasingly become part of the work culture and of the daily practice[37] of the producers, of the workers, of the other subjects who have understood the efficiency and the 'plasticity' of this tool."[38] "We have fulfilled our dreams," Antonella said during that phone call. "Vecchie Uve" 2017 was bottled that June after fifteen months in autoclave.[39] "We'll taste it," said Paola "and see how much time it will take before we can release it." During my visit to the Vigna dei Confinanti in late 2018, she said there may be Boschèra[40] in the mix as well.

Epilogue

Da vino diventa spumante.

—Francesco Drusian

We still have machines made of steel, but they now obey bits that are weightless.

—Italo Calvino, *Six Memos for the Next Millennium*[1]

Memory of another fitful twinkling light at a rainy early autumn lunch in Solighetto: sopa coada (in Italian: zuppa covata) is a Trevisian layered bread and pigeon baked dry together over slow hours, served in a shallow bowl flooded heartily and murkily with glimmering hot broth the pigeon had been cooked in first. Against the usual ban on hot and cold together at table, it is matched by these hills' most famous thing to drink: the delicate lemony, blossom-, jasmine-, ginger-scented, tart, and classically bitter-tinged sparkling Prosecco wines that are salve in hot sunny summers and snow-covered winters alike and, wine's coefficient of absorption for carbon dioxide increasing with the lowering of the temperature to which the liquid is subjected,[2] always served chilled. These were my first hours in the Veneto's Trevisian hills.

"We think our wines are the best wines in the world," consorzio director Innocente Nardi said to a small group of journalists that evening as we tasted several bottles of Prosecco Superiore. What, I wondered, would lead someone to say this of these subtle wines? How could *these* be the best wines of the world? But even in their shynesses were tastes and aromas for which I needed many more references than what carried me through other wines. What world that can understand them in such a way? I could note only that there were things to see, then wonder about what those were. What did I need to understand and how would I get to whatever that place was? The day was a glinting, beckoning and focused bubble which perhaps itself did not know yet how to talk to me. It was maybe also a matter of codes, "a necessity, or at least a desire," as Poe had written, "of transmitting information from one individual to another, in such manner as to elude general comprehension [and so that] analytic ability is very forcibly called into action," for which reason, he added, "cryptographical solutions might with great propriety be introduced into academies."

"Is there such a thing that we call "the Veneto" and as such remains fixed in the memory of who frequents novels and cultivates themselves?" writes Isnenghi. "Probably yes. The Veneto is also a literary dream."[3] Its hills and their wine are a narrative accumulation, a working of its writers in chemistry, in dialect to be read out loud, in homage to sound and light and softly speaking complicating and proud wines. I hope that my writing into some kinds of understanding is seen as the act of admiration that it is. By now I have moved complete circle, back to a beginning that looks very different in the overlapping end. I started out for the contadini, lost sight of them or perhaps more accurately lost the sound of their voices, found them again in descendants, photographs, words, fractions, and unexpectedly in the work of Carpenè and some of the other nineteenth-century scientists and growers. There was other help finding them in Lumley's pragmatic organizational image of my project as a beach: there is sand, and people, and pebbles and lunches and there is the sea, he described to me, a far greater share but which is another world (71 percent of the earth's surface, I wrote out later, is water, 97 percent of which is ocean,[4] 80 percent of which unknown to who is writing about it[5]) about which there is far less information. Sea of Treviso's contadini who of these nineteenth- and twentieth-century hills would have until 1964 been small landowners or renter-sharecroppers or day laborers in farms or rice fields in the flats below and whose voices both need to be a part of my story and are notoriously hard to find. I have included bits and pieces of what I could, and I have thought of them throughout. I hope there are happier stories to tell of them than what most others tell. I think often of the sounds of those hazy collaborations. Sometimes I can see the stable and hear the sounds of voices and the wind and glasses, hear an evening's flickering shine.

Theirs on feast days was that first meal I ate here. The first thing I ate at all soon after landing in Venice's airport from New York and then driving into the hills around the city of Conegliano on a stormy mid-September afternoon, first of the early three-day cold spell that would lead to the area's latest harvest in decades. Hearty and brothy and something complicatedly but decidedly grape around those cool bubbles poured with it, the meal was my first thought of the striking hills I was about to learn are Prosecco's best, a foundational taste just beyond reach of what Prosecco is fused with the moment I also learned that sopa coada existed. I go back to that tension each time I return. It is a lunch, though the Prosecco bottles change, that is my way to again and again and never less suddenly know that I am here once more to taste and think. But on that first visit, three days shimmying between a lot of mist and spells of pouring rain, what

That first sopa coada
Susan H. Gordon

was presented in generally forbidden hot and cold was my first glimpse of something that seemed ungraspably untold and ancient, a Trevisian story glistening in the frank fog of fairytale and in damp steep vineyards where rows of large and clustered indented three-lobe leaves protected loose bunches of daffodil-yellow Prosecco grapes that cascaded longer than a hand's reach into a pyramidal berried point. Around me the subdued diffused light that Conegliano, Valdobbiadene, and Asolo are just as famous for was showing Prosecco as something possible to see beyond the lucrative frenzy and confusion that the outside has about its wines. I have not stopped wanting to look since, not stopped wondering, finding something, wondering more, gathering disparate and concurrent stories of the hills' people, history, character, flavor, of the current unthinkable success of the land of sopa coada and polenta now shining gold as well. What does Prosecco, in taste, and story and look and whereness, sound like?

"Judicious analysis, scrupulous hygiene in the winemaking areas, the application of all the other good norms of theory and practice in winemaking," I first read Montanari and Ceccanelli's words three years later in a reading room of the New York Public Library. "Instead in the medium and small private cantine, winemaking and wine aging always leave much to be desired; and on top of that, in some cases, it is conducted in an absolutely irrational way." I needed to read something else. I spent the rest of that day finding Veronelli, Pablo

Echaurren, Marcello Baraghini in *Le parole della terra "Manuale per enodissidenti e gastroribell,"* Soldati in *Vino al Vino*, to connect the conversation between the contadini and the industry that has become industrial and to lean back heavily toward empiricism, poetry, philosophy, and the kinds of wines I love. By the time I came back to Montanari it was easier to see what my problem was. By the standards of many, including me, today, the paragraph just after reads as a compliment to the smaller producers said to be *natural* and an indictment of the larger considered *industrial*. "In fact, especially in the small cantine of the grape-growers, the areas, used often as promiscuous growing spaces, are more than not the negation of the most elementary hygienic practices, and all the procedures adopted are still too empirical; almost never, finally, in these cantine are sulfited salts used during vinification and in aging wines, sulfited salts which are instead widely and judiciously adopted in the other winemaking means mentioned above."[6] "Buying some good vino sfuso," Baraghini wrote restoratively in the introduction to *Le parole della terra*, "allowed me to spend time in cantine in a way that was otherwise impossible, made of tasting, chatting, stories, sharing secrets, and sometimes a snack or a meal. That is a little piece of contadino civilization that survived the violence of the consumerism that arrived here."[7]

In media and drinker talk about winemaking everywhere, in response to that consumerism and industrialization a kind of ruralism has arisen today, a simplified idea of nature and countryside in which it is hard to not hear echoes of fascist movements, *natural, pure, farmer* wines whose context is *untouched landscape and winemaking*, flanked by twenty-first-century soil and rock studies, and above all *not industrial*. "The discourse of the authentic as culturally pure, rural, and traditional produced myriad exclusions, not only of all that was urban, but also of the culturally hybrid and cosmopolitan," writes Patriarca of fascist times. The intense consumerism of today is bound with telltale words like *authentic*, and also *discover, hidden, unknown*, words that make little sense of who is being spoken of this way but are tools for the kind of control and taking that colonizing and newsmaking forces enact. Writing as process, listening, trusting, never ever pronouncing: at the end of my book I found my thesis, or at least my reason for writing it. "Finirò," Carpenè wrote in 1893 in defense of his sparkling method, "con una mia opinione (che taluno considerà ardita)."[8] This book is also my proposal that there is another Italian moment beholden to neither impulse, and that is the nineteenth century just before and the few decades after the Risorgimento and exemplified by the Prosecco hills, through Carpenè, through Conegliano, by the work to create an Italian spumante defined by everything that is the tumbling reaches of the Valdobbiadenese (where the cooperative cantina sociale Val d'Oca, too, has now begun to make a Prosecco Superiore of *those* hills' old varieties: "Uvaggio Storico" is Glera, Verdiso, Perera, Bianchetta Trevigiana, in sweeter, softer, old-fashioned Dry) and the overall gentler rise of the Coneglianese and Asolo with its own sharp suddennesses. The forests aren't the only salvationally impure part of this story.

Purity isn't even true about vineyards, is equally impossible to be so about the most hands-off winemaking. Contrary to accusations of technology, Carpenè's dream was of his machines and what grew not *naturally* as no vinifera does but *viticulturally* and *healthfully*, his books and writing to teach and guide, into custodian, rallier, inheritors of this land. Throughout twenty-first-century Italy, it is common to hear wine-growers talk of their wines as made of tradition, nature, and technology, the right measures of which I believe are realized reflections of what, for example, Carpenè was after. Many, as he foresaw, have become wealthy. In the hills, restaurants from humble to more elevated serve equally perfect sopa coada. And everywhere, Veronelli and Soldati's spirits live on in wine-drinkers' new fascination with Italy's indigenous grapes and once-upon-a-time wines. But the world those two writers worried about is different now, cities no longer mean supermarkets and genetically modified foods, since the rise of industrial fast food in the 1980s, farmers markets and urban agriculture have been arriving as have the supermarkets to the supposed-idyllic non-city. This time, more hope, empiricism includes data and digital measuring, a cyborg[9]-minded *return to nature*, weather stations monitored on smartphones in biodynamic vineyards, same for soil readers so that vines can be left only as untouched as possible, growing hybridization that includes also the "growing complexity of the traditional antinomy of urban versus rural with the additional expansion of social hybridity where different existential visions collide."[10] Outside the mutually dependent tug-of-war between industrialists and artists today as in that late-nineteenth-century point, lie the far more connected scientists and farmers. Between ancient systems of past modernnesses and technology that seems modern to us today, a fusion or alliance between what is imagined as two things, for the time being. "You've got a new generation, in the United States anyway," McIlhenny said to me in response to that. "They read about ancient traditions and plug into that, but they also have a mastery of what science and tech can do to make up for what they don't have of the intuition of makers in the past; and technology is accessible and inexpensive." A new generation that also brings a new appreciation and market for the hills' traditional style, long made in various degrees of meaning to, gently fizzy wines that taste of fermenting yeast along with grapes, frizzante from the change of seasons trapping some CO_2 in the fermentation vessel for bubbles. As the autoclaves[11] now do on purpose, controllably and stronger, earning the hills that DOCG label in 2009, and carrying words like *Prosecco, Treviso, Valdobbiadene, Conegliano, Asolo* and slowly all of their multiplicities into the rest of the world.

"After gathering the documents of so much history and of so many sweet testimonies of good living, I nevertheless have the feeling of not having said everything," Maffioli wrote at the end of his *Cucina Trevigiana*. "I realize that there is another dimension in the Joyous Marca, which is also the basis of its gastronomic fortunes. Hearths and domestic and public tables are altars for celebrating the ancient ritual of the taste of life, the taste of friendship, the art of conversation, and of remembering, commemorating, transforming the past into present . . . which the light intoxication of wine transforms into legendary tales. We hope this book of ours will be useful to those who read it, not only as a volume of recipes, but also as memories of a city and its territory. . . . As if I had told the story of a large family of mine, exemplary in many respects, which in a healthy desire to equal and surpass others, with agreement among all, has made itself superior [in food and wine], which all of Italy now recognizes."[12] As a grateful guest, I agree.

What in that first sip I tasted, I don't remember. But I know what must have reached me first. Opening a wine is to confront it, to pour and look, smell, then taste. After writing something like *pale lemon gold*, I would have had to move to the next step, smelling and then breaking what happened next into components. I would have had to notice first how reticent yet insistent the wine's aromas. The emotional effect first, and then what were the ways to get to it. Haunting, there, but quietly. Then, thinking a split second before writing, *ginger, lilac, lime*, and what parts of that; the louder oils present in the peel would come to mind first, then maybe *pith* for the telltale bitterness that's such a part of these wines and a taste so present as to be smellable. As in all sparkling wines the bubbles would have worked to deliver flavors more emphatically, giving these graceful aromas greater boldness upon taste. I wanted to become a wine drinker who could hear, anticipate, be worthy of these wines of such distinctive determination and home and waiting. "In the autumn," wrote Comisso, "on all these hills from Asolo to Onigo, to Cartizze, to Pieve di Soligo, to Conegliano, to Vittorio Veneto, [the] final feast[13] is that of the harvest."[14]

In 1949, the spumantizzatore Bortolomiol was founded by Giuliano Bortolomiol, son of a long family line of grape-growers and who had graduated from the Scuola Enologica. First, he had founded Valdobbiadene's Prosecco Fraternity to rebuild more than a thousand hectares of vineyards destroyed by World War II's bombs, then unlike many farmers here stayed with a mission to improve the wines and develop the market, with a familiar thought in mind. Enthralled with Valdobbiadene's vineyards, with spumante and the Martinotti method now available —"with our Italian method we can reach the highest levels, I have no doubts"[15]—in 1960 he released Valdobbiadene's first official Brut typology, less remaining sugar still than the traditional Extra-Dry made throughout the hills and a push as strong for the grapes of Valdobbiadene as it was for what Conegliano's technology made of them. Until Nino Franco's 1979 autoclave, *that* family transformed their wine into their spumante here. But Bortolomiol was unseeing in one regard: with four children all daughters, he once thought it would be necessary to sell the winery when he retired. Today, all four sisters run the azienda—previously a silk production mill where work had been women's, too, and guaranteed a woman a role within her family. In the summer of 2021, Bortolomiol eldest Elvira became the first woman elected president of the Conegliano-Valdobbiadene Consorzio. It is she who pours the wine at my visit in 2018, who began to tell us who her father was.

Glossary

Terms and cross-references are in cursive; incidental new terms are, too, with definitions where they appear.

GRAPE VARIETIES[1] (NAMES SHARED BY FRUIT AND VINE)

Bianchetta Trevigiana Average, spheroid grapes; thick yellowish-green skin with a whitish bloom (pruinose) and slightly astringent flavor; loose pulp, juicy with simple flavor; in average-size pyramidal compact bunches with one wing which is sometimes large enough to make the bunch seem doubled; below-average-size pentagonal, three-lobed and sometimes even five-lobed leaves with gray-green underside.[2]

Glera (Tonda) Name as of 2009. Notably firm average-size spheroid grapes of dotted yellowish-golden thin skin pruinose and a little tannic; juicy pulp, simple, sweet flavor; in larger-than-average elongated, pyramidal, winged bunches under average-to-large pentagonal trilobal and at times pentalobal leaves.[3] Vigorous, late ripening (early October), early bud-burst and flowering.

Glera Balbi Name as of 2009. Prized Glera *clone*, loose bunches for best ripening; can have grapes of different sizes.

Glera Lunga Name as of 2009. Average-sized, dotted elliptical grapes with not-very-thick skin of a deep golden color tending to pink, pulp with a special herbaceous flavor; in average-size truncated pyramidal, basically compact bunches under large, pentagonal, three-lobed leaves.[4]

Perera Average-size spheroidal grape with golden-yellow skin, heavily pruinose; pulp with a characteristic, aromatic flavor, in medium-to-large conical or pyramidal bunches with one or two wings, under medium-large pentagonal, three-lobed leaves. Late ripening, somewhat late bud-burst.[5]

Prosecco (Tondo) Name of *Glera (Tonda)* before 2009.

Prosecco Balbi Name of *Glera Balbi* before 2009.

Prosecco Lungo Name of *Glera Lunga* before 2009.

Rabbiosa Trevisian synonym for veronese Durella, and therefore likely a local *biotype*. Average-size ovoid grape with thick yellowish or golden-green skin, heavily pruinose; loose pulp, with a simple, sourish taste; under slightly trilobed or whole leaves, smooth light-green underside. Late bud-break, somewhat late ripening.[6]

Verdiso Above-average size, slightly ellipsoidal grapes with thin slightly dotted yellowish-green skin, pruinose; juicy, soft pulp with simple flavor; in long, pyramidal (and in the coolest and most fertile soils, compact) bunches with or without one wing; under average-size pentagonal leaves almost as wide as they are long, trilobed or no lobes, and turn yellow in the autumn. Somewhat late bud-break, late ripening (first half of October).[7]

ORGANIZATIONS

cantina sociale Winemaking cooperative.

Carboneria Secret patriotic and risorgimentale society founded in the Kingdom of Naples during the first days of the nineteenth century.

Consorzio A producer membership organization charged with guiding, promoting, and protecting a wine (for this book's purposes) zone's denomination-covered wines, including viticulture, winemaking, labeling, and disseminating information to producers, media, and drinkers.

Giovine Italia Out of the *Carboneria*; secret risorgimentale society founded by Giuseppe Mazzini in 1831 in Marsiglia to promote, make happen, and plan for one unified republic of the Italian people and therefore against the piemontese Savoias heading a conglomerate Italian kingdom, as did instead happen.

partigiani The covert armed resistant movement of women and men in Italy against the fascists during World War II; hard to count but estimated to be at least 250,000 Italians, plus other nationals, including Italian Americans and Russians. Among the partigiane are Clara Lollini, married to partigiano and chemist Michele Giua, and their daughter Lisa.[8]

Società Enologica Trevigiana Membership-based winemaking and teaching society founded in 1868 in Conegliano by Antonio Carpenè and Abbot Felice Benedetti.

PEOPLE

agricoltore Agriculturalist, with a somewhat more relaxed meaning.

Count Amedeo Carlo Avogadro Piemontese noble family scientist, 1776–1856; in 1811 wrote correctly about atoms' role in compound structure; risorgimentale in 1821 uprising.

Conte Marco Giulio Balbi Valier Experimented with Prosecco *clones* in the hills in the mid 19th century. The resulting Prosecco Balbi biotype is considered a top-quality vine.

Pietro Bembo Writer, member of court in Rinascimento Asolo; first non-tuscan to write in the vulgar/common tuscan that would become Italian, following the work of tuscans Dante Alighieri and Giovanni Bocaccio.

bracciante Day laborer.

Brandolini Trevisian noble family of Cison del Valmarino area in the Marca Trevigiana.
Tullio Brugnatelli Chemist, Antonio Carpenè's chemistry teacher; risorgimentale in 1848 uprising.
Stanislao Cannizzaro Sicilian chemist (1826–1910), brought Avogadro's work to a larger audience; risorgimentale in 1848 uprising.
Antonio Carpenè Chemist, winemaker, writer, inventor, cofounder of the Scuola di Viticoltura ed Enologia in Conegliano in 1876; risorgimentale in 1860.
Etile Carpenè Antonio Carpenè's son, continued his father's business and further developed Prosecco spumante
C. B. Cerletti First director of the Scuola di Viticoltura ed Enologia in Conegliano
colono An *agricoltore*, lavoratore della terra, *contadino*
contadina Woman in the *contadino* culture; whose roles were quite different at home, in the vineyard, and in day-laboring.
contadini Plural for either all men or men and women. *Contadine* for all women, but not often seen.
contadino Man who farmed the land for a living: *peasant* certainly doesn't cover it, neither does *farmer*. It is a term with historical and cultural meanings, referring to a man who is part of Italy's centuries-old agricultural traditions and with a variety of classes within that: from landowners to renters to sharecroppers to day-laboring *braccianti*. Largely oral and conservative, traditional.
Caterina Cornaro Queen of Cyprus when Venice adds that territory to its Stato da Mar. In return for losing her kingdom, she is given a court in the Trevisian town of Asolo.
Contessa Marta Gradenigo Count Balbi's widow, listed as a winemaker in 1892 *Annuario*.
Tullio De Rosa Mid-twentieth-century enologist in Conegliano; particular focus on spumanti and tasting.
Angelo Malvolti Great-great-grandson of Francesco Maria; viticulturist growing Balbi Prosecco; his vineyards and financing were one-half of founding Carpenè Malvolti.
Francesco Maria Malvolti Of noble Trevisian family, academic of agrarian studies, his 1772 written mention of quality Prosecco is earliest, so far.
Federico Martinotti Later-nineteenth-century enologist in Asti, Piemonte; director of Regia Stazione Enologica Sperimentale di Asti; his patented method and apparatus become the Italian method of spumantizzazione.
Paron Stefano Massarioto Illuminato Checchini, wrote in a Trevisian dialect for the contadini after unification; an ally of the *contadini*.
popolo Populace, the people, the citizenry.
viticoltore One who grows the vines and by inference the wines. *Viticulturist* is a direct and technical translation. *Winegrower* is a nice substitute.

PLACES

Asolo Town and comune in Treviso province; on right bank of the Piave river.

Asolo Prosecco DOCG Spans 18 comuni of the Treviso province, on the right bank of the Piave river; producing Prosecco Superiore.

casera Small structure in the Trevisian hills where people who brought their flocks to higher pastures in the summertime could stay, sleep, cook, and eat.

chiusura Small farmhouse with a small amount of land attached.

Conegliano City and comune, on the left bank of the Piave river; location of Italy's first winemaking school.

Conegliano-Valdobbiadene Prosecco DOCG Spans 15 comuni of the Treviso province on the left bank of the Piave river; producing Prosecco Superiore.

Marca Trevigiana The countryside, both flat and hills, of Treviso. The Alta Marca is the upper Marca.

masseria Farmhouse with a larger amount of land attached than what a *chiusura* has; the structure, too, is then larger than a chiusura.

Tre Venezie Collective name for the Venezia Euganea (covering today's Veneto region and part of Friuli), Venezie Tridentina (covering Trentino-Alto Adige region), Venezia Giulia (today parts of Friuli-Venezia Giulia, parts of Slovenia and Croatia); originating in 18th century to indicate cultural and geographical Italy under foreign rule.

Treviso Province (also city and comune) in the Veneto region.

VITICULTURE AND ENOLOGY

botte I translate this as *cask* instead of barrel, in many sizes, roughly 10 hectoliters to 500; for storing and/or raising wine. I understand *barrel* as a French barrique, about 228 liters. Dalmasso and Dell'Olio note that for a while a botte was a unit of measurement in the Conegliano area and accounted for in today's terms about 8 hectoliters of wine, and in contemporary ones 20 mastelletti coneglianesi (each of these: 18 boccali, or 40 liters).

biotype Subvariety, or member of a variety.[9] Mutations of a variety (also called a *cultivar*) as it adapts over time to the different local conditions in which it is planted, genetically identical but with differences in appearance and most importantly aromas and taste. Having been around longer, older varieties have more biotypes. See *clone* for process and difference.

cantina The room in which wine is made; *wine cellar* and perhaps *winery* are the closest English words.

cantina sociale See *Organizations*

clone A cutting from an individual grapevine chosen for desirable traits including disease resistance, weather hardiness, yields, and fruit quality; genetically and morphologically identical to source plant.

Consorzio The membership organization charged with protecting and promoting a wine zone's producers, growers, and winemakers (and when applicable, spumante bottlers).

DOC Denominazione di origine controllata (denomination of verified origin): legal name assigned to a wine for specific qualities, including grape varieties, place, growing and winemaking techniques, historical role or presence; a label indicator of quality.

DOCG Denominazione di origine controllata e garantita (denomination of verified and guaranteed origin) as per above; wine quality is verified by outside official; generally a label indication of strictest and most historical winemaking.

DOP European Union catchall term for Italy's protected wines and foods; Denominazione di Origine Protetta (denomination of protected origin).

enochimico Wine chemist, or chemistry as related to wines

enologo Enologist; winemaker with understandings of a technical focus reflective of the late nineteenth century and later.

fecce The solids that precipitate from wine. Generally this term is equivalent to *gross lees*; larger particles in fermenting *must* made up of the accumulation of grape parts, dead and satisfied yeast cells, and other such-sized particles all settled to fermentation vessel bottom, and often harbingers of harmful bacteria so new wine is racked off of them (different from *fine lees*, though sometimes a synonym, too: smaller particles that remain in new wine, adding flavors and body and generally unwanted in Prosecco spumante, but usually a component of the ancestral method, or *sui lieviti, on the fine lees*).

filare First, vines trained as festoons draped through live trees; later, row in specialized vineyards.

Millesimato A bottle of Prosecco Superiore with the vintage noted, by law at least 85 percent all from that one vintage. Wines without this label can also be from one vintage and generally are, but for various reasons denomination rules may not allow including the vintage on the label.

mosto Or *must*: the liquid-and-solids result of crushing the grapes, juice complete with skins and sometimes stems.

Prosecco Superiore Labeling term for fully sparkling Prosecco wine from the Conegliano-Valdobbiadene or Asolo hills under specific viticultural and winemaking conditions, since 2009.

spumante Unlike frizzante wines (in modern-day legal wine terms: no more than 2.5 bars of pressure), a number reflection of technology: from Champagne's stronger-glass bottle-based re-fermentation method to Prosecco's re-fermentation apparatus, both of which can produce the minimum of 3.5 bars of pressure EU regulations require of Quality Sparkling Wine, and in practice generally 5 to 6 atm. There are typologies by *sweetness level* to consider, too, and the well-made spumanti will be a question of such in balance with acidity, carbon-dioxide-granted pressure and saltiness, and other taste components.

tino Vat: vessel built to hold the crushed grapes while they undergo their alcoholic fermentation, or fermentation into wine, additionally in the form of a truncated upright cone, open at the top and fitted at the bottom with a little opening from which to extract the pomace; once exclusively of wood (bent staves of chestnut, oak, and so on, bound together by iron bands), today often of cement, stainless steel, or fiberglass.[10] A *botte* is a tino built instead to store and raise wine.

tinaia Literally, place where the *tini* are; term for *cantina*, today sometimes identified as a tuscan term.

tipicity Typical, usual, representative of the area it comes from.

travasare Common winemaking term, taught in international wine education by its French name. A verb form of travasamento, the process by which wine is transferred from one botte to another, leaving most of the *fecce* behind. At bottle (drinking) level and therefore with sediment instead from precipitating tannins, colors, or *lieviti*, this would be something like decanting.

vinacce Pomace: remainder after grapes are pressed, of stems, skins, seeds, some fermented *must*. Made from vinacce, the distillate grappa is an Italian habit, and Carpenè Malvolti has long made grappa from Prosecco (and then Glera) vinacce.

variety The more common term for a *cultivar*: a grapevine with fruit selected and propagated for specific attributes such as flavor, color, texture, identifiable also by fruit and leaf appearance and since the twentieth century corresponding to specific DNA patterns. A member of the set of *Vitis vinifera sativa*, cultivated *vitis vinifera.*

vigna Piece of land on which vines are cultivated; can include many plots and subdivisions. *A vigna* is the cultivation of vines. Also symbol of the Roman Catholic Church, as community of the faithful, "the vigna of the Lord."[11]

vigneto Small vineyard, connotation of *plot*. Size is less important though than that it be of a vineyard of similar soil and climate characteristics and as such produces wine with specific tastes and characteristics tied to it. Such a site can also define a variety's most *itself* form.

vite Grapevine (for connections and importance, the Italian term for *life* is *vita*).

vitigno Cultivar, or a grapevine variety that is cultivated, which is circular thinking since in order to not be a wild vine with unpredictable fruit appearance and traits, a variety is propagated—cultivated—by planting cuttings of existing and desirable vines. *Vitigno*, however, is used in everyday wine language whereas *cultivar* has a technical sound. Sharing space with *vite*, it is clearly a grapevine, an instance, a changing, or domesticating of one.

Vitis vinifera One of about thirty species of climbing vine genus *Vitis*, and thought native to at least two centers, "one in the near East and another in the western Mediterranean region."[12] All quality wine grape varieties belong to this species (interspecies hybrid grapes made by research centers exist and can make very good wine, but for many reasons with varying degrees of validity are not generally accepted as top-quality wine materials). *Vitis vinifera sativa* is the subspecies of cultivars, or domesticated agricultural varieties (*Vitis vinifera silvestris* are the species' wild vines, which therefore cannot be categorized into varieties).[13]

PROSECCO DOCG TYPOLOGIES, IN ORDER OF INCREASING PRESSURE

Still (Tranquillo)
Frizzante (also *Sui lieviti* if made in bottle and not autoclave)
Spumante

PROSECCO SUPERIORE TYPOLOGIES, PER EU CLASSIFICATIONS

Zero dosage No residual sugar. A little misleading since Prosecco is never *dosaged*, the French term for sugar added at bottling as per Champagne practices.

Extra-Brut 0 to 6 grams of sugar per liter, remaining after second fermentation.

Brut 0 to 12 grams of sugar per liter, remaining after second fermentation. Technically Extra-Brut falls within Brut but this newer labeling term (since 2014 in Asolo; since 2019 in Conegliano-Valdobbiadene) alerts you to the lower amount of residual sugar. In *Tecnica*, De Rosa points out that *Brut* is from Champagne, and means that when there is no dosage, it is a sparkling wine that is "not finished," from which habit the term has come to be used to mean a content of sugar not "more than 1% (10 grams), and usually 0.5 to 0.8%," even for wines like Prosecco which use a different production and residual-sugar-level process.

Extra-Dry 12 to 17 g/l, traditional.

Dry 17-32 g/l (a trail from when sparkling wines—precisely, Champagne—were much sweeter than today. *Dry* indicated one that was less sweet. Cartizze, where the grapes are sweetest, traditionally had a hard time fermenting until the end (no sugars left), and so its traditional style is Dry. There is a handful of producers, including the brazen Duca del Dolle, fermenting these to *Zero dosage*, deliciously.

NON-EUROPEAN UNION TERMS FOR SWEETNESS

Amabile A light wine with noticeable sweetness.
Asciutto Dry (lowercase *d*) still wine.

WATERS

acqua
canale
rio
roggia
rosta
ruscello
torrente
torrentello
val
valle

TREES

acacia
cherry
chestnut
fico
frassino
gelso
nespolo
oak
poplar (talpon)
susinèr
walnut

RIVE, OR LABEL DESIGNATION FOR FORTY-THREE ESPECIALLY STEEP HISTORICAL AREAS OF THE CONEGLIANO-VALDOBBIADENE DOCG

Twelve are names of a comune:
Cison di Valmarino, Colle Umberto, Farra di Soligo, Follina, Miane, Pieve di Soligo, Refrontolo, San Pietro di Feletto, San Vendemiano, Susegana, Tarzo, Vidor.
Thirty-one are names of a frazione of those comuni:
Arfanta, Bagnolo, Barbisano, Bigolino, Campea, Carpesica, Col San Martino, Colbertaldo, Colfosco, Collalto, Combai, Corbanese, Cozzuolo, Farrò, Formeniga, Guia, Manzana, Ogliano, Premaor, Resera, Rolle, Rua, San Giovanni, San Michele, San Pietro di Barbozza, San Vito, Santa Maria, Santo Stefano, Scomigo, Solighetto, Soligo.

Timeline

1603 Cultural-studies Accademia degli Aspiranti founded in Conegliano.
1753 Accademia dei Georgofil founded in Firenze.
1765 Reggio-Emilian Jesuit biologist Lazzaro Spallanzani publishes *Saggio di osservazioni microscopiche.*
1769 Accademia degli Aspiranti adds agrarian classes.
1775 Garbellotto botti-maker launches in Treviso.
1796 Napoleon arrives in Italy.
1797 Conegliano, devoted to Venice since 1339, renews its oath of loyalty on March 29.
1797 Venetian Signoria meets for the last time, headed by 120th and last Doge Manin, who dissolves the republic on May 12; signed on May 17. Napoleon enters the city.
1797 Napoleon cedes Venice and territories to Austria on 17 October.
1798 Austrian troops enter Venice, ceremony with Manin on 18 January.
1798 to 1805 Kriegskarte military maps created; Carboneria forms in Naples (first anti-French focus, then decisively anti-Austrian).
1799 Adelfi forms in Paris, Jacobin, anti-Napoleon, extends to Piemonte and Milan in early nineteenth century; by end of eighteenth century, the Filadelfi, democratic, founded in France, spreads in northern Italy.
1805 Terraferma back under Napoleon, Kingdom of Italy.
1805 Giuseppe Mazzini is born on June 22 in Genova.
1807 Giuseppe Garibaldi is born on July 4 in Nizza.
1811 Avogadro publishes *Essai d'une manière de déterminer les masses relatives des molécules élémentaires des corps, et les proportions selon lesquelles elles entrent dans ces combinaisons.*
1815 Venice and its Terraferma back to Austria, now as part of its Lombardo-Veneto Kingdom.
1821 Uprising in Sicily, Avogadro participates, uprising fails.
1831 Giuseppe Mazzini founds Giovine Italia in Marsiglia, for Italian republic, against kingdom and (therefore, later) Piemonte rule.
1838 Antonio Carpenè is born on August 17 in Brugnera di Sacile.
1845 Trevisian contadini sign petition to protect their rights to the woods.

1848 Italy's first war of independence, Sicilian chemist Stanislao Cannizzaro fights; so does Trevisian agriculturalist Angelo Vianello.

1851 French Rousseau's first attempts at a large sealed pressurized vessel for wine.

1855 Conegliano railway station opens.

1857 French chemist Louis Pasteur publishes *Mémoire sur la fermentation appelée lactique*, based in part on Spallanzani's work.

1858 Pasteur publishes *Mémoire sur la fermentation alcoolique*.

1858 Sicilian chemist Cannizzaro publishes *Sunto di un corso di filosofia chimica*.

1858 French chemist Edme Jules Maumené experiments with Afroforo and afrometro.

1859 Against Austria and somewhat allied with France, Italy's Second War of Independence.

1860 Mazzini publishes *Doveri dell'uomo*.

1860 Cannizzaro presents Avogadro's results at the first International Congress of Chemists, in Karlsruhe, September 3–5.

1860 Carpenè leaves school to fight for Italy with Piedmont army, leaves it to fight with Garibaldi.

1861 There is an Italian nation; Garibaldi hands control to King Vittorio Emanuele II of Piemonte's Savioa family; the Kingdom of Italy is proclaimed on March 17.

1861 Tullio Brugnatelli publishes *Trattato di Chimica Generale* textbook.

1861 Carpenè earns Diploma in Farmacia at the R. Università di Pavia, with a thesis on aluminum.

1862 Carpenè earns Dottore in Chimica at the R. Università di Pavia, with a thesis on lactic acid.

1866 Much of the old Venetian empire joins Italy, after Third War of Independence against Austria.

1867 Garibaldi addresses Conegliano on March 5.

1868 Carpenè and Vianello's viaggio in Italia to see the states of Italy's wines.

1868 Carpenè cofounds the Trevisian province's Società Enologica in Conegliano with agricultural abbot Benedetti.

1869 A group of Italian chemists publishes the *Dizionario Generale di Chimica*.

1870 Italian army takes Rome on September 20.

1870 *Ampelografia Generale della Provincia di Treviso* is published.

1871 Carpenè publishes *Sunto teorico e pratico di enologia pei vinificatori della provincia trivigiana*.

1872 Mazzini dies on March 10 in Pisa.

1874 Carpenè and Vianello publish *La Vite* based on *Sunto* and viaggio and on the occasion of the 1873 exposition in Vienna.
1876 The R. Scuola di Viticoltura e di Enologia is created in Conegliano by royal decree.
1877 The School opens.
1878 Carpenè publishes third edition of *Sunto*, now called *Sunto teorico-pratico di enologia*, with Vol. 1 on vinification and Vol. II (for the first time) on enochemical analysis.
1880 Bertolaso, maker of enological machines, opens in Verona.
1881 Carpenè spumante system wins a gold medal at the Esposizione internazionale di macchine enologiche in Conegliano.
1882 A patent is issued for the sistema Carpenè per la fabbricazione dei vini spumanti.
1883 Carpenè and nobleman viticulturist Angelo Malvolti convert the Società from a multimember enterprise riddled with debt and disagreement into the Carpenè Malvolti company.
1888 Carpenè publishes fifth edition of *Sunto*, vols. 1, 2; thirty-two pages and fifteen pages longer than 3rd edition, respectively.
1888 Montello contadini sign petition of protest to protect their rights to the forest.
1895 Banca di credito agricolo ed industriale di Conegliano opens.
1895 Piemontese enologist Federico Martinotti patents his spumante method and apparatus.
1897 *Ampelografia Italiana* is published.
1902 Carpenè dies on March 23 in Conegliano.
1912 In England, metallurgist Harry Brearley makes stainless steel of chromium and iron.
1915 Italy enters World War I on May 24 in part to regain Venetian lands from Austria.
1917 After Italy's terrible loss at the Battle of Caporetto (now Slovenia), Austria lays seige to Conegliano and hills.
1918 Italy defeats Austria at Vittorio Veneto in September.
1918 A badly battered city of Conegliano is liberated on October 28.
1918 Villa Giusti armistice sanctioning Austrian empire's surrender goes into effect on November 4.
1918 World War I ends on November 11; Italy gains Trento, Trieste, Istria, and Alto Adige.
1923 The Istituto Sperimentale per la Viticoltura di Conegliano opens. Tullio De Rosa is born in Spilimbergo.
1924 The words *Prosecco* and *Conegliano* first appear on a spumante label, from Carpenè Malvolti.

1924	The School opens at its second and current location.
1925	Benito Mussolini installs fascist dictatorship throughout Italy in January.
1927	Bottega del Vino of Conegliano's School opens.
1937	First Prosecco exhibition: vini di tavola, vini di lusso.
1930s to 1950s	Etile (son) and Antonio (grandson) Carpenè fine-tune the system of spumantizzazione in autoclave for Prosecco.
1945	Mussolini is executed by partigiani on April 28.
1945	World War II ends on May 8 in Europe; on September 2, overall. Fascist Italy surrenders, including colonies on Adriatic, in Libya, on the Horn of Africa. Partigiana Italy is victorious. Italy loses some Venetian territories.
1946	Referendum ends the Kingdom of Italy and creates the Repubblica Italiana on June 2.
1947	The republic's constitution is approved in December; article 31 creates the regions as first-level subdivision of the nation.
1948	First iron autoclave, enameled on the inside, is made in Canelli, Piemonte, for Asti's wines.
1949	Bortolomiol founded as Italian-method Prosecco producer.
1953	Autoclave maker Sirio Aliberti is founded in Canelli.
1958	The beginnings of the Ristorante da Brun.
1959	First festival of Trevisian cuisine.
1961	Luigi Veronelli publishes *I vini d'Italia*.
1962	Consorzio for Conegliano and Valdobbiadene created.
1963	Italy creates DOC system.
1964	Italy's mezzadria sharecropping system is ended by law.
1964	Osteria Dei Colli is opened.
1964	Tullio De Rosa publishes *Tecnica dei vini spumanti*.
1966	Strada del Vino Bianco is created.
1968	Mario Soldati's first viaggio in Italia, in autumn.
1969	Conegliano-Valdobbiadene DOC created on April 2.
1970	Prosecco Tondo, Verdiso, and Bianchetta Trevigiana are added to the Registro nazionale delle varietà di Vite (under the Ministero delle politiche agricole alimentari e forestali) on May 25.
1970	Mario Soldati's second viaggio in Italia, including Conegliano and the hills, in autumn.
1975	Mario Soldati's third viaggio in Italia, in autumn.
1977	Montello and Colli Asolani DOC created.

1977 Veronelli uses legal means to stop distribution of Coca-Cola in Italy by claiming that the beverage was not "produced with substances coming from the fruits or plants referred to in the denomination"; for one day.

1979 Nino Franco buys its first autoclave.

1985 Consorzio Vini Asolo Montello created.

1986 Slow Food founded in Piemonte, after demonstration against a McDonald's opening in Rome.

1987 Gambero Rosso wine guide is launched.

1992 Boschera is added to the Registro nazionale delle varietà di Vite (under the Ministero delle politiche agricole alimentari e forestali) on October 30.

1994 De Rosa dies in Conegliano. Perera and Marzemina Bianca are added to the Registro nazionale delle varietà di Vite (under the Ministero delle politiche agricole alimentari e forestali) on November 24.

1999 Slow Wine guide is launched.

2000 Prosecco Lungo is added to Italy's Ministero delle politiche agricole alimentari e forestali's Registro nazionale delle varietà di vite on December 6.

2003 Osteria Dei Colli closes.

2009 Change and expansion of first DOC, creation of Conegliano-Valdobbiadene and Asolo DOCGs, Prosecco grapes become Glera; on July 17.

2010 Almost 66 million bottles of Prosecco spumante are produced under the Conegliano-Valdobbiadene Prosecco DOCG label.

2012 1 million bottles of Prosecco spumante are produced under the Asolo Prosecco DOCG label.

2019 One-fermentation style Prosecco frizzante (sui lieviti, col fondo, ancestral) included in DOCG.

2020 WHO declares the novel coronavirus (COVID-19) outbreak a pandemic on March 11.

2020 Large-scale protests against police brutality begin throughout the United States on May 26.

2021 Prosecco DOCG bottles set a record (and just shy of capacity) of producing more than 100 million Conegliano-Valdobbiadene and more than 21 million Asolo; including the DOC bottles, more than 650 million bottles of Prosecco spumante are made and sold.

Notes

INTRODUCTION

1. From *The One-Straw Revolution*, by Masanobu Fukuoka, Larry Korn, ed. New York: New York Review Books, 2009. Used by permission of the Estate of Masanobu Fukuoka.
2. These smoothed-out numbers are from a more complex story: the two DOCGs (about to be defined in this introduction) that make Prosecco Superiore are both historical regions but with differing legislative, cultural, and territorial compositions and time lines. The Asolo DOCG, created as a legal entity in 2009, has been catching up to its production capacity (through making this wine instead of selling grapes or base wines or producing other styles of Prosecco), with 17 million bottles produced in 2019, almost 19 million in 2020, and 21 million in 2021. While the Conegliano-Valdobbiadene DOCG, longer its own entity and a larger territory than its neighbor as well as differing uses and organization of that, produced almost 90 million bottles of Prosecco Superiore in 2019, backed up to about 88 million in 2020, and reached 100 million in 2021. In addition, the export focus of each differs: the Conegliano Valdobbiadene DOCG exports around 40 percent of its production, while Asolo exports 60 percent. All numbers are approximated from economic reports and press materials issued by the two DOCGs' consorzi (see Glossary, or throughout this book). Both make Prosecco styles that are not the sparkling Superiore, but in very small numbers. Both are now nearing production capacity, and as Giulia Pussini of the Conegliano-Valdobbiadene consorzio points out, their "medium- and long-term objective remains the consolidation of the intangible value and quality of our wine, rather than numerical growth."
3. This is generally true throughout Europe, to varying degrees in other winemaking regions, and mostly untrue in the United States, where wines are named by the wine-grape variety/ies they've been made of.
4. Not sparkling or fizzy. Tranquillo, in Italian.
5. In 2022, under the Giorgia Meloni–led administration, this government agency's name was changed to Ministero dell'agricoltura, della sovranità alimentare e delle foreste, a change reflecting in at least two ways an inward, disconnecting outlook.
6. Visentin and Vallerani, "A Countryside to Sip," 3.
7. "Prosecco is believed by some to come from the homonymous village in eastern Friuli; others believe that Prosecco, which we call Coneglianese, is very different from that of Friuli," See: Ministero d'Agricoltura, Industria e Commercio, *Ampelografia Italiana*, 93.

8. In what is now Italy's autonomous region of Friuli-Venezia-Giulia, the nation's easternmost and a traditional borderland of Italian, Austrian, and Slovenian language food and people.
9. *Baedeker's Southern Germany and Austria,* 276.
10. See Ferdinando Abbri, "L'edizione veneta del Lärbok i kemien di J. J. Berzelius," in *La chimica e le tecnologie chimiche*, ed. Bassani, 73–86. For example, "Nel panorama degli antichi Stati Italiani la Repubblica veneta costitui nel Settecento un contesto di grande rilievo per lo sviluppo della chimica come scienza. È noto che nel corso della prima età moderna la chimica dovette faticare non poco per affermarsi come disciplina sia sul piano culturale sia su quello istituzionale. Nei diversi contesti politici si ritrovano situazioni molto specifiche e lo storico della scienza italiana deve tenere conto delle peculiarità regionali." ("On the panorama of the ancient Italian states in the eighteenth century the Venetian Republic constituted a context of great importance for the development of chemistry as a science. It is known that during the early modern age chemistry had to work hard to establish itself as a discipline on the cultural level as well as on the institutional level. There are very specific situations to be found in the various political contexts and the historian of Italian science must take into account the regional particularities.")
11. "Denominazione di Origine Protetta," which covers both DOC and DOCG under European Union standards, created in 2010.
12. Sold "loose," that is has not been bottled and instead is drawn from barrel or other vessel into bottles, often plastic, or damigiane brought by the customer.
13. Schmitt, "It's official."
14. "Raccomandosi alla misericordia di Dio Signor ed alla Madre sua santissima." Molmenti, *La storia di Venezia*, 506.
15. With a brief return to France, from roughly 1805 until 1815, under France's Regno d'Italia.
16. "France, Austria and Italy," 4.
17. "France, Austria and Italy," 4.
18. Duggan, *Force of Destiny.*
19. The poor state of Veneto's agriculture has been attributed in nineteenth-century writings and up through to its abolition in 1964, the Italian sharecropping system often called mezzadria. See C. B. Cerletti, "La questione sociale fra i mezzadri e i vigneti," *Annali*, 1878.
20. Questlove and Greenman, *Creative Quest.*
21. Soldati, *Vino*, 17.
22. Roughly *tasting journeys*. Soldati, *Vino,* 17.
23. His book's subtitle is *Alla ricerca dei vini genuini*, or *In search of the genuine wines*.
24. Who then headed Conegliano's viticultural research center.
25. *In the short term*, *soon*, with intimations of *short-lived*, De Rosa, *Andar*, 125.
26. The people and the winemakers of Champagne.
27. Such wines, by unnamed producers, would be included in the area's Strada del Vino Bianco in the 1960s and '70s. Schiratti, *Invito*, 64–65.
28. The most famous of the many is the fifteen-volume *Atti della Giunta per la inchiesta agraria e sulle condizioni della classe agricola*, called by everyone the *Inchiesta Jacini,* conducted from 1877 to 1886 to examine the state of the new nation's agriculture

and the contadini who ran it. It begins with a section headed, "La difficoltà di un'Inchiesta agraria in Italia" (*The difficulty of an agricultural inquiry in Italy*).

29. And defined in *Grande dizionario italiano Hoepli* as "1. Investigation or research aimed at obtaining an exhaustive amount of data relating to a fact or phenomenon. 2. Extraordinary investigation of a transitory nature, ordered by the competent authority to a body specifically created or designated, to obtain a series of information," translation mine.
30. Danilo Gasparini, "Disegnare le campagne . . . disegnare le vigne: note di storie dell'agrimensura," Gasparini and Rossi, *Disegnare*, 23.
31. Taken as an average in growing seasons (from previous year's October to September of harvest/vintage) from 1993 to 2018: for the 2019 growing season, far more rain in October, February, and especially April and May; far less rain in December, January, March, June, July, and September. *Bolle*, 16.
32. In the December 16, 1986 *Il Manifesto* (with a "Communist daily newspaper" header), the weekly insert *Gambero Rosso* (name of the osteria in Carlo Lorenzini's *Pinocchio*) included a "Manifesto" signed by writers and food thinkers and called "Slow Food."
33. For far more on food and origins in Italy, see Montanari, *Il mito.*
34. Vianello and Carpenè, *La vite*, 33.
35. An especially wonderfully detailed one describes one of my favorite wines and places: "Col Credas," a Prosecco Superiore DOCG made by the Adami house from grapes grown on one of the denomination's steepest hillsides:

    Style: Sparkling wine
    Grape variety: 100 percent Glera
    Viticultural area: Credazzo Farra di Soligo
    Average vineyard altitude: 350–400 meters
    Work hours per year/hectare: Approx. seven hundred
    Soils profile: Clayey, dry, nutrient-poor soils, fairly shallow, bedded in calcareous rock
    Site and aspect: Steep hillslope, south- and southeast-facing vineyards oriented along relief contours
    Climate: Mild, with cold winters and warm, but not humid, summers. Average rainfall: ca.1250mm with peaks in June and November, and lows in January and August. Solar radiation, April–October: 92745 cal/cm2
    Training systems: Double-arched cane (Doppio capovolto, Cappuccina)
    Vineyard density: 2,500–3,500 vines/ha
    Vineyard yield: Not exceeding 130 q/ha
    Harvest period: 20 September–10 October
    Vinification: Light pressing with bladder press, gravity settling of must, fermentation at controlled temperature (17–19°C) with cultured yeasts. Maturation on the fine lees in steel for three months.
    Second fermentation: "Metodo Italiano" in steel pressure tanks. Second fermentation at 15–17°C. Cold tartrate stabilisation (-4°C). Tight filtration before bottling to remove spent yeast from sparkling wine.
    Cycle length: Approx. forty days
    Technical data:

Alcohol	11 percent vol.
Sugar	4 g/l

Total acidity 3.20
pH 6 g/l
Atmospheres 4.80/5bar

Sensory profile:

Presentation: Straw yellow. Dense mousse and delicate, long-lasting bead.

Bouquet: Broad and generous, intense and well balanced, releasing delicate floral notes, such as wisteria and acacia blossoms, and crisp impressions of fruit, particularly of apple and pear.

Palate: "Extreme" brut in style, with classic tangy acidity, very dry yet richly flavored; always strikingly full-volumed and remarkably refreshing, with a lengthy progression in the mouth. Impressively mirrors the fragrances on the nose. Exhibits superb harmony and elegance, crowned with a notably dry finish.

36. "I carri che conducono le uve si ricevono sotto una tettoia. Si riempiono dei recipienti appositi di uva, e questi mediante un semplice macchinismo, si portano all'altezza del primo piano del fabbricato, dove trovasi una porta per la quale due operai li ricevono, li adagiano sopra il piano di una bilancia e tengono nota del peso. I recipienti pesati si vuotano dall'uva, la quale viene versata in appositi cassoni di deposito, posti in prossimità ai pigiatoi. Ricevuta l'uva si procede alla pigiatura." Vianello and Carpenè, *La vite*, 150–53.
37. "È il tartrato acido potassico il sale più importante che il vino contiene; esso influisce non solo sulla sapidezza e finezza dei vini, ma ben anche assai sulle loro proprietà igieniche." Carpenè, *Sunto*, vol. 2, 2nd ed., 63.
38. "Offriamo ai vinificatori questa piccola guida all'analisi chimica dei mosti e dei vini, nella lusinga che tornerà loro opportuna, lontani però dall'intendimento di coprire una mancanza, che si verifica in Italia, di un trattato cioè di enochimica, il quale, oltre alle nozioni di chimica applicata alla viticoltura ed all'enotecnica, contenga la descrizione dei processi più rigorosi e consigliati per l'analisi completa di quanto sta in diretta relazione colla viticoltura e con l'arte di fare e conservare il vino." Carpenè, *Sunto*, vol. 2, 2nd ed., 1.
39. Some producers in New York and in California were grandfathered in, and without a tradition otherwise, accepted: you can still buy those producers' bottles, labeled with *Champagne, New York, California.*
40. Of venetan, UNESCO notes that it is spoken by 3,800,000 people, "according to Ethnologue. 69.9 percent of population of Veneto (National Bureau of Statistics of Italy)." UNESCO Atlas of the World's Languages in Danger at http://www.unesco.org/languages-atlas/index.php
41. "Senza la pretesa di fare del vinificatore un enochimico, riconosciamo l'utilità grandissima che quello può conseguire dalla conoscenza dei processi chimici usati alla determinazione della quantità dei principali componenti dei mosti e dei vini, mentre questo deve dedicarsi, oltre a ciò, allo studio ed alla soluzione di problemi più difficili e delicati, studio che richiede assai maggior corredo di cognizioni, di suppellettili scientifiche e di tempo." Carpenè, *Sunto*, vol. 2, 2nd ed., 1.
42. As explained to me by Domenico Scimone, general manager of the Carpenè Malvolti winery.
43. Which was often seasonal or otherwise temporary, unlike other zones especially south.
44. Patriarca, *Italian Vices*, 160.

45. "Pochi mesi dopo, a Mestre, operai dimostrano sotto la sede del sindacato gridando: 'pane, lavoro, morte ai contadini, cosa fa il governo Mussolini.'" Casellato, "Sestieri."
46. Patriarca, *Italian Vices*, 159.
47. Patriarca, *Italian Vices*, 160.
48. Jefford, "Wine."
49. See John D'Agata's *The Lifespan of a Fact* for seven years of his proofreader's never understanding what an essay is, a problem most (based on various reviews and online comments) of the book's readers shared.
50. Wood, *Everything*, 13.
51. Indebted to director Daniele Luchetti's words on Italian public life in the 1970s, unavoidably through an interpreter: "It's amazing to consider how patient we were. Patient enough to sit through a political meeting and listen to everyone speak. Patient enough to read books that set out monumental ideologies that today we find indecipherable. We were reading obscure writers, watching obscure movies, simply because we had patience, and we knew that from that patience something good would come." Luchetti, "Our Brilliant Friends: Conversation & Watch Party," online event, February 28, 2022, McNally Jackson Independent Booksellers, https://www.mcnallyjackson.com/.
52. For more, see, for example, Venuti, *The Scandals.*
53. Rockmore, "Prove It!"
54. Ording, *99 Variations.*
55. Haraway, *Staying.*
56. Which he might not have actually said. He did, though, just as helpfully for storytelling, write that, "Indeed, any attempt to trace the detailed course of a transition process would involve an uncontrollable exchange of energy between the atom and the measuring instruments which would completely disturb the very energy balance we set out to investigate." *Atomic Physics and Human Knowledge* (New York: Science Editions, Inc. 1961) 6–7. Open Library.
57. "The Ptolemies were uncommonly cultured rulers and established as a paradise for scholars a lavishly endowed temple to the Muses, the Museum." McNeely and Wolverton, *Reinventing Knowledge*, 15.
58. McNeely and Wolverton, *Reinventing Knowledge.*
59. "What comes first is always primal, typically causal. Roland Barthes said, 'Everything suggests, indeed, that the mainspring of narrative is precisely the confusion of consecution and consequence, what comes *after* being read in narrative as what is *caused by*,' and in atlases this is generally the case." Wood, *Everything Sings*, 13.

CHAPTER 1

1. From *An Unfinished Woman: A Memoir* by Lillian Hellman, introduction by Wendy Wasserstein copyright © 1999. Reprinted by permission of Back Bay Books, an imprint of Hachette Book Group, Inc.
2. See Glossary.
3. Richard Powers, *The Overstory: A Novel* (New York: W. W. Norton, 2019), 6.
4. Merriam-Webster: "an increase in degree or intensity / exaltation of virulence of a virus," https://www.merriam-webster.com.

5. Carbonic acid.
6. If Carpenè could see them today, his chest would swell with the familiar pride that drove him then but he might not recognize the walks he'd take now, nor would he be likely to know the grape's new name, changed from a word so identified with these hills to protect against an outside world now eager to cash in on the success found more than one hundred years after his dream. "Nel paesaggio troviamo scritte la nostra identità e la nostra storia," wrote president of the Veneto region Luca Zaia in the introduction to *Esercizi di Paesaggio* and in a similar sense of pride and practical. "Tutelarlo e valorizzarlo significa difendere il nostro futuro." ("In the paesaggio we find written our identity and our history. Protecting it and exploiting [not in the American-English sense] it means defending our future.")
7. Corresponding to the lagoon city of Venice, capital of the Republic, its seat of power, and one of its three administrative divisions at the time of its fall (with the inland Domini di Terraferma reaching from the eastern Alps to the top of Lake Como and the maritime Domini da Mar along the Adriatic sea down to Cyprus)
8. From Conegliano was, for example, eighteenth-century agrarian philosopher Pietro Caronelli, who lamented the state of local viticulture after a local 1809 frost forced hurried replanting and shoddy care in both vineyard and cellar so that the contadini could quickly recoup the much-needed income they had lost. He wrote heavily on the need to expand and improve the area's viticulture, and on science as showing the ways to do so. The street on which Italy's first winemaking school originally was is now named for him. His annotated works are listed in a heavily opinionated *Dizionario Ragionato ad uso degli amatori delle cose agrarie, e della gioventù* published in Venice in 1808 and findable on Google Books.
9. In Pavia, Lombardy: during the Austrians' last period of occupation, they had created the Lombardy-Veneto state by including Venice's Lombardian Terraferma.
10. See, for example, Orosi, *Farmacologia.*
11. Kernot, *Storia della Farmacia*, Table of contents.
12. Loosely translates to *village*, and also carries the notion of a self-contained culture/place, as in the land one comes from. *Il Paese* will also mean one's country, equally understandable as *locality* and *nation.*
13. Kernot, *Storia della Farmacia*, 138.
14. Translated soon after into English as *Duties of Man* authored by *Joseph Mazzini*
15. Mazzini, *Doveri*, 136–38.
16. The Medici battalion; see, for example, Piero Corbellini, *Diario di un garibaldino: della Spedizione Medici in Sicilia* (Como: Riccardo Gagliardi, 1911). HathiTrust.
17. See, for example, Franca Giansoldati, "Le donne del Risorgimento che hanno fatto l'Italia, la storia al maschile va riscritta," *Il Messaggero*, November 15, 2019. https://www.ilmessaggero.it/mind_the_gap/donne_rivoluzione_clima_risorgimento_italia_anita_garibaldi_ravenna_mind_the_gap-4864505.html
18. Taught by Tullio Brugnatelli. See Luigi Cerruti "*Concordia Discors.* I chimici italiani dell'Ottocento," in *La chimica e le tecnologie chimiche*, ed. Bassani, 34-35.
19. And within the province of Treviso, Conegliano seems to have been especially loyal: After more than three centuries, notes an unnamed writer in 1927 in the following abbreviated translation, the comune of Conegliano suddenly wanted to belong to Venice. This devotion began in 1336–1339 during Venice's victorious war against the Scaligeri, and from 1389 the city pledged its "faithful love" to Venice, to

which it remained devoted. On March 29, 1797, overlooking the dangerousness of the republic's imminent collapse, Conegliano renewed its oath of loyalty by Comunal vote. "A Venezia—dopo oltre tre secoli di ribellioni e di guerre contro Padova, Verona e Treviso—il Comune di Conegliano spontaneamente volle appartenere. La dedizione avvenne negli anni dal 1336 al 1339, durante la guerra vittoriosamente combattuta dalla Serenissima contro gli Scaligeri. Tra il 1356 e il 1389 Conegliano subì altre vicissitudini: se ne impadronirono il re d'Ungheria e quindi i Carraresi signori di Padova, ma dal 1389 la bella città fece ritorno con fedele amore alla bella Venezia. A Venezia Conegliano restò sempre devota, solo per brevissimo tempo dal 1509 al 1513 avulsa dalla repubblica marinara e invasa dai tedeschi dell'imperatore Massimigliano. Nel 1797, sovrastando i pericoli della caduta imminente, Conegliano rinnovò per voto Comunale del 29 marzo il giuramento di fedeltà." "Conegliano Vinipara," *Sorrisi*, 1, in Spriveri, S., and M. Coradazzi with Domenica Scocchia, eds. *Anniversario 130*.

20. Kernot, *Storia della Farmacia,* 153.
21. So said his son-in-law, Vallery-Radot, in *Louis Pasteur.*
22. "It should be noted that before Louis Pasteur's revolutionary discovery, the addition of the liqueur de tirage was a rather ad hoc process that invariably resulted in broken bottles (up to 80 percent in 1828). This was not helped by the fact that glass at the time was hand-blown and bottles varied widely in quality. It was not until the early nineteenth century that advances in bottle technology removed much of the uncertainty from second fermentation." "{Tirage} or bottling the blended wine," Union Des Maisons De Champagne, https://maisons-champagne.com/en/appellation/stages-in-winemaking/from-grape-must-to-still-wines/article/tirage-or-bottling-the-blended-wine.
23. Carpenè, "Guerra alle nostre abitudini," *Rivista* 1878, 34–35.
24. Under Venice, agriculture in the area had existed and been destined for an internal Venetian market, composed of the needs and desires of the nobility and landowners and the needs of an unusually dense contadino population. See Berengo, *L'Agricoltura veneta.*
25. Though several are now being repurposed as vineyard-side tasting rooms.
26. He may have: in his own sweeping geographical introduction in *La Vite*, Dr. Torquato Taramelli notes that to his own observations he has added measurements from topographical military maps.
27. "Is the purpose of geography to make war? An exhibition of maps and art," Fondazione Benetton Studi Ricerche. http://www.fbsr.it/en/cultural-heritage/projects-and-researches/is-the-purpose-of-geography-to-make-war/
28. Mazzini, *Doveri*, v.
29. Mazzini, *Doveri*, 139.
30. (More wine and less pellagra).
31. In *Rivista* 1880.
32. Introduced into the Veneto area in the early seventeenth century. Berengo, *L'Agricoltura veneta*, 10–11.
33. Vianello and Carpenè, *La vite*, 42.
34. The word he uses is "cammino," or "walk": walking, camminare, inherent in what otherwise means a path or journey.
35. Carpenè, "Guerra alle nostre abitudini," *Rivista* 1878, 3.

36. See Berengo, *L'Agricoltura Veneta*, 297.
37. Vianello and Carpenè, *La vite*, 102.
38. See Tomasi and Gaiotti, *The Terroirs.*
39. Under whom, at least in theory, as the behavior of a paesaggio, Italian landscape of land and people, was a reflection from outside and above, and not the habits, practice, traditions, responsive wisdom of its people. See Giuseppe Barbera, "Il paesaggio viti-culturale dell'Antropocene," Fondazione Benetton, online seminar, April 21, 2021.
40. There are Russians, too: see, for example, one grave in Cison del Valmarino, TV.
41. Vianello and Carpenè, *Rassegna* 1887, 482.
42. Ulliana, *La scuola enologica*, 13.
43. *Planted*, in dialect.
44. *Scimitar.*
45. *Bunch.*
46. He was also fond of the "Piccolit" grape. All of which he said during a session at the Pubblica Accademia of Conegliano on February 26, 1772 to address the problems of the hills' rapidly spoiling wines. As published in *Giornale d'Italia: spettante alla scienza naturale e principalmente all'agricoltura, alle arti, ed al commercio*, vol. 8 (Venezia, 1772), 388–89. Google Books.
47. See Tomasi and Gaiotti, *The Terroirs.*
48. Brock, *The History of Chemistry*, 1.
49. Vianello and Carpenè, *La vite*, 58.
50. One square millimeter holding at least fifty hectoliters.
51. "The growers love the nation." Vianello and Carpenè, *La vite*, 39.
52. This wasn't lost on Cima of Conegliano, either. He painted this missed place into an existence he could see from Venice, in now archaeologically meticulous architectural, flora, and fauna details.
53. Where they would become targets for Mussolini's black-shirted supporters; but this story stretches there only as horizon.
54. Carpenè, "Guerra alle nostre abitudini," *Rivista* 1878, 33.
55. Ibid., 34–35.
56. The word used, as it is today, is *vite*: one vine, with its close resemblance to *vita*, *life*, and up to definite article equal in its singular form to Italian *life* in plural—*vite*.
57. Carpenè, "Sui vini d'incerta conservazione," *Annali* 1872, 176.
58. Calculated as an average from ten measurements of my own (by now relatively uniform muscle memory) glass lifting and tasting, at a 22 inch/55.8 cm arc, each calculation itself figured as a constant speed, without the acceleration especially at the end which surely there is—thanks to Buzz Spector, who proposed that all producers lift their glasses at the same rate.
59. Letter from Venetian academician and politician Luigi Luzzatti, in De Leo and Favero, *Ca' Foscari e Carpenè Malvolti*, 50. You could say such a phrase drove Carpenè, too.
60. Kitchen credenzas and hope chests.
61. For more on which trees were grown here and for which uses, see Berengo, *L'Agricoltura Veneta.*
62. See "Inventario d'un piccolo laboratorio enotecnico," *Sunto*, vol. 2, 2nd ed., 3–6.

63. This particular place is in many ways invented. I drew from a visit to the Carpenè Malvolti winery, illustrations and directions in Carpenè's books, research on nineteenth-century chemistry laboratories, research on Treviso's landscape, and natural and cultural resources.
64. Lithography for multicolor prints.
65. For the reproduction of an advertisement for it ("among injection systems, the one which with the most simple instruments gives the best results") with a sale price of 2,000 lire, see Gregoletto, *Vite Ambulante*, 363.
66. In 1875.
67. De Leo and Favero, *Ca' Foscari e Carpené Malvolti*, 77.
68. Carpenè, *Sunto*, vol. 1, 5th ed., 119–20.
69. Ibid.
70. Ibid.
71. Mazzini, *Doveri*, 10.
72. In 1920, this bell tower would be fitted with a new bell, made from the artillery left behind by Austria, defeated by the Italians in the Battle of Vittorio Veneto. Luca Nardi, "Giù le campane del campanile di Col San Martino, saranno restaurate. La più antica fu collocata dopo la Grande Guerra," *Quotidiano del Piave*, March 3, 2021. https://www.qdpnews.it/comuni/farra-di-soligo/giu-le-campane-del-campanile-di-col-san-martino-saranno-restaurate-la-piu-antica-fu-collocata-dopo-la-grande-guerra/.
73. See Laura Tripaldi, *Parallel Minds*, 25.
74. Carpenè, *Sunto*, vol. 1, 5th ed., 117–30.
75. "Production of spumante wines."
76. Inspired by Levi, *The Periodic Table.*
77. Montanari, *La cultura*, 17.
78. The website gives no source or date; I am using this as memory or family tale.
79. The title translates to "Our daily polenta."
80. Called braccianti, a derivative of *braccia*, or arms as main asset, these contadini worked for whoever would hire them, doing a variety of physical tasks.
81. "Carne di bue," she tells him: beef, specifically ox meat
82. "Non si celebra l'alleluia senza tagliatelle carne di bue e vino tondo, quello di monte s'intende, si andava a comperarlo nelle osterie, ma non tutte ne tenevano, c'era tanto di insegna sul muro con la scritta "Qui si vende vino di monte." Coltro, *La Nostra Polenta Quotidiana*, 63.
83. "Gerle di vimini," she says.
84. Vianello and Carpenè, *La vite*, 126.
85. According to the *OED*, *scientist* first appears in 1834: "W. Whewell in Q. Rev. 51 59 Science . . . loses all traces of unity. A curious illustration of this result may be observed in the want of any name by which we can designate the students of the knowledge of the material world collectively. We are informed that this difficulty was felt very oppressively by the members of the British Association for the Advancement of Science, at their meetings . . . in the last three summers. . . . Philosophers was felt to be too wide and too lofty a term . . . ; savans was rather assuming . . . ; some ingenious gentleman proposed that, by analogy with artist, they might form scientist, and added that there could be no scruple in making free with this termination when we have such words as sciolist, economist, and atheist,—but this was not generally palatable."

86. Coltro, *Paese perduto*, vol. 3, Appendix ("Appunti per una storia sociale e linguistica del mondo contadino").
87. Carpenè, *Sunto*, vol. 1, 5th ed., iv.
88. The *OED*, which offers the additional mathematics definition of "the property of neither increasing nor decreasing" gives 1901 as its first instance of the word in English.
89. Unification is the first time you can talk about an Italian economy, writes oral historian Dino Coltro; before that, the economies of the various states, various political systems, were all based on agriculture. See *Paese perduto*, vol. 3. And, Nicola Zitara, opening his book on unification with thoughts of whether to consider it as a liberal and national theme or as the Italian borghesia's European assumptions, finds that "what interests him is only that whichever approach you choose, unification coincided with the formation of a national market." *L'unità d'Italia*, 19.
90. "In a manner of speaking," he writes in colloquial form; "so to say."
91. In occasion of Paris's 1878 world exposition. Carpenè, *Rivista* 1877, 213.
92. Carpenè, *Rivista* 1877, 279.
93. Letter from Carpenè, October 1872; published in *Annali* 1872, 100.
94. Tomasi, "Arriva il Progresso!" 13.
95. *Kriegskarte*, vol. 1, 385.
96. There is another Rujo in Cison.
97. "La località è suggestiva, soffusa di quella austera ma gentile bellezza che è la caratteristica dei colli coneglianesi." Luzzatto, "La comunità ebraica," 42.
98. Vianello and Carpenè, *La vite*, 34-35.
99. "Che storicamente si riferisce ai giardini all'interno dei monasteri e che è poi stato ripreso per indicare i vigneti pianeggianti vicini alle case," Giulia Pussini and Innocente Nardi of the Consorzio Conegliano-Valdobbiadene in an email in 2020.
100. Many traditional winemakers follow a lunar bottling schedule: during the waning moon referred to here, sweet wines or those meant for aging are bottled. Frizzante, and presumably spumante, wines are to be bottled during a waxing moon.
101. Caccianiga, Antonio, "Il Marsala spumante dello Stabilimento Carpenè-Malvolti," *Rassegna* 1891, 3-5. Caccianiga was also one of the original four or five founding members of the Società Enologica.
102. Or, primary materials: used in Italian food and drink to mean the ingredients, with a focus on having been grown well and correctly and of top quality.
103. He writes, "piantare."
104. Patented in 1895 to solve new commercial needs of another aromatic Italian grape: Moscato Bianco.
105. Application for "Installation for the continuous production of sparkling wines" filed by Federico Martinotti on July 12, 1895. CH10711A granted on January 31, 1896. Google Patents.
106. De Leo and Favero, *Ca' Foscari e Carpenè Malvolti*, 80.
107. "Potassium sulfate" or K_2SO_4 : once used to protect against harmful bacteria; as metabisolfito di potassio ("potassium metabisulfite" or $K_2S_2O_5$) is used today.
108. Carpenè, "Sull'origine dei solfati nel vino," *Rassegna* 1891, 389.
109. Veronelli, *I vini d'Italia*, 186.
110. De Rosa, *Tecnica*, 8.
111. Lanaro, "Genealogia di un modello, *Il Veneto*, 22.

112. Launched by Ligurian Ferdinando Swift in Venice 1875, three years after Mazzini's death in Tuscany.
113. See the comune's website, https://www.comune.pievedisoligo.tv.it.
114. For more, read about the Trattato di Campoformido.
115. Soldati, *Vino*, 301.
116. Slogans of Mazzini's secret revolutionary society Giovine Italia
117. Edgar Allan Poe, "A Few Words on Secret Writing," *Graham's Magazine* (July 1841), 33. Google Books.

CHAPTER 2

1. Red grape grown throughout central Italy; areas in Toscana and Emilia-Romagna are its most classic zones. "In Toscana è una varietà che ha una grandissima importanza," reads 1879's *Ampelografia italiana*.
2. According to Great Britain's *Reports on the Vienna Universal Exhibition of 1873*: "The veritable wine, which possesses remarkable finesse and an agreeable sub-acidity, is not unlike the best Beaujolais growths with, however, more colour, body, and force." In the 1850s, Ricasoli—risorgimentale and obsessed with *his* area's wines—had redefined Chianti to be mostly made of Sangiovese grapes, with the previously starring Canaiolo variety now playing a secondary role, and requiring local white grapes Malvasia Bianca Lunga and, later, Trebbiano Toscano to be in the mix, too. Those last two have been controversial in the Chianti Classico area since the mid-twentieth century, and for better or, possibly, worse since 2006 the zone's denomination no longer allows their inclusion.
3. Boschiero, *Relazione sulla industria dei vini*, 30-31. The highest Medal of Progress awards included Bettino Ricasoli's Chianti di Brolio 1861, 1855, and 1865. An 1870 still "vino bianco prosecco" by Ortolan Giovanni was given an honorable mention. Three Italian spumanti were awarded: a Medal of Progress to one from Napoli; an honorable mention to another from Napoli and to one from Asti.
4. Carpenè and Pasteur corresponded on the use of these, with the former's views the accepted notion today. As reported in *Annali* 1876, 189-90: Carpenè wrote to "Il Pasteur, autorità ben conosciuta in fatto di fermenti e fermentazioni, e ciò allo scopo di avere una conferma alle mie idee esposte in base ad argomenti ed una maggior luce in argomento di grande importanza per l'enotecnica. Riporto qui tradotta la lettera del Pasteur, colla quale, pur mostandosi riservato nell'esprimere le proprie idee, dà delle nozioni che devono tornare utili per quanti si applicano alla pratica delle fermentazioni," introducing Pasteur's reply, translated into Italian: "Signore! / Arbois (Jura), 2 agosto 1876. / Sgraziatamente io non ho istruzione alcuna da darvi al soggetto della solforazione, e ciò perchè, per così dire, non ho mai fatto esperimenti diretti col gaz solforoso e coi soliti per modificare le fermentazioni. Del loro impiego non ho se non che delle idee preconcette, risultanti dallo studio approfondito dei principali fenomeni fermentativi. Presumo adunque che l'acido solforoso in dosi convenienti debba arrestare oltre che la fermentazione principale, tutte le fermentazioni secondarie e le malattie (1)." *That* note leading to: "(1) Il Pasteur qui mostrasi contrario alle mie vedute. Probabilmente con solforazioni eccessive, in modo da saturare il liquido, si potranno avere effetti contrarii

a quelli da me ottenuti. Ma io ho sempre trattato dell'efficacia dell'acido solforoso in dose praticabile per vini ed altre bibite, come ad esempio la birra."

C: "Pasteur, a well-known authority in terms of fermentation and fermentation, in order to have a confirmation of my ideas presented on the basis of reasoning and a greater light on the subjects of great importance for winemaking. I present here translated a letter from Pasteur, who, while showing himself reserved in expressing his ideas, gives some notions that could be useful for those who apply themselves to the practice of fermentations."

P: "Unfortunately, I have no instruction to give you on the subject of sulfurization, and this is because, so to speak, I have never done direct experiments with sulfurous gas and with sulfites to modify fermentations. Of their use I have only preconceived ideas, resulting from the in-depth study of the main fermentation phenomena. I therefore assume that the sulfurous acid in suitable doses would stop all secondary fermentations and diseases as well as the main fermentation (1)."

C: "(1) Pasteur's view here is contrary to mine. Probably with excessive sulfurations, in order to saturate the liquid, it will be possible to have effects contrary to those obtained by me. But I have always dealt with the effectiveness of sulfurous acid in a practicable dose for wines and other drinks, such as beer."

5. Doppio capovolto; one kind of vine training, widely used in the hills today, in a specialized vineyard, or one planted only to vines. It is a two-armed modification of Guyot (see Index), and as such is cane (secondary, two-year-old wood) pruned.
6. From an online presentation on winter pruning, by the Consorzio Conegliano-Valdobbiadene, on December 9, 2021.
7. Pasteurization, Pasteur's technique of heating foods like wine and milk, relies on this principle to stop yeasts and bacteria.
8. See McGee, *Nose Dive.*
9. Bitter and pleasing. There is no equivalent in English, except to learn of the kinds of *bitter* that are *pleasing*.
10. The Champagne method of creating and finishing sparkling wine is, at its most basic: still wine in a bottle, to which is added a *tirage* (a mixture of sugars and yeasts) to transform that into sparkling wine, and then that outcome finished with a *dosage*, which is sugars and perhaps older wine or something else to balance and create the final wine. No-dosage Champagne wines skip this last step, obviously. Prosecco Italian-method spumante does not use dosage; any final sugars are the result of calculations made during the *tirage* phase in autoclave.
11. "L'acido carbonico, anidride carbonica, è un gas senza colore dotato d'un sapore leggermente acidulo. È solubile nell'acqua e maggiormente nell' alcoole e nel vino. Alla temper. di gr. 10 ed alla pressione normale (0.76 cent.) un litro d'acido carbonico discioglіesi in un litro d'acqua, od in mezzo litro di alcoole circa, a norma della sua densità, ossia del suo stato più o meno anidro. Un litro di vino alle condizioni citate di temperatura e pressione può disciogliere poco più di un uguale volume di detto gàs." *Sunto*, vol. 2, 2nd ed., 69.

12. "La tensione interna si porta al pari di quella di 6 atmosfere, essendovi oltre alle cinque di acido carbonico, anche quella dell'aria rimasta imprigionata all'atto del tappamento della bottiglia." Carpenè, *Sunto*, vol. 1, 5th ed., 71.
13. In the Pescara province, inland so near Abruzzo's only un-sea-touched province—L'Aquila, where my paternal family line is from.
14. Biodynamic viticulture based on the work of Austrian philosopher Rudolf Steiner, and spontaneous fermentation by way of yeasts present on grape skins and in the winery rather than adding bought, "commercial," yeasts are precious concepts and tenets of "natural" winemakers and drinkers.
15. Carlo De Cesare, 1864, as quoted in Giuseppe Cappelleri, "L'evoluzione delle tecnologie chimiche in ambito vitivinicolo," in *La chimica e le technologie chimiche*, Bassani, ed., 381–82.
16. See, for example, Scott Byington, "Spontaneously Generating Life in Your Classroom? Pasteur, Spallanzani & Science Process," *The American Biology Teacher* 63, no. 5 (May, 2001), 340. https://doi-org.i.ezproxy.nypl.org/4451121.
17. With *Mémoire sur la fermentation appelée lactique*, this forms the basis for his culminating work on germ theory twenty years later.
18. Pasteur is also a teller of scientific stories, and understands that for his finding of fermentation and germs to stick now, the still lingering spontaneous generation would have to go. In his move from materiality and into interaction with unseen living forces, he is seen as coming close to spiritual, religious. With him, fermentation moves from self-contained interior decomposition and disruption to a lively conversation with a living organism in the world, that is everywhere in fact. From a breaking apart in conversation with the living world: "Pasteur established that alcoholic fermentation invariably produces not only carbonic acid and ethyl alcohol, as was well known, but also appreciable quantities" of other substances, and "emphasized the complexity of alcoholic fermentation," Gerald L. Geison, *The Private Science of Louis Pasteur* (Princeton: Princeton University Press, 2016), 107, ACLS Humanities e-book, https://hdl-handle-net.i.ezproxy.nypl.org/2027/heb.05758. "La présence de la glycérine dans le vin, où elle est associée à des matières albuminoïdcs et à des phosphates, mérite l'attention sérieuse des physiologistes" recommended Pasteur in *Memoire sur la fermentation alcoolique*. And, "De cette convention relative à la nomenclature adoptée, il résulte que l'expression de *fermentation alcoolique* ne pour pas désigner tout phénomene de fermentation où il se produirait de l'alcool; car il peut y en avoir de diverses sorts ayant ce caractère commun." This complexity of process is a matter of difference that could be tasted, supported, solid ground for comparing wines and, when they merited, showing Italy's greatness. In the same year Pasteur confirms fermentation—1858 (and it is a "final blow," he writes in his *Mémoire sur la fermentation appelée lactique* report on it)—Stanislao Cannizzaro finishes writing his *Sunto*.
19. Carpenè, *Sull'acido lattico*, 15.
20. Carpenè, *Sull'acido lattico*, 9. The word *origine* appears eight times.
21. Unlike today's use of subscripts to denote how many of each element, he writes the numbers as superscripts. In just a few decades elements' ions will be understood, and superscripts will be reserved for writing those atoms with an electric charge.
22. Carpenè, *Sull'acido lattico*, 9–10.

23. Measured this way, water is one part hydrogen to eight parts oxygen. It is also the smallest amount of oxygen observed in a compound substance, and in addition the other H and O compounds were observed to contain oxygen in multiples of that relative weight, so that hydrogen dioxide, which is 1 part hydrogen and 16 parts (or 2 equivalents) oxygen, was written HO^2. See Brugnatelli, *Trattato*.
24. The measure of elements.
25. For more on why he held to this model, see Cannizzaro, *Sunto*.
26. Where, like a lot of the South and from which came some of Italy's best wines, vineyards were often part of latifondi, large commercial farms, and contadini components of a protective social and economic system.
27. Where, if under different circumstances, private or sharecropped small vineyards were the model as they were in Treviso.
28. For more on Cannizzaro's presentation based on concepts and terminology, see "Stanislao Cannizzaro," *Accademia nazionale delle scienze dette dei XL*, https://www.accademiaxl.it/pubblicazioni-2/ipertesti-tematici/biblioteca-virtuale/percorsi-della-chimica/stanislao-cannizzaro/.
29. See Cerruti, *Bella e Potente*, for more on the philosophical difference between materials and goods and chemistry's role therein.
30. Notes, the *OED* points out, the *Cyclopædia: An Universal Dictionary of Arts and Sciences* in 1728. https://www.oed.com.
31. *OED* on *manipulation*: "*Chemistry*. The action of handling apparatus, reagents, etc., in experiments; experimental procedure; an instance or example of this. Now *archaic*. Michael Faraday, *Chemical manipulation*, vol.1 (London: W. Phillips, George-Yard, Lombard-Street, 1827)." https://www.oed.com.
32. Montanari, *Il mito*, 5.
33. The old moat that ran along the gardened homes on modern Conegliano's main street was called the Refosso; the word means basically *double moat*. Fosso is a moat, or a ditch, and is preserved in names of, for example, a local restaurant.
34. A quick note on Mussolini's recasting (an outsider's view of) rural and contadini identity as Italian identity, agrarian politics, and funding (for an up-close personal look at the latter, see for example, Iris Origo's diary-based trilogy): drawing a firm line between rural and city were the 1931 and 1939 "Provvedimenti contro l'urbanesimo" laws to prevent intra-Italy migration from countryside to cities, or even between cities, and which were repealed in 1961 (*Gazzetta Ufficiale della Repubblica Italiana* 102, no. 43, 714. https://www.gazzettaufficiale.it/eli/gu/1961/02/18/43/sg/pdf), three years before the end of the mezzadria sharecropping system (*Gazzetta Ufficiale della Repubblica Italiana* 105, no. 233, 4100. https://www.gazzettaufficiale.it/eli/gu/1964/09/22/233/sg/pdf).
35. *Fosso*, too.
36. *Kriegskarte*, vol. 1, 389.
37. In the 1940s, for an example of recasting such thinking by way of quantities, for the Province of Treviso: two pages of 120 "Stabilimenti vinicoli di proprietari-viticoltori," with productions from 300 hectoliters to 2,500 hl and altogether 100,550hl, while a half-page of 27 "Stabilimenti vinicoli di commercianti-industriali" with a total of 38,650 hl included Carpenè Malvoti with 300 hl, most such producers in the hundreds, a few up to 2,000 hl and one with 8,000 hl. Three cantine sociali totaled 106,00 hl. Montanari and Ceccarelli, *La viticoltura e la enologia*, 455–57. For

now, about half of the producers of Conegliano-Valdobbiadene DOCG are small and account for 1.4 percent of bottles, while the forty-six largest (22.5 percent of total producers) make about 90 percent of the bottles; *Rapporto economico* 2021.

38. Indeed in tying together an Italy, one culture that beyond environmental differences was recognizable from "the "jurnatari of Sicily to the brazente de giornada of Veneto," was that of the contadini, writes Dino Coltro. "Substantially the culture is the same. We must not mistake certain 'cultural islands' of the north and south as the demonstration of peculiarities native to the cultura contadina, but as particular situations in a wider socio-cultural context." Even, later, the "new motives of an agricultural proletariat seemingly different, graft onto the ancient and primitive rootstock of the contadino culture, without changing its substantial monolithism, ethics, philosophy." Dino Coltro, "Appendice," *Paese perduto*, vol. 3, 423.
39. Carpenè, *Sull'acido lattico*, 16. "Studj" uses the Venetian *j* instead of the Italian *i*.
40. Carpenè, *Sunto*, vol. 1, 5th ed., 25.
41. See Glossary.
42. Negri and Petrini, *Roma Caput Vini*, chapter 3.
43. Ulliana, *La scuola enologica*, 5.
44. "For example, according to one Greek alchemical lexicon, mercury was 'the seed of the dragon,' or 'dew,' or 'milk of the black cow,' or 'Scythian water,' or 'water of silver,' 'water of the moon,' 'river water,' and 'divine water.' A Persian source lists over fifty synonyms for the philosophers' stone." Brock, *The History of Chemistry*, 17.
45. Brock, *The History of Chemistry*, 17.
46. Carbonelli, *Sulle fonti storiche*.
47. Brock, *The History of Chemistry*, 17.
48. "To evade the police and the censors . . . by coding their dissent: for example, by disguising opposition in cultural debates (about language, literary styles, or the study of history), or in poetic, artistic and musical symbols, or even in fashion symbols (hair and beards were a particularly rich source of political expression from the 1790s)" (Duggan, *Force of Destiny*, 57) or "a hand placed on the forehead and an index finger struck on the palm or wrist six times to signify each of the letters of 'Italia.'" (Duggan, *Force of Destiny*, 62.)
49. Carpene, *Sunto*, vol. 1, 5th ed. 35–38. Poetic: "Le bolle di gas che, rare ai primordi della fermentazione, vanno poi ad impossessarsi di tutta la massa liquida, agitandola e sconvolgendola in tutti i sensi, sono appunto di acido carbonico, il quale si estrica, non potendo disciogliersi, se non in proporzioni tanto minori, quanto più alta è la temperatura del liquido fermentante. E siccome questo gas è più pesante dell' aria, così ne consegue, che, invece di espandersi e frammischiarsi a questa man mano che si sprigiona dalla massa del mosto, si agglomera nella parte vuota del tino, e quando lo ha riempiuto, lo sormonta pe' suoi bordi, ed a guisa d'un liquido che non può più capire in un vaso, trabocca lambendo le pareti esterne del tino e va a depositarsi nel suolo della tinaia (t). Col proseguire delle reazioni chimico-fisiologiche citate, il mosto va via via trasformandosi in vino."
50. All the same element.
51. "Era avviata nell'ambito di una cultura chimica Italiana impermeabile alla storia della scienza." ("It was done in the environment of an Italian chemistry culture impermeable to the history of science.") From, more than a century later, same

conditions, "Leonello Paoloni," *Treccani*, on the man who published four volumes of Cannizzaro's letters in the later twentieth century. https://treccani.com.

52. di Meo, "La storia della chimica," 181.
53. He actually writes "of Avogadro, of Ampère and of Dumas," but as the former's work arrived three years after Avogadro's, within the first pages Cannizzaro identifies Avogadro as having come to this conclusion first.
54. "Oltre al fastidio dei cattivi odori, vi era il pericolo delle esplosioni. Molti chimici subirono gravi danni a causa dei loro esperimenti. . . . andavano giustamente fieri di appartenere a un corpo di persone coraggiose e risolute." David M. Knight, "L'Ottocento Chimica, Una nuova scienza," *Storia della Scienza* (2003), https://www.treccani.it.
55. Brock, *The History of Chemistry*, 90.
56. Brock, *The History of Chemistry*, 4.
57. In the long jostling between chemistry and pharmacy, from the Paracelsians' insistence that chemistry serve medicine, to the iatrochemists and apothecaries, the former rose in prestige and functions in Italy during these years.
58. Trevisian countryside.
59. Coltro, *Paese perduto*, vol. 1, 40.
60. On November 4; but this date would be wrongly celebrated by the fascists.
61. Caniato and Palugan, *Conegliano*, 96.
62. From American Jesuit priest and historian Walter J. Ong's work on oral and literate knowledges: "Oral cultures appropriate actuality in recurrent, formulaic agglomerates, communally generated and shared. Formulas are communally fixed ways of organizing simultaneously object and response-to-object. They can be typified by epithets, that is, standard, expected qualifiers or surrogates—the sturdy oak, the rosy-fingered dawn, the brave soldier, the noble chief, the hated foe, Son of Peleus (for Achilles). It should be noted, however, that to say that oral cultures use and need formulas is not to say that their verbalization processes, either in formal prose or poetry or in other discourse, consists simply of stringing together in various orders items from a catalog of fixed phrases. The use of formulas is far more complex than this. The surface actualizations in ancient Greek formulaic poetry, for example, overlie what Michael M. Nagler has well characterized as 'an inheritance of habits, tendencies, and techniques rather than of completed entities,' which make oral poetry 'spontaneous-traditional' art in which 'the oral poet . . . at the moment of performance makes spontaneous, and therefore original realizations of inherited, traditional impulses.' What Nagler says of poetry would apply, mutatis mutandis, to other forms of discourse in oral cultures, where the entire noetic economy is dominated by formulas, so that poets simply maximize, in often exquisite ways, processes of thought and discourse endemic through the entire culture. The point here is simply that the elements figuring in the discourse are basically formulary to an extent radically greater than what highly literate cultures can ordinarily tolerate." *Interfaces of the Word*, 19.
63. To oversimplify. One official vehicle for this were the public schools to be established, which for generations insisted on only Italian being taught and allowed to be spoken, and written, in classrooms.
64. See Duggan, *Force of Destiny*, 91, 220.
65. Latifondi.

66. Wine has been made throughout Italy for so long that there are hundreds of identified grape varieties throughout its twenty regions, most of which grow nowhere else and which are thought to be perhaps a quarter of how many are actually there, in small vineyards growing rogue until the later twentieth century's ongoing project of restoration—when "new" varieties are identified by a mix of DNA readings and ampelography, they are added to Italy's national registry of officially recognized grapes; see, for example, D'Agata, *Native Wine Grapes*. Through poverty, remoteness, communication issues, hard-to-reach hillsides, the contadini acted as protectors of Italy's wine patrimonies for centuries, preserving minor grape varieties from devastations like planting something else after vine diseases came with increased contact with the Americas and, a little later, global commercial pressures to sell better-known wines made of French-become-international varieties like Chardonnay and Cabernet Sauvignon.
67. "The Philadelphia exposition has ended in the best way possible in regard to the work of the Comitato Enologico Italiano," reported Torino delegate Canavesio to the international fair in 1876. "The first step has been taken, the road is plotted; a brand, a very essential thing in these paesi, has been established," *Rivista* 1878, 25.
68. There are ideas that came up in a conversation with abruzzese winemaker Cristiana Tiberio.
69. Coltro, *Paese perduto*, vol. 1, 59.
70. Ibid.
71. To ponder in this context, too, "the ideas that obsessed and still obsess chemistry—principles, elements, atoms, the problem of differentiation, the relationship between one and the many, generation interpreted as an ephemeral transgression of a static order or as the result of a perpetual conflict," Bensaude-Vincent and Stengers, *A History of Chemistry*, 11.
72. Ong, *Interfaces of the Word*, 20.
73. Ibid., 21–22.
74. To paraphrase Levi in *The Periodic Table*: "Like Moses, from that cloud I expected my love, the principle of order in me, around me, and in the world. I was fed up with books . . ."
75. Carpenè, *Sunto*, vol. 1, 5th ed., 121.
76. Ibid.
77. Named in 1814 by finder Sir Humphry Davy. See *OED*. https://www.oed.com.
78. In 1807 by Sir Humphry Davy who named it, too. See *OED*. https://www.oed.com.
79. See Brugnatelli, *Trattato di Chimica Generale*. He was Carpenè's teacher, and due to date and physical proximity, I am proposing that this was the pupil's textbook.
80. *Annals of Philosophy*, vols. 2, 3. London: 1813. Google Books. As summarized and cited by Le Moyne College's Dept. of Chemistry's Carmen Giunta at https://web.lemoyne.edu/~giunta/berzelius.html.
81. And, from that summary: "If the first two letters be common to two metals, I shall, in that case, add to the initial letter the first consonant which they have not in common: for example, S = sulphur, Si = silicium."
82. This offers a temporary confusion: is everything made of hydrogen?
83. "L'emancipazione della donne dovrebb'essere continuamente accoppiata per voi coll'*emancipazione dell'operaio* e darà al vostro lavora la consacrazione d'una verità universale," Mazzini, *Doveri*, 142.

84. "Only starting around 1870 is it possible to speak correctly of an Italian economy; before then existed the economies of the states that 'formed Italy,' and which were subjected to suffocating tariffs and customs procedures on both external and internal markets": see Coltro, *Paese perduto,* vol. 3, Appendix for an overview of changes, development, influx of Russian and US grain, 1883 strikes of the mondine, women rice workers, and the great strikes of the rice-growing valle padana in 1884.
85. "Carbonic acid," https://www.britannica.com.
86. Giuseppe Cappelleri, "L'evoluzione delle tecnologie chimiche in ambito vitivinicolo," *La chimica e le tecnologie chimiche*, ed. Bassani, 391–92.
87. Which will be awarded a gold medal at the Esposizione internationale di macchine enologiche in Conegliano. "Il sistema Carpené per la fabbricazione dei vini spumanti venne premiato con diploma d'onore dal Reale Istituto Veneto di scienze, lettere ed arti in Venezia, con medaglia d'oro all'Esposizione Nazionale di Milano (premio speciale per le invenzioni del R. Ministero d'agricoltura, industria e commercio) e con una medaglia d'oro all'Esposizione internazionale di macchine enologiche in Conegliano." Carpenè, *Sunto*, vol. 1, 5th ed., 118. While the year isn't mentioned, a guess is 1881, when such an event was held.
88. Cannizzaro, *Sunto*, 22.
89. Brugnatelli, *Trattato*, 27.
90. The suffixes -ic and -ous in English.
91. Cerruti, *Bella e potente*, 15–16.
92. Ibid., 23.
93. Ibid., 681.
94. Ibid.
95. Carpenè, letter to Italy's Ministry of Agriculture, Industry, and Commerce printed as "Sulle condizioni dell'Industria vinicola nel Veneto," in *Rassegna di agricoltura*, 135.
96. Attributed to Antonio Carpenè in Gregoletto, *Vite Ambulante*.
97. Their foods, also, though that is, in an inverse of John Dickie's observation in his *Delizia* that it is a book about Italy's foods only since beverages worked differently, another story: no foods are offered up, he notes, as symbols of Patria, but there is at least one toast, with wine (170–71).
98. Giuseppe Cappelleri, "L'evoluzione delle technologie chimiche in ambito vitivinicolo," in *La chimica e le tecnologie chimiche*, ed. Bassani, 382.
99. Gaspari, *Grande guerra*.
100. Traditional agricultural sites of a certain measure.
101. Carpenè, "Pensiamo ai pericoli," *Rivista* 1881, 513–19.
102. "Vini da pasto," Carpenè writes. "Pensiamo ai pericoli," ibid.
103. "È pure evidente per tutti che se il consumo di vini bianchi era maggiore un tempo di quello dei rossi, oggidì s'è invertita la cosa, ed i vini bianchi neppure in queste località si possono più considerare veri vini da pasto. Ciò devesi attribuire ai gusti modificati dei consumatori, ma molto alle condizioni nervose della popolazione, per le quali preferiscono i vini rossi tonici ai bianchi eccitanti." Carpenè, "Pensiamo ai pericoli," ibid.

104. Which he has isolated from red grape skins in a collaboration with Enrico Comboni, with whom he will promote the substance. It is also a name he hopes to give to his first daughter.
105. "L'enocianina, materia colorante rossa, potrà dare il carattere tonico e meno eccitante ai vini bianchi, provvedendoli di quanto loro manca, con la stessa sostanza colorante dell'uva e convertendoli in rossi, e potrà essa facilitare lo smercio dei vini bianchi, come che sotto forma di vini spumanti si potrà agevolarne lo smaltimento, essendo questi ultimi vini eminentemente igienici, per il gas acido carbonico che contengono." Carpenè, "Pensiamo ai pericoli," *Rivista* 1881, 513–19.
106. Cerruti, *Bella e Potente*, 13.
107. "Sulla bilancia del consumo," he writes. Carpenè, *Sunto*, vol. 1, 5th ed., 121.
108. See writings on Henry Louis Le Chatelier's 1884 presentations on chemical equilibrium; https://www.britannica.com and others.
109. Cerruti, *Bella e Potente*, 37.
110. In a letter to G. B. Cerletti dated 19 November 1875, addressed to "Carrissimo Cerletti" ("dearest") and signed *L'Amico affez.*[mo] *A. Carpenè* and addressing him informally in second-person. It is published in *Annali* 1875, 157–60. Cerletti is a fellow Garibaldian and director of the winemaking school they would cofound the following year.
111. Cannizzaro, *Sunto*, 27.
112. Comboni, *Trattato di Enochimica*, 341.
113. The guided defining description is: "D'azzurro, alla fascia d'argento carica della parola TARVISIUM, sostenente una torre merlata di cinque, accostata da due torri merlate di tre, con due bandiere svolazzanti uscenti dalle basi delle torri laterali, quella destra posta in banda, e quella sinistra posta in sbarra. Nella parte inferiore, nel mezzo, sopra un terreno lastricato, un portone con una finestra armata e con due torri basse senza merli; a ciascuno dei lati stanno due piccole torri a quattro merli, e tra questa altre due torri di vedetta con il tetto rosso rotondo cimato da una croce." http://www.comuni-italiani.it/026/stemma.html.
114. De Leo and Favero, *Ca' Foscari e Carpene Malvolti*, 58–60.
115. Carpenè, *Sunto*, vol. 1, 5th ed., 120.
116. Carpenè, *Sunto*, vol. 2, 2nd ed., 71.
117. Caniato and Palugan, *Conegliano*, 24.
118. See Cerruti, *Bella e Potente.*
119. Montanari and Ceccarelli, *La viticoltura e la enologia*, 128–33.
120. Agreed-upon top quality. Esteemed, valuable.

CHAPTER 3

1. See *razionale*, https://www/treccani.it.
2. "Streets."
3. Literally: iron-ed streets.
4. See in the *OED*, "way," which for the historical meanings it carries is a decent English place to use in place of *vie*, today more accurately translated as *street*. https://www.oed.com.
5. Anna Lowenhaupt Tsing, *The Mushroom*, 205.

6. *Ampelografia Generale della Provincia di Treviso.*
7. "Volgarmente detta."
8. "Sopranominata."
9. "Corrispondente a."
10. "Forse."
11. *Ampelografia Italiana.* Rovasenda is head editor.
12. Rovasenda, *Saggio.*
13. "La maggior facilità delle comunicazioni si letterarie che commerciali procurate dalla stampa e dalle ferrovie, hanno di molto favorito gli scambii di vitigni e l'esame dei medesimi, cosicchè non è infondato il presumere che il movimento ampelografico che si manifesta riuscirà a serie migliorie nella viticoltura." Rovasenda, *Saggio*, 6.
14. "Le commissioni si eseguiscono contro assegno ferroviario," *Storie di vite* no. 1, 2018, Carpenè Malvolti, PDF.
15. See Glossary.
16. *Là dove*, 274.
17. See Glossary.
18. Gasparini, "La viticoltura nel Nord Italia," 26.
19. "Enological establishments."
20. *The Atheist.*
21. See: *L'Ateo*, no. 4, 2005. https://docplayer.it/968294-L-ateo-di-fine-ottocento.html.
22. Vianello and Carpenè, *La vite*, 125.
23. Carpenè, *Sunto*, vol. 1, 5th ed., 2.
24. See Glossary.
25. Carpenè, *Sunto*, vol. 1, 5th ed., 117.
26. Ibid., 87.
27. See Glossary.
28. Carpenè, *Sunto*, vol. 1, 5th ed., 131.
29. Ibid., 117.
30. Vianello and Carpenè, *La vite*, 129.
31. *Rows*, sort of. The word is used for this system of festoons, then later for the measured rows in vines-only vineyards.
32. Vianello and Carpenè, *La vite*, 41.
33. Carpenè, *Sunto*, vol. 1, 5th ed., 41.
34. "In a more rational shape and construction"; Vianello and Carpenè, *La vite*, 122.
35. *Storie di vite*, no. 1, 2018.
36. See Dalmasso and Dell'Olio, "Vini bianchi tipici."
37. See Glossary.
38. Carpenè, *Sunto*, vol. 1, 5th ed., 24.
39. Vianello and Carpenè, *La vite*, 45, 49.
40. "Un'ora con Bepi Mazzotti: testimonianza di Bepi Mazzotti e Andrea Zanzotto, in compagnia di Toti Dal Monte e Valter Balliana," *Incontri*, Rai, 1972.
41. Meaning, *at the stalls.* Alle Stalle is also the name of an area in Rolle, steep town in the northwest of the Conegliano-Valdobbiadene DOCG and one of the last towns to hold a stable in the piazza, a contadini center town. For more on Rolle viticulturally, see Tomasi and Gaiotti, *The Terroirs*, 216.

42. A Venetian vernacular glossary on the blog *Passa parola nel Veneto orientale* notes of the word *Massariòt*: "etymology: pejorative derivative of 'Massaro.' Contadino, owner of a small piece of land. By extension: uncouth and ignorant man. Personal observations: unlike *Bisnent* (etymology: he who has nothing) or *Pisnent*, who was the simple bracciante agricolo, the massariot was a 'chel podea dir a sua' who could express his opinions, in the osteria. . . . The massariotti [the spelling varies across sources] called themselves 'paroni' and so they wanted to be called by the laborers and family members." See, "Dialetto/vernacolo del Veneto orientale," http://www.passaparolanelvenetoorientale.it/.
43. *Tino*, or *vat*, in a dialect of the Marca Trevigiana.
44. *Must*, in a dialect of the Marca Trevigiana.
45. Carpenè, *Sunto*, vol. 1, 5th ed., 9.
46. Ibid., 10.
47. Vianello and Carpenè, *La Vite*, 41.
48. Carpenè, *Sunto*, vol. 1, 5th ed., 24.
49. "Le bottiglie saranno conservate in cantina fresca e caricate una sopra l'altra a formare una catasta per risparmio di spazio." Carpenè, *Sunto*, vol. 1, 5th ed., 129.
50. Ibid., 136.
51. Loosely: *traveling teacher posts*.
52. As Italianist Robert Lumley pointed out to me in conversation.
53. Carpenè, *Sunto*, vol. 1, 5th ed., 41.
54. By way of Hugh of Saint-Victor in his *Didascalicon* encyclopedia. See *Lapham's Quarterly*, *Technology*, 14, no. 4 (Winter 2021), 129.
55. Carpenè, *Sunto*, vol. 1, 5th ed., 55.
56. Ibid., 35.
57. *L'enotecnico*, 38.
58. Carpenè, *Sunto*, vol. 1, 5th ed., 11.
59. Vianello and Carpenè, *La Vite*, 144.
60. A red wine grape throughout the Venetian Terraferma, especially today's Friuli Venezia Giulia.
61. Carpenè, *Sunto*, vol. 1, 5th ed., 3.
62. Ibid., 28.
63. Ibid., 32.
64. Ibid., 34.
65. *Ampelografia Italiana*, 94.
66. Carpenè, *Rivista* 1875, 158.
67. Carpenè, *Sunto*, vol. 1, 5th ed., 122.
68. By way of acido solforico (sufuric acid) plus bicarbonato di soda (soda bicarbonate).
69. Carpenè, *Sunto*, vol. 1, 5th ed., 37.
70. Patriarca, *Numbers*, 74.
71. *Dizionario Enciclopedico Italiano*, vol. 10, Q-SCH (Roma: Istituto della Enciclopedia Italiana fondata da Giovanni Treccani, 1959).
72. Carpenè, *Sunto*, vol. 1, 5th ed., 1.

CHAPTER 4

1. Entirely in uppercase lettering and roughly, *Agitator and prophet / roused the third Italy / and across the centuries / he joined / with a pure soul, Socrates / with an austere intellect, Dante / In the first centenary of his birth / the citizens / with the kindly patronage and guidance of the scholastic alliance.*
2. *Baedeker's Italy*, xvi.
3. The city of Treviso, official head of the province. Same for the *Belluno* in this sentence.
4. *Kriegskarte*, vol. 1, 387, map XIII.12.
5. *Kriegskarte*, vol. 1, 387.
6. Caniato and Palugan, *Conegliano*, 104.
7. Ibid., 11.
8. *Annuario d'Italia*, 1221.
9. See, for example, https://archiviopossessori.it/archivio/54-sbarra-pultio.
10. See emblem of Conegliano's Accademia Agraria degli Aspiranti, Dalmasso and Dell'Olio, "Vini bianchi tipici," table 1.
11. Dalmasso and Dell'Olio, "Vini bianchi tipici," table 9.
12. There are ciglioni; *Là dove*, 48.
13. Massimo Rossi, "Per una geografia storica del paesaggio viticolo collinare trevigiano. Un esempio: Colbertaldo," *Là dove*, 46–53.
14. Caniato and Palugan, *Conegliano*, 104.
15. Ibid., 21.
16. Vianello, *Annali* 1872, 370.
17. Casellato, "I sestieri popolari."
18. "Many, many," from Caniato and Palugan, *Conegliano*, 33. "Le numerosissime osterie che caratterizzavano la Conegliano dei pionieri e ancora la caratterizzano. Ristoro, utero, rifugio, luogo di assorbimento dei disagi, da sempre l'osteria ha sostituito in città la piazza, fabbricandola in disseminate particole all'ombra dei portici."
19. Caniato and Palugan, *Conegliano*, 33.
20. Ibid.
21. Ibid.
22. *Là dove*, 68.
23. Interview with Wilma and Walter Balliana, July 11, 2021. In the hills' town of Col San Martino in the mid-twentieth century, they ran an osteria named Dei Colli and famously loved by local card-playing men and celebrity Italians alike.
24. Anna Pizzati, *Conegliano: Una "quasi città" e il suo territorio nel secolo XVI* (Treviso: Fondazione Benetton Studi Ricerche, 1994), blurb.
25. Casellato, "I sestieri popolari."
26. "The *peota casse* (peota is in the Venetian language) are widespread associations in the Veneto, traditional organizations in nonprofit noncorporate form, of ancient or of recent constitution, which collect their members' small impromptu savings and grant modest consumer loans based on mutuality." Complete definition at the website of the Associazione Nazionale Enciclopedia della Banca e della Borsa, https://www.bankpedia.org.
27. Casellato, "I sestieri popolari."

28. Wilma and Walter Balliana, interview on July 11, 2021.
29. Still no women doctors.
30. Brunello, *Ribelli*, 2.
31. Roughly, *osteria-frequenting participants in rallies.*
32. Brunello, *Ribelli*, 130.
33. "Singolare caso di sineddoche che deriva dall'antica consuetudine, oggi desueta, di appendere proteso un ramo di abete sulla facciata delle case come insegna in occasione della vendita diretta di prodotti dell'attività agricola, con particolare riguardo al vino" ("Singular case of synecdoche that derives from the ancient custom, now obsolete, of hanging a spruce branch on the facade of the houses as a sign on the occasion of the direct sale of agricultural products, with particular regard to wine"), from *Ossigeno nascente,* Alberto Bertoni and Marco Marangoni, Università di Bologna, https://site.unibo.it/atlante-poeti/it/poeti-nord-est/luciano-cecchinel.
34. In Italian, according to *Ossigeno nascente*, Alberto Bertoni and Marco Marangoni, Università di Bologna, https://site.unibo.it/atlante-poeti/it/poeti-nord-est/luciano-cecchinel: "rame vecchie che non avevano maniglia / a dire delle porche già di ferro" from Luciano Cecchinel, "Tóvena via là sote Nàdega," *Al tràgol jért: 1972–1984* (Pederobba: ISCO, c1988).
35. Orefice, *Altre Tempi.*
36. Danilo Gasparini, in conversation.
37. Caniato and Palugan, *Conegliano*, 33.
38. Ibid., 9.
39. Ibid., 13.
40. Ibid.
41. Ibid., 21.
42. *Of.*
43. An abbreviation of *domiciliato*, or *domiciled*
44. Potočnik and Villa, *Il paesaggio di Cima*, 75.
45. Ulliana, *La scuola enologica,* 9.
46. *Annuario d'Italia*, 1222.
47. Ulliana, *La scuola enologica,* 9.
48. As yet another hidden distortion, during the fascist Ventennio, *avanguardista* was distorted to mean instead teenagers who were fascist paramilitary; this is of course not what Gasparini is saying.
49. Private conversation, July 12, 2021.
50. Antonio Caccianiga, *Ricordo della Provincia di Treviso* (Treviso: Zoppelli, 1874), from Ulliana, *La scuola* enologica, 9.
51. Sergio Cella, *Dizionario Biografico degli Italiani*, vol. 2, 1970.
52. "But in my opinion, the conferenze ambulanti, in the present state of general culture and agricultural education, called practical, are palliative," Carpenè, *Annali* 1875, 158.
53. "Catalyst," *OED.* https://www.oed.com.
54. Ibid.
55. Caniato and Palugan. *Conegliano.*
56. See "paesaggio," Treccani. https://www.treccani.com.
57. Tsing, *The Mushroom*, 152.

58. The Bordeaux wine zone in France.
59. Antonio Caccianiga, *Ricordo della Provincia di Treviso* (Treviso: Zoppelli, 1874), from Ulliana, *La scuola* enologica, 10.
60. Dalmasso and Dell'Olio, "Vini bianchi tipici," 16.
61. Ulliana, *La scuola* enologica, 7.
62. Roughly and beautifully usefully, this is an adjective form of *put in good order.*
63. Quoted in Ulliana, *La scuola enologica*, 5. See also Francesco Maria Malvolti in *Giornale d'Italia: spettante alla scienza naturale e principalmente all'agricoltura, alle arti, ed al commercio*, vol. 8 (Venezia, 1772), 388–89. Google Books.
64. Ulliana, *La scuola enologica*, 4.
65. "Ma—strano a dirsi—alla ricchezza di studi e notizie sulla viticoltura ed enologia della Marca Trevigiana, che riscontrammo nell'ultimo periodo del secolo XVIII, una deplorevole povertà, per non dire un silenzio assoluto, seguì per tutta la prima metà del secolo XIX, tanto nelle memorie locali, quanto nelle suddette pubblicazioni regionali" ("But—strange to say—the wealth of research and news on the viticulture and enology of the Marca Trevigiana, which we encountered in the last period of the eighteenth century, is followed by a deplorable poverty, not to mention an absolute silence, for all of the first half of the nineteenth century, as much in local memories as in the aforementioned regional publications"). Dalmasso and Dell'Olio, "Vini bianchi tipici," 16.
66. Allusion to an October 1872 letter by Carpenè, published in *Annali* 1872.
67. *Rassegna di agricoltura*, 135.
68. "Sulla necessità dell'istituzione di scuole viticole-vinicole, teorico-pratiche" as listed in *Annali* 1875, 114. Google Books.
69. *Annali* 1875, 114.
70. Ulliana, *La scuola enologica,* 42.
71. "The last and first name and the address of all government functionaries and workers," the *Annuario* promised. "The mayors, municipal secretaries and notaries, the professionals, the industrialists, the merchants, the representatives, the commissioners." The rules that governed the how and where of the administrative, electoral, judicial, communications, distances, fairs, products of the soil, and of industry, specialties, monuments, etc., of each municipality.
72. Ulliana, *La scuola enologica,* 45.
73. Roughly, on March 5, 1867, from this balcony Giuseppe Garibaldi spoke to the people. Today, on via Camillo Benso Conte di Cavour.
74. For the chemical effects of movement and CO_2, see Mark J. Mitchell, et al., "A model of carbon dioxide dissolution and mineral carbonation kinetics," Proceedings of the Royal Society. A. 466, 2009, 1265–90. http://doi.org/10.1098/rspa.2009.0349.
75. Roughly: To who does not not drink wine, God denies also water. Benedetti, Federica, "Aspetti antropologici dell'opera di Luciano Cecchinel," *La parola scoscesa: Poesia e paesaggi di Luciano Cecchinel*, edited by Alessandro Scarsella (Venezia, Marsilio Editori, 2012), 58.
76. Caniato and Palugan, *Conegliano*, 104.
77. Partially, a walk made of slippery stone, smooth slab, frightened hornbeams and oaks, to which "no one arrives anymore." In Italian, from *Ossigeno nascente*, Alberto Bertoni and Marco Marangoni, Università di Bologna, https://site.unibo.it

/atlante-poeti/it/poeti-nord-est/luciano-cecchine: *dietro il sasso sdrucciolato sulla lastra levigata / dove nessuno più ormai arriva / . . . di là dei carpini e le roveri spauriti.* From Luciano Cecchinel, "de là del sanjut sofegà del stran" ("di là del singhiozzo soffocato dello strame"), *Sanjut de stran* (Venezia: Marsilio Editore: 2012).

78. "Un'ora con Bepi Mazzotti," *Incontri*, Rai.
79. In Italian, from *Ossigeno nascente*: *ma sulle svolte lisce e le stalle / di quella notte tenera di Tovena / vegliano deboli tante lacrime gialle / da su sopra i prati erti di Nadega.* From "Tóvena via là sote Nàdega," in *Al tràgol jért: 1972–1984* (Pederobba: I.S.C.O., 1988).
80. Similar to Austria's in Klosterneuburg. See Ulliana, *La scuola enologica.*
81. From photo of document "Contributi degli Enti consorziati e frequenza ai due corsi della Scuola. Conegliano, 27 novembre," *Anniversario 130.*
82. By the Lindemann-Weierstrass theorem.
83. Or, a possible *x*.
84. Angelo Vianello, "Due professori stipendiati da sei ettari di Vigna: studii sperimentali," *Annali* 1872, 246-53, 364–71.
85. Vianello and Carpenè, *La vite*, 82–99.
86. Vianello, "Due professori stipendiati da sei ettari di Vigna: studii sperimentali," *Annali* 1872, 246.
87. *Vines as vineyards.* Vianello, "Due professori stipendiati da sei ettari di Vigna: studii sperimentali," *Annali* 1872, 248.
88. Mulberry trees.
89. Vianello, "Due professori stipendiati da sei ettari di Vigna: studii sperimentali," *Annali* 1872, 366.
90. Ibid., 368.
91. Ibid., 371.
92. *Kriegskarte*, vol. 1, 384.
93. Crucial, if hard to find today, the writings of Jewish professor Adolfo Vital of and on Conegliano: 1908's *Piccola guida storico-artistica di Conegliano*, sponsored by among others, Carpenè Malvolti (see Luciano Caniato and Giuseppe Palugan, *Conegliano*, 24]; 1913's *Effemeride di storia coneglianese*; 1928's *Conegliano: gemma della Marca Trevigiana.* Giovanni Tomasi and Silvia Tomas write that in 1855, twenty-nine Jewish people lived in the district of Conegliano and seventy-three in Ceneda. The census of 1881 showed thirty-nine people in Conegliano and forty-nine in Vittorio (today, Vittorio Veneto and the result of Ceneda and Serravalle being combined), "in total 88 people of Jewish religion." In 1919, seven people remained in Conegliano and six in "Vittorio Veneto." After 1944, three Jewish natives of Vittorio Veneto and Conegliano, residents of other northern Italian zones, as well as a Jewish Venetian living in San Vendemiano were deported to Germany and "all perished in the Shoah. One of the last Jewish people of Conegliano, the professor Adolfo Vital, died in hiding in 1944," while those living in Vittorio Veneto survived. "Today in [Vittorio Veneto] there are a few Jewish people of ancient origin." See, Giovanni Tomasi and Silvia Tomasi, *Ebrei nel Veneto orientale: Conegliano, Ceneda e insediamenti minori* (Firenze: Casa Editrice Giuntina, 2012).
94. Per Vianello and Carpenè, *La vite,* 48: 7,614.
95. From the two-year old *Annali.*
96. Particularly steep hills.

97. Ulliana, *La scuola enologica*, 64.
98. Piovene, *Viaggio*, 40.
99. As I saw at the school. Also noted in Ulliana, *La scuola enologica*.
100. Ulliana, *La scuola enologica*, 51.
101. *Rivista* 1880, 165.
102. Ibid.
103. Ibid.
104. Ibid.
105. Ulliana, *La scuola enologica*, 45.
106. *Rivista* 1880, 164.
107. Ulliana, *La scuola enologica*, 44.
108. Dalmasso and Dell'Olio, "Vini bianchi tipici," table 9.
109. "Da Conegliano, come raggi di una ruota ["rays of a wheel"] si spanderanno nella Veneta Regione e più in là i benefici lumi della scienza applicata alla più bella e più lucrosa fra le produzioni del suolo." Carpenè, *Rivista* 1877, 40.
110. "Gli attrezzi che usavano i bottai di Aquileia (presenti in una lapide funeraria) sono gli stessi usati ancora oggi a Conegliano, le cui fabbriche sono forse le prime nel mondo, per la qualità del manufatto, capacità di produzione totale e gamma di produzione, con botti da 1 a 150.000 litri. . . . Le botti di Conegliano, costruite usando il rovere di Slavonia, alimentano anche una fiorente esportazione in tutto il mondo vinicolo. Spesso girando il mondo ho trovato con soddisfazione botti di ditte di Conegliano, scelte per una ideale conservazione dei vini. Ad esempio certi ottimi vini israeliani diventano tali in botti della Ditta Garbellotto di Conegliano. Sempre a Conegliano, si vedono anche botticelle uso famiglia, da 10-15-25-40-50-100-150 litri in su, per l'invecchiamento di vini, liquori, distillati e aceti fini da tavola." ("The tools used by the coopers of Aquileia (pictured on a tombstone) are the same still used today in Conegliano, whose factories are perhaps the first in the world, for the quality of the product, total production capacity and production range, with botti from 1 to 150,000 liters. . . . The Conegliano barrels, built using Slavonian oak, also fuel a thriving export throughout the wine world. Often traveling the world I have found with satisfaction barrels of companies from Conegliano, chosen for ideal conservation of wines. For example, some excellent Israeli wines become so in barrels made by the Ditta Garbellotto di Conegliano. And in Conegliano, there are also barrels for family use, from 10-15-25-40-50-100-150 liters upwards, for the aging of wines, liqueurs, spirits and fine table vinegars.") Maffioli, *La cucina trevigiana*, "Le botti di Conegliano" section.
111. Ulliana, *La scuola enologica*, 62.
112. King of the Kingdom of Italy from 1878 to 1900. This had been the viale dei Passeggi.
113. Carta dei vini at the school's fiftieth anniversary in 1927, *Anniversario 130.*
114. Vianello and Carpenè, *La vite*, 34.
115. See the "Elenchi dei comuni fillosserati o sospetti d'infezione fillosserica," in Montanari and Ceccarelli. *La viticoltura*, 128–34.
116. Montanari and Ceccarelli. *La viticoltura*, 134.
117. Caniato, *Conegliano*, 55.

CHAPTER 5

1. Organized by School cofounder and director C. B. Cerletti. *Annuario del Ministero di agricoltura, industria e commercio* (Roma: G. Bertero, 1904), 463. Google Books.
2. Antonio Caccianiga, "Introduzione," *Nuova rassegna* 1887, 4. "Portava, he continued of the defunct (though the name would return in some years) *Rivista*, "la voce dell'insegnamento oltre i ristretti confini della scuola, era l'apostolato della scienza, metteva in comunicazione i suoi lavori cogli studi degli altri istituti, colla propaganda dei giornali agricoli ed enologici, serviva di eccitamento a continue indagini, a nuove applicazioni delle scoperte scientifiche alle pratiche agricole ed industriali." ("It carried the voice of what was taught beyond the narrow confines of the school, was the apostolate of science, put its works into communication with the studies of other institutes, with the propaganda of agricoltural and enological journals, served to incite ongoing research, the new application of scientific discoveries to agricultural and industrial practices.")

   To which four pages later, Carpenè added about the *Rivista*, "E si fu a merito di queste persone se la R. Scuola diede splendidi risultati cosi da farla rifulgere come astro luminoso sull'orizzonte enologico, astro salito tant'alto che se qualche nube riuscì un momento ad intercettarne i raggi, non ne seppe affievolire la potenza luminosa e ciò perchè il suo nucleo è costituito di materiali vivi, in piena energia dinamica da lasciarci ritenere lontano quel raffreddamento cui sottostarà per legge dell'universo ogni massa cosmica calorifica e luminosa. Ritorno volentieri tra le fila di operai che lavorarono tanto proficuamente nel cessato periodico e vi ritorno con la piena fiducia che molti altri servigi renderà all'enologia questa nuova Rassegna." ("And it was to the merit of such persons if the R. Scuola gave such splendid results as to set it ashine like a luminous star on the enological horizon, star risen so high that if some cloud succeeded for a moment in intercepting its rays, it did not know how to weaken the luminous power because its core is constituted of living materials, in full dynamic energy that lets us keep far away the cooling which by laws of the universe every calorific and luminous cosmic mass will undergo. I return gladly among the ranks of operai who worked so profitably for the ceased periodical and I return with full faith in the many other services to enology that will be rendered by this new Rassegna.") "Ai lettori della 'nuova rassegna," *Nuova rassegna* 1887, 8. Google Books.
3. Of 1871, which seems to have been a poor vintage: "Ma l'anno 1871 essendo stato, specialmente nel veneto [sic], sfavorerole alquanto alla buona maturazione dell'uva, somministrò buona quantità di vini di incerta conservazione. In alcuni dipartimenti della Francia l'aspetto dei vini desta pure inquietudine ai vinificatori." ("But the year 1871, having been, especially in the Veneto, unfavorable in regard to proper maturation of the grapes, dispensed a large quantity of wines of uncertain lastingness. In some departments of France the appearance of the wines is also causing concern for winemakers.") Carpenè, "Sui vini d'incerta conservazione," *Annali* 1872, 174.
4. "Ora che la calda stagione incomincia, molti vinificatori mostransi inquieti perchè certa quantità di vino specialmente bianco dell'ultima raccolta minaccia alterarsi. Ve ne sono infatti quest'anno di quelli che non hanno subita la completa fermentazione nell'autunno e che perciò rimasero dolcigni." That last word a compound

meaning of *sweetish* and *unpleasant*. Carpenè, "Sui vini d'incerta conservazione," *Annali* 1872, 174.

5. See Dalmasso and Dell'Olio, "Vini bianchi tipici."
6. Also professor at the Scuola. See, for example, Ulliana, *La scuola enologica* and https://treccani.it.
7. The Consiglio per la ricerca in agricoltura e l'analisi dell'economia agraria (CREA), founded in Rome, Italy, in 1999.
8. Drying is an ancient Italian technique, used as much to concentrate grapes of a too-cold vintage as for style and preservation.
9. Dalmasso and Dell'Olio, "Vini bianchi tipici," 41.
10. In wine terminology, *travasato* in Italian, *racked* in English: the wine is always *racked* a little early. See Glossary.
11. No residual grape sugar remaining; in practice can sometimes be, unnoticeably, <1%.
12. *We are in the era of racking*. Carpenè, "Siamo all'epoca de travasi," *Rivista* 1878, 16.
13. As prescribed by the two DOCGs' disciplinari.
14. Sannino, "Sui caratteri," *Rivista* 1907, 251–54.
15. Vitruvius in "Control Group," *Lapham's Quarterly: Technology*, 122.
16. "The Papadopoli counts, the Corinaldi counts of Monselice . . . the old aristocratic and noble class that was beginning to avail itself of the new tecnicians: precisely, the enologi," Danilo Gasparini, "Nascita del distretto enologico," *Là dove*, 58.
17. Vianello and Carpenè, *La Vite*, 126.
18. "Gli spumanti artificiali o di saturazione, che di tali odori sono privi [which lack the secondary flavors of the first production method], migliorano sensibilmente se prodotti con vini che prima della saturazione abbiano odore gradevole com'è quello dei vini bianchi di Prosecco giovani e un po' dolci," Sannino, "Sui caratteri," *Rivista* 1907, 254.
19. Carpenè, *Sunto* vol. 1, 1888, 123–25.
20. See, for example, the explanation by the Ministero dell'agricoltura, della sovranità alimentare e delle foreste: https://www.politicheagricole.it/flex/files/2/3/6/D.86a647c9965e8dd53e1b/ConosciamoVinoItaliano.pdf.
21. Or, an *atmosphere*: a unit of measurement equal to that average air pressure at sea level at a temperature of 15°C (59°F).
22. Probably *silvered*: "Due bottiglie di rame della contenenza di un litro circa vennero una stagnata molto diligentemente con stagno del più puro del commercio, l'altra *argentata* dalla Casa Cristofle et C. Vennero riempite allo stesso momento d'un vino preparato per divenire spumante e, insieme a queste, anche un certo numero di bottiglie di vetro da Champagne con lo stesso vino. Chiuse con turacciolo fissato con filo di ferro, furono portate ed accatastate nella medesima cantina. Dopo sette mesi, le due bottiglie metalliche vennero aperte ed esaminatone il contenuto; la bottiglia stagnata offri un risultato pari a quello che si consegue conservando il vino in bottiglie di vetro entro cui si dimenticarono dei grani di stagno, cioè assoluta assenza di spuma, fermento completamente morto, il vino conteneva in dissoluzione gr. 0.36 per litro di stagno, esalava infine un odore fetido, dovuto probabilmente a traccie di solfostanetile e non aveva più dell' 8.5% d'alcole, con un residuo solido di gr. 47,5 per litro. La bottiglia argentata, al contrario, offrì la più fortunata compensazione, il vino fu trovato chiaro, la spuma abbondante, il gusto buono ed il deposito ben formato; l'analisi non isvelò la benchè minima traccia d'argento o

di rame, la superficie dell'argento si riscontrò perfettamente inalterata." Comboni, "Vagoni Serbatoi," *Rassegna* 1888, 407.

23. Cobb, *The History of Stainless Steel*, 2.
24. Though *tinaia* appears a few times in these Veneto agriculturalists' and enologists' writings, the word today is noted as a tuscan one, hinting that it either receded back into local use from its nineteenth-century inclusion in Italian or that as the new nation's local languages became officially replaced with the tuscan dialect chosen to be "Italian," some overcompensated.
25. Interview with Franco Adami on January 29, 2021.
26. Just a guess.
27. Tomasi and Gaiotti, *The Terroirs*, 75.
28. See Tomasi and Gaiotti, *The Terroirs*.
29. Only as seen from a distance. "Dalla pianura," wrote Mazzotti in 1957, "il Montello sembra un lungo dorso piatto e unito. In realtà è tutto un susseguirsi di collinette e doline, con un paesaggio che varia in modo incredibile. Percorrere le sue strade interne è una continua grata sorpresa. Molte doline (di natura carsica, a cui corrispondono nel sottosuolo numerose grotte) sono coltivate a vigneto, e danno buonissimi vini." ("From the plain, the Montello seems like a long flat and united slope. In reality it is all a series of little hills and depressions, with a landscape that varies in incredible ways. Traveling along its interior roads is a continuous gratifying surprise. Many depressions (of karstic nature, to which correspond below ground many caves) are planted with vineyards, and give superlative wines.). *Immagini.*
30. Dalmasso and Dell'Olio, "Vini bianchi tipici," 25.
31. On October 21 and 22.
32. Brunello, *Ribelli*, 85.
33. Ibid., 96.
34. Ibid., 70.
35. Ibid., 74.
36. Gasparini, "La viticoltura nel Nord Italia," 6.
37. Brunello, *Ribelli*, 88.
38. Ibid., 94.
39. Ibid., 65.
40. See Brunello, *Ribelli.*
41. Brunello, *Ribelli*, 69.
42. In the 1850s, when a collection of mildew diseases ran through the Venetian provinces, writes Berengo in *L'agricoltura veneta*, 303. They radiated out from neighboring Veronese countryside, landing in Treviso, too. "Nel 1851–52, e poi con forti ondate sino al '56 e a; '57, la crittogama investì le province venete , trovando il suo centro nel Vicentino ma irradiandosi nel Veronese, nel Padovano, nel Trevisano e poi nel Friuli; solo alcune piccole zone, come quella collinare di Fonzaso nel Feltrino che vantava buoni vigneti, riuscirono a salvarsi."
43. "Ci sono anche le cattive abitudini, l'ignoranza dei contadini e più spesso dei proprietari. Si invoca che prima che al contadino l'istruzione dovrebbe essere impartita al proprietario. È il sistema agricolo dominante che molto si oppone all'estensione della vigna bassa nel Veneto," Carpenè, "Più vino e meno pellagra," *Rivista* 1880, as quoted in Ulliana, *Scuola* 28–30. On page 674 of a scanned copy of the article from a bound volume of the 1880 *Rivista* publications, a more detailed

relationship (see Notes for chapter 7, "Promised Lands"): "D'ordinario il proprietario non gode prestigio verso il contadino perché quasi come questo e talvolta anche più, ignora l'arte dei campi, cioè non possiede la scienza agronomica e la pratica agraria necessarie per imporsi all'operajo dei campi, il quale quantunque pieno di pregiudizi non ignora l'ignoranza del padrone e spesso se ne ride di lui." (Roughly: Ordinarily the owner does not enjoy prestige in the eyes of the contadino because often like him and sometime even more so, the former ignores the art of the campo, that is he does not have the agronomic science and the agrarian practical training necessary to dictate methods to the worker of the campo, who already filled with prejudice does not ignore the proprietor's ignorance and often laughs at him.)

44. "I nostri contadini esercitano l'agricoltura colle cognizioni tradizionali alle quali sono tenacemente avvinti. Se il proprietario loro impone una innovazione, non vi si rifiutano, ma tendono ad esimersene con ogni mezzo; si persuadono difficilmente e solo dopo ripetuti esempi." ("Our contadini practice agricolture with the traditional knowledge which they are tenaciously enmeshed. If the proprietor imposes an innovation, they don't refuse it, but they tend to avoid it by any means; they are persuaded with difficulty and only after repeated examples.") Vianello and Carpenè, *La vite*, 40. It was a problem with city dwellers and politicians, too: "Or bene, in tutta Italia, diciamolo pur francamente, vi ha fra moltissimi il mal vezzo di accettare con diffidenza le nuove istituzioni e si è inclinati a criticare senza ragione più che ad incoraggiare giustamente." ("OK, in all of Italy, let's speak frankly, many, many people have the bad habit of accepting new institutions with diffidence and are inclined to criticize without reason more than they are to justly encourage.") Carpenè, *Rivista* 1877, 42.

45. Compare with, as one example, an 1872 article by one Paolo Antoniotti: in "L'agricoltura e la scienza": "Il contadino, il pratico agricoltore, limitato nella sua intelligenza, non istruito ed appoggiato unicamente all'empirica sua attività, si contenta di contemplare collo sguardo della maraviglia le periodiche rinnovazioni della natura senza punto darsi pensiero dei migliori metodi d'agricoltura investigare gli agenti della fertilità, e conoscere gli elementi delle sostanze organiche e minerali. Ei zappa e semina, e colla profonda sua compiacenza vede il miracola della nascita e crescenza della pianta nutrendosi felicemente della speranza del buon risultato della produzione, la quale però non di rado, riescendo scarsa, gli porta l'inganno e con esso il cattivo umore e la miseria. Non è quindi lodevole nè prudente consiglio l'abbandonarsi in braccio alla Provvidenza; l'accorto agricoltore deve rintracciare la miglior ricompensa delle annuali sue fatiche anche nella direzione più intelligente de' suoi lavori campestri." *Annali* 1872, 48.

46. "Si insegni all'agricoltore una buona pratica agraria, lo si convinca pure sulla bontà di questa, ma esso non potrà ricevere chiara in mente la conoscenza delle condizioni svariatissime per le quali la pratica insegnata è utile in dati luoghi e somministra risultati negativi in altri; perciò adottandola, non sempre conseguirà gli effetti desiderati. I concimi chimici, ad esempio, non godono la fiducia che si meritano ed è poco generalizzato il loro uso. Ciò devesi perchè vennero sparsi di spesso su terreni e per colture che non li esigevano. Si dira anche che i concimi chimici vennero fatti cadere in iscredito dalla poco onestà dei fabbricatori; gli inganni saranno

sempre frequenti per l'agricoltore privo delle fondamenta della scienza." Carpenè, *Annali* 1875, 158.

47. I read with surprise a footnote condemnation of the Montello boscaioli in *La vite*, 102. It is not clear which coauthor wrote it; it does not come up again in Carpenè's other writings about woods.
48. "É spettacolo che veramente rattrista ogni cuor ben nato il veder delle frotte di contadini, colle mogli, figlioletti e vecchi cadenti, abbandonare con pochi stracci la terra, che essi e I loro padri han bagnato dei sudori della loro fronte, per andare in cerca di una fortuna che pur troppo il più delle volte sta solo nell'immaginazione di una mente debole e soffrente e sulla bocca di melliflui arrolatori, che perfidamente speculano sulla credulità del povero lavoratore della terra." Cerletti, "La questione sociale fra I mezzadri e I vigneti," *Rivista* 1878, 742.
49. "Fealty"; to science? The word means also *loyalty*, made up, too, of *faith* and *honesty.*
50. "Elevates, ennobles the industrialist who works with their own intelligence and their *arms*," an echo of *braccianti*. Carpenè, "La verità sui vini spumanti," *L'Enotecnico*, 37.
51. "Non accadrebbe se I proprietari avessero saputo diriger bene ed il vignaiolo fosse più esperto nell'eseguire le molteplici operazioni viticole." Carpenè, *Annali* 1875, 159.
52. Italy's north.
53. "Se in quelle vallate e versanti ameni tendesse la coltura al miglioramento delle qualità, se si dedicassero alla vite in vigneti chiusi tutte quelle cure ed attenzioni che noi nel settentrione siamo costretti di usare, allora quel vino acquisterebbe 'una rinomanza tale come verun altro nel mondo'" Carpenè, *Rivista* 1877, 41.
54. Based on Gasparini, "La viticoltura nel Nord Italia," 24.
55. Those in the know.
56. Carpenè, "Sul legname per la costruzione delle botti," *Rivista* 1878, 448.
57. "Roba," he writes.
58. Carpenè, "Sul legname per la costruzione delle botti," *Rivista* 1878, 448.
59. Carpenè, "La verità sui vini spumanti," *L'Enotecnico*, 37.
60. *Erariale* is the word he uses.
61. Carpenè, "Sul legname per la costruzione delle botti," *Rivista* 1878, 448.
62. The first volume is in 1893. The title page reads: "Un gruppo di enotecnici residenti in Conegliano sotto la Presidenza del Prof. Cav . Dott . Antonio Carpenè con la collaborazione degli enotecnici licenziati dalle R. Scuole Enologiche del Regno; Redattore—Capo Arturo Marescalchi." The latter a graduate of the school in 1889, viti-vini publisher, including during the fascist years (https://www.treccani.it).
63. "Se i mezzi chimici d'investigazione fossero maggiori e bastassero a scoprire nettamente i diversi composti che si generano durante la fermentazione, a seconda della temperatura sotto cui viene condotta, saremmo in possesso di cognizioni preziose per la tecnologia enologica. La pratica insegna che una fermentazione condotta lentamente dà una divergenza nei risultati in confronto di un'altra accelerata col mezzo d'una temperatura più favorevole." Carpenè, *Sunto*, vol. 1, 3rd ed., 26.
64. Defined as "their ability to readily take up and retain moisture," Merriam-Webster, https://www.merriam-webster.com. From review of *Dei recipienti vinari* in *Rassegna* 1887.
65. Carpenè, "Dei recipienti vinari e della cantina," *Sunto*, vol. 1, 5th ed., 52. Elaboration on Carpenè, "Dei recipienti vinari e del miglior modo di conservali," *Annali* 1874,

15–20. This work would be republished several times in growing booklet form as part of the Bilbioteca agraria Ottavi, and in 1905 republished as *I recipienti vinari: ed il modo di depurarli e conservali*, 4th ed., "entirely revised and enhanced" by Marescalchi. Google Books.

66. "Since Aristotle's time the properties of a body had been rooted in a 'principle,' a substance that acted and explained," Bensaude-Vincent and Stengers, *A History of Chemistry*, 54.
67. Carpenè, "La verità sui vini spumanti," *L'Enotecnico*, 33.
68. Ibid., 37.
69. Wine fermented in a closed bottle—the Champagne method or the French method.
70. Wine fermented in an "open," as in unbottled, receptacle.
71. Since its earliest years, the Carpenè Malvolti winery has also made brandy, which until it was disallowed much like the name *Prosecco* for wines made of these grapes elsewhere, would have been called, after the region in France which made this traditional style of brandy, *Cognac*.
72. Carpenè, "La verità sui vini spumanti," *L'Enotecnico*, 36.
73. Ibid., 33.
74. *Bollettino del comizio agrario di Mantova e dei distretti riuniti di Asola*, 1893, 237. Google Books.
75. Or, the somewhat vague and heavy-handed English word *mousse*.
76. Carpenè, "La verità sui vini spumanti," *L'Enotecnico*, 36.
77. Ibid., 35.
78. Dalmasso, "Vino Spumante," *Enciclopedia italiano*, 1936, https://treccani.com.
79. Ibid.
80. "Autoclave" in *OED*, https://www.oed.com.
81. See De Rosa, *Tecnica*, chapter 17.
82. The words reflecting a changing world. Working with Massimo Pivetti and Grazia Gabrielli, eds, *Grande Dizionario Italiano* di Aldo Gabrielli, 4th ed. (Milano: Ulrico Hoepli Editore, 2020): *Tecnica* is *technique*; technology, skill, "complesso di norme su cui si fonda l'esercizio pratico di una determinata attività manuale, intellettuale, o artistica" ("complex of norms upon which is based the practice of a determined manual, intellectual or artistic activity"); and a "procedure to follow in the execution of a determined work." While *tecnologia* is technology, "studio dei vari procedimenti tecnici applicabili a un'arte o un'undustria" ("study of the various technique procedures applicable to an art or an industry"); the technical methodology related to a particular art or industry; study of the processes of working on raw materials to transform them into finished products; in chemistry, "tecnica applied to an industrial process, to scientific research, and similar."
83. De Rosa, *Tecnica*, 449.
84. See Pirovano, *Trattato di vinificazione*.
85. Pirovano, *Trattato di vinificazione*, 282.
86. "From the greek aphros = schiuma and phoreys = portatore," Lorenzo Tablino, "La storia degli spumanti in autoclave," *Lorenzo Tablino - Enologo: Consulenza, Curiosità e Formazione sul Mondo del Vino* January, 27, 2005, http://tablino.it.

87. "Instrument used for measuring the pressure in sparkling wines." Translated from Massimo Pivetti and Grazia Gabrielli, eds, *Grande Dizionario Italiano* di Aldo Gabrielli, 4th ed. (Milano: Ulrico Hoepli Editore, 2020).
88. Lorenzo Tablino, "La storia degli spumanti in autoclave," *Lorenzo Tablino—Enologo: Consulenza, Curiosità e Formazione sul Mondo del Vino* January, 27, 2005, http://tablino.it.
89. Antonio Carpenè's son and grandson.
90. De Rosa, *Tecnica*, 449.
91. Lorenzo Tablino, "La storia degli spumanti in autoclave," *Lorenzo Tablino—Enologo: Consulenza, Curiosità e Formazione sul Mondo del Vino* January, 27, 2005, http://tablino.it.
92. "Cr," periodic table at The Royal Society of Chemistry website. https://www.rsc.org.
93. "Il materiale che viene primo alla mente suggerito, per la costruzione dei vagoni serbatoi, è il ferro in lamiera di spessore conveniente, oppure la lamiera di ferro stagnato, o di rame stagnato. Ma i metalli, come si disse, sono più o meno tutti intaccati dall'anidride carbonica contenuta in buona quantità nei vini dell'annata specialmente e dagli acidi liberi costantemente presenti nel vino, come facenti parte della naturale sua costituzione. In conseguenza di detta azione chimica, viene alterata la composizione del vino, i suoi caratteri, le suo igieniche proprietà, Il radicale metallico, entrando in combinazione tra i componenti del vino forma dei composti metallici che agiscono alla lor volta provocando precipitazione di ossido metallico e di materia colorante, di enocianina sotto forma di lacca, e così i vini da taglio perdono della loro intensità colorante e quindi il loro valore viene ad essere diminuito. Se il recipiente è ferro, i sali ferrosi che si formano, per quanto restino per sole traccie nel vino, al primo contatto con l'aria, per ossiddazione, convertonsi in sali ferrici che alla lor volta si decompongono per formare del tannato ferrico che brunisce il vino e, se è bianco, lo rende incommerciabile, se non venga eliminato con opportune successive operazioni." Comboni, "Vagoni Serbatoi," *Rassegna* 1888, 402–3.
94. Such as malleability and temperature conduction.
95. Cobb. *History of Stainless Steel*, 2.
96. "A Non-Rusting Steel.; Sheffield Invention Especially Good for Table Cutlery," *The New York Times*, Jan. 31, 1915. https://nytimes.com.
97. But which is also held to be a term to which producer Francesco Drusian holds legal rights.
98. Carpenè, "E son dei secoli che si predica!" *Rivista* 1878, 98.
99. Robert Lumley, conversation on July 2, 2021.
100. Dalmasso and Dell'Olio, "Vini bianchi tipici," 75.
101. In the comune Cison del Valmarino, including its two frazioni, Rolle and Tóvena.
102. "Mentre racconta dei lavori agricoli invernali comincia a piovere, come annunciato dal cielo plumbeo fin dalla mattina. Angelo e Roberta ci invitano ad entrare in casa e ci fanno accomodare al grande tavolo della cucina. Si apre una bottiglia di prosecco di cui chiaramente chiediamo l'origine; è così che scopriamo che, mentre la produzione per il mercato esterno si arresta alla fase del conferimento dell'uva, per l'uso domestico Angelo prosegue la tradizione vitivinicola del nonno utilizzando i suoi stessi macchinari: «A: sì è un po' di vino che faccio io in casa.

F: ah! Fai del vino in casa?

A: sì, due o tre ettolitri, così . . .

B: per uso famigliare?

A: sì.

B: come si fa a farlo in casa, scusami . . . botti?

A: ho le macchine d'acciaio, ancora quelle di mio nonno. Io pigio l'uva, gli acini vengono separati dal graspo e anche dalle bucce, sì, nella vasca va a finire solo il succo e poi da lì faccio i diversi travasi e basta»." AISO 2020 interview, Angelo Nardi by Francesca Poggetti and Bianca Pastori.

103. See Glossary.
104. As, if outside of the scope of this book, it could be argued that the iron-carbon alloy of steel is to the new nation: Since its opening day in 1884 in the central landlocked Italian region of Umbria the industrial complex called the Società degli Alti Forni, Fonderie ed Acciaierie di Terni has supplied the greater part of Italy's production of steel. At the end of World War II during which would come, and go more quickly than it was built for, the Pact of Steel, Italy would need to once again search for itself in the best terms possible. The partigiani's fight against Mussolini's regime is sometimes referred to as the latest risorgimentale movement, the rebuilding of places and structure a continuation of national struggle to find itself through its own components and methods. Steel would form the basis of the nation's postwar boom economico, also called the Italian miracle, a newer industry for an Italy.
105. For this last word, there is no good word in English for *stoppering*, *closing*, or *plugging* specifically a wine-filled bottle. "L'ingegnere Bortolo Bertolaso fonda nella campagna veronese la sua azienda per il settore vitivinicolo. La direzione è chiara: la forte propensione allo sviluppo ingegneristico porta la società a concentrarsi sulla produzione di macchine tappatrici. Attraverso quattro generazioni, la famiglia Bertolaso ha ampliato e diversificato la produzione di impianti per l'imbottigliamento, senza tradire lo spirito pionieristico del fondatore: l'impegno nella ricerca di soluzioni tecniche sempre più avanzate." Zimella, 1880. https://www.bertolaso.com.
106. The well-known inquiry "into the conditions of the agricultural class in Italy," decreed by law on March 15, 1877, and led by senator and economist Stefano Jacini "represents the most complete documentation on the state of the agricultural economy of post-unification Italy." The results were published from 1881 to 1890 and summarized in Jacini's final report. Sistema Bibliotecario di Ateneo of the Università degli studi, Firenze, https://www.sba.unifi.it/. See archives at Italy's Ministero per i beni e le attività culturali, http://www.archivi.beniculturali.it
107. Dalmasso and Dell'Olio, "Vini bianchi tipici," 19.
108. In mathematics also, *having an infinite number of solutions*
109. Like Friuili's beloved white Friuliano wine, made of the eponymous grape but also to indicate a desired white wine, *Prosecco* was at times used here, too, to indicate, say, the best Ribolla ("la Ribolla migliore, o una Ribolla con caratteristiche che la distinguevano dalla massa dei vini che andavano sotto questo nome."). Costantini, "Chel Prossec," *Tiere furlane / terra friulana*, 9–10.

110. What is now the bordering regions of Veneto (then, Venezia Euganea), Friuli-Venezia-Giulia (Venezia Giulia), and Alto-Adige (Venezia Retica) in northeastern Italy, and that included some territories lost to Italy after World War II.
111. Compiled by Montanari and Ceccarelli.
112. Costantini, "Chel Prossec,"*Tiere furlane / terra friulana*, 13.
113. Subtitled "Rivista di cultura del territorio," agriculture presupposing culture and vice versa.
114. To quickly situate: the Lombard-Venetian Kingdom was created in 1815 by the Congress of Vienna following the end of Napoleon's empire and was under Austrian rule. After the Second War of Independence, most of the Lombardy part was annexed to the piemontese Savoy-ruled Kingdom of Sardinia in 1859 (whose king, Vittorio Emanuele II, two years later became the first king of the new nation of Italy). The remainder was mostly the old Venetian lands, including what is now Veneto and Friuli, annexed to the new nation in 1866.
115. "Prosecco" on *Bioversità del Veneto* website of the Provincia di Vicenza, Istituto di Genetica e Sperimentazione Agraria. http://biodiversita.provincia.vicenza.it.
116. Advertisements from the Società Enologica Trevigiana, selling cuttings of Prosecco Bianco, appear in the school's journals; on, for example, the last page of *Annali* 1875.
117. For the vineyards of the Rhine valley, or German Riesling.
118. Sannino, "Note ampelografiche," *Rivista* 1913, 125.
119. *Là dove*, 274.
120. See *Glera* entry, D'Agata, *Native Wine Grapes.*
121. See Glossary.
122. *Annuario d'Italia*, 1222.
123. See "Nelle auspicatissime nozze della nobile contessa Marta Gradenigo col nobile conte Giulio Balbi-Valier," in catalog of Veneto-wide library network, Biblioteche in Polo, https://binp.regione.veneto.it/SebinaOpac/.do.
124. De Leo and Favero, *Ca' Foscari e Carpenè Malvolti*, 57.
125. Sannino, "Note ampelografiche," *Rivista* 1913, 125.
126. Carpenè, *Sunto*, vol. 1, 5th ed., 4.
127. Ibid., 5.
128. Ibid., 55.
129. Ibid., 79–80.
130. Ibid., 6.
131. Ibid., 76.
132. Ibid., 88.
133. Ibid., 87.
134. Carpenè, "I campioni dei vini," *Rassegna* 1889, 5.
135. Of the two most usual bottle shapes, one with sloped shoulders (the Burgundy-style bottles), this is the other one, with square shoulders.
136. A tall slender bottle as those used for German, and now by extension other areas', Rieslings.
137. "Procurino i produttori di fare miglior scelta di bottiglie per spedire campioni di vini. Queste dovrebbero essere della capacità di mezzo litro poco meno, quantità sufficiente per un assaggio ed anche per una analisi occorrendo. La bottiglia sia di forma bordolese a tinta chiara pei vini rossi e di forma renana pei vini bianchi. Sia

ben pulita, e dopo riempiuta si chiuda con buon turacciolo nuovo, che sarà poi coperto con una capsula di stagno. Non si consiglia l'uso di ceralacca chè facilmente comunica al vino il sapore e l'odore di trementina. Si applichi alla bottiglia un'etichetta elegante sulla quale vi sarà il nome del produttore o della ditta venditrice, quello del vino, il numero d'ettolitri disponibile, il prezzo etc. La bottiglia sia involta, in un foglio di carta fina bianca, o colorata e posta poi entro ad un cono di paglia e, infine, coricata in una cassettina elegante.

Sono tutte cose le descritte che sembreranno per taluni miticolosità di secondaria importanza, ma realmente ne hanno molta. L'eleganza, la grazia, il buon gusto, anche in dette preparazioni, dinotano la perfezione e danno la misura della civiltà e dei costumi di un paese," Carpenè, "I campioni dei vini," *Rassegna* 1889, 5.

138. Dalmasso and Dell'Olio, "Vini bianchi tipici," 20.
139. Carpenè, "Pensiamo ai pericoli, *Rivista* 1880, 518.
140. He writes *lustri*, a literary term for *five years*; "parecchi lustri"
141. Roughly *needed new blood*.
142. "Sebbene nella primavera promettesse abbondante il prodotto in vino per quelle viti che superarono incolumi i rigori eccezionali del verno, perchè erano fiorite di grappoli da far dimenticare la primavera precedente, il prodotto venne meno al declinare della vita vegetativa, si da toccare il quinto appena delle produzioni normali. E sì che l'Italia avea bisogno di rissanguarsi dai dani patiti pel 1879 e piena di speranza al sorgere del 1880 davasi al lavoro, onde riguadagnare il perduto! ma l'oidio, l'antracnosi, gl'insetti e più che altro le grandini e la peronospora viticola menarono tanto danno, da lasciare il viticoltore deluso nelle sue più liete speranze." Carpenè, "La viticoltura nel 1880," *Rassegna* 1880, 738.
143. Ibid.
144. *Today we walk with the light of science.* Carpenè, "E son dei secoli che si predica," *Rivista* 1878, 97.
145. Symptoms are taken from those listed online at the National Ataxia Foundation and the Mayo Clinic.
146. Carpenè, "La verità sui vini spumanti," *L'Enotecnico,* 37.
147. Mechanically, "lever." "Simple machine, made of a rigid bar that turns around a fulcrum [fixed point], and in which a power exerted on one end of the bar overcomes a resistance wielded at the opposite end," translated from Massimo Pivetti and Grazia Gabrielli, eds, *Grande Dizionario Italiano* di Aldo Gabrielli, 4th ed. (Milano: Ulrico Hoepli Editore, 2020).
148. See Tomasi and Gaiotti, *The Terroirs.*
149. The nature of these often membership-based or cooperative establishments (called also cantine sociali) had changed under Mussolini and then again after the war as their book waited to be published. For more on fascism and cantine sociali, see "Durante il ventennio fascista: declino o metamorsforsi della cooperazione? (1925–1944)," in *Il movimento cooperativo in Italia,* Fornasari and Zamagni. E.g., "Dopo i violenti attacchi al movimento cooperativo cattolico e socialista, culminati nello scioglimento della Confederazione delle cooperative italiane e della Lega nazionale delle cooperative, il ceto dirigente fascista avviò una profonda ristrutturazione del settore cooperativo, trasformandolo in un organismo di quel 'regime reazionario di massa' che secondo alcuni studiosi rappresentò la novità storica del fascismo, la cui relativa longevità si basò non solo sul ricorso ad istituzioni illiber-

ali e antidemocrathiche, ma anche sulla ricerca e l'organizzazione del consenso. Questa trasformazione avvenne tramite l'inserimento del movimento cooperative nelle strutture dello Stato corporativo fascista," 121. ("After violent attacks on the Catholic and socialist cooperative movement, which culminated in the dissolution of the Confederation of Italian cooperatives and the National League of Cooperatives, the fascist ruling class began a profound restructuring of the cooperative sector, transforming it into an organism of that 'reactionary regime based on masses' which according to some scholars represented the historical novelty of fascism, whose relative longevity was based not only on the recourse to illiberal and anti-democratic institutions, but also on the search for and organization of consensus. This transformation occurred through the inclusion of the cooperative movement in the structures of the fascist corporate state.")

CHAPTER 6

1. From *Material and Mind*, by Christopher Bardt. Copyright © 2019 MIT Press. Reprinted with permission of MIT Press.
2. "Prosecco," *OED*. https://www.oed.com.
2. *Baedeker's Southern Germany*, 277.

CHAPTER 7

1. From *Epistenology: Wine as Experience*, by Nicola Perullo. Copyright © 2020 Columbia University Press. Reprinted with permission of Columbia University Press.
2. Dalmasso and Dell'Olio, "Vini bianchi tipici," 31. They identified four subzones, adding Vittorio to the three that make up this book.
3. "The Song of the Piave," written in 1918.
4. See Glossary.
5. "Climateric" but in context this reads *climatic inclemencies*, or inclement weather; I'm assuming a misspelling. *Ampelografia italiana*, 93.
6. Once an official administrative unit.
7. "Il Prosecco era tempo addietro il principale vitigno del Coneglianese, non solo, ma di Valdobbiadene, Asola [sic] e Vittorio. In piano ha pochissima importanza essendo stato quasi completamente sostituito dal Verdiso e nel Coneglianese perde ogni di terreno scomparendo sotto il flagello dell'oidium, della peronospora e delle inclemenze climateriche. Ora il massimo di coltura lo si ha solo nel distretto di Valdobbiadene e nei comuni di Pieve di Soligo, Soligo, Solighetto, Farra, Follina, Colsammartino che costituiscono una serie di colli che partendo dalla sinistra del Piave si estendono da Valdobbiadene a Vittorio." "Prosecco Bianco," *Ampelografia Italiana*, 93.
8. Which is 106 million bottles. See, for example, Davide Bortone, "Prosecco Superiore Conegliano Valdobbiadene, 100 milioni di bottiglie: «raggiunto limite fisico»," *Wine Mag: Giornale italiano online, guida vine e news*, December 21, 2021. https://www.winemag.it.
9. Carpenè, *Sunto*, vol. 1, 5th ed., 118.
10. Berengo, *L'agricoltura veneta*, 235.

11. Ibid., 242–43.
12. Like the earlier *grass-terraced*, these are the UNESCO-cited ciglioni: "The hillsides arranged in embankments; another rare example of human endeavours aimed at rendering the steep slopes cultivable. In this hostile hilly terrain, over the course of centuries, humans shaped the slopes to be able to farm, by creating narrow terraces which, thanks to the solid soil types (marls and conglomerates) and to the grassing over of the terrain did not require supporting walls to be built." Tomasi and Gaiotti, *The Terroirs*, 153.
13. Or casei.
14. To be elaborated later in this chapter.
15. Franceso Drusian, interview on October 13, 2019.
16. Zoom interview with Innocente Nardi on June 4, 2020. Vittorio Veneto is a compound of place and ideas, made of Ceneda and Serravalle and then *Veneto* added after World War I.
17. "2019 Addendum, Evaluations of Nominations of Cultural and Mixed Properties," *ICOMOS report for the World Heritage Committee* 43 (2019), 17. http://whc.unesco.org.
18. For example, of the zone of San Michele di Feletto. See *Kriegskarte*, vol. 1.
19. "Volendo noi esaminare questa regione anche solo dal lato della coltivazione della vite e della produzione del vino, troviamo necessario il premettere alcuni cenni sulla natura del suolo e sulle condizioni climatologiche, quali fattori precipui di ogni produzione agraria." Taramelli, "Cenni sul suolo e sul clima della provincia di Treviso," *La vite*, 7.
20. Gasparini, "Disegnare le campagne . . . disegnare le vigne: note di storie dell'agrimensura," Gasparini and Rossi, *Disegnare*, 7.
21. "Ridente," is written: "which gives joy; happy; pleasant. Favorable, propitious," definition translated from *Grande dizionario italiano Hoepli.*
22. Carpenè, discussion reported in Antonio Caccianiga, chapter 16 of "La R. Scuola di Viticoltura ed enologia," *Feste e funerali*, 1897, 226–27. "Conegliano s'è meritamente guadagnato una fortuna accogliendo in suo seno il R. Istituto viticolo. Favorita oltre modo dalla natura per la sua ridente posizione, pel suo clima e terreno fecondi alla vite, devesi ritenerla destinata a farsi centro nel Veneto della produzione vinicola razionale."
23. "Francesco Maria . . . per la prima volta aveva citato nel 1772 il termine prosecco davanti agli accademici dell'agricoltura . . ." Lugato, *I nobili Malvolti*, 36.
24. Dalmasso and Dell'Olio, "Vini bianchi tipici," 20.
25. Istituto Nazionale di Statistica (ISTAT), "Population," *Serie Storiche,* http://seriestoriche.istat.it
26. *Accurately recognized* and a *position exactly determined.*
27. Not *communal.*
28. *Gathering place.*
29. These last two can also officially describe a "località," word used also more often and less officially with the English meaning of "locality."
30. Bringing, the story goes, cultivars and winemaking technology to the north. In the Treviso hills, south of the Piave anyway, the Atestina people, around 950 BC, were viti-vini precursors. Iris Fontanari Martinatti, *La vite e il vino nella farmacia di Plinio il Vecchio* (Spini di Gardolo: Edizioni Arca, 2001), 30. Google Books.

31. From 2021 Annual Report of Consorzio di Tutela Conegliano-Valdobbiadene DOCG, 92, plus Asolo DOCG's 60 percent export rate, per Asolo Consorzio media materials.
32. *Protection.* of the "quality and image" of this territory's wines, "the possibility [of this Prosecco] to make itself known beyond regional borders and abroad." Ferena Lenzi, "Nascita e sviluppo della Denominazione," *Là dove*, 21. To prevent wines being sold with the name of 'Prosecco dei colli di Conegliano e Valdobbiadene' but that have neither that origin nor those characteristics, writes Giuseppe Schiratti of the Consorzio's role: "Vigilare affinchè non siano posti in vendita con il nome di 'Prosecco dei colli di Conegliano e Valdobbiadene' vini che non ne abbiano l'origine nè le caratteristiche." Schiratti, *Invito*, 66.
33. Rossi, "Per una geografia storica," *Là dove*, 52.
34. *Territories* but add to the English word a sense of historical legitimacy. For more on the meanings, discussion, and possibilities of this word, see Gordon "What a Little Hilltop," *Gastronomica*.
35. Rossi, "Per una geografia storica," *Là dove*, 52-53.
36. Perhaps, speculates Fabio Giavedoni for Slow Wine, leading to Italy's use of 72 cl bottles once exporting arrived in the 1970s and demanded smaller formats: exactly 75 of them from each damigiana, easy enough to reverse number of bottles and number of centiliters as the 750 milliter bottles turned standard format. "Perché le bottiglie di vino più utilizzate misurano 0,75 litri di capacità?" https://www.slowfood.it, December 9, 2020.
37. Also: *sweet.*
38. Miro Graziotin, "Archeologia rurale e storia contadina," *Là dove*, 82.
39. In this case, *the town.*
40. Piovene, *Viaggio*, 39.
41. "Cittadine," he says. Interview on January 20, 2021.
42. *La viticoltura e la enologia nelle tre Venezie.*
43. Montanari and Ceccarelli, *La viticoltura*, 78–79.
44. Emigration, mostly relatively local and temporary.
45. "2019 Addendum, Evaluations of Nominations of Cultural and Mixed Properties," *ICOMOS report for the World Heritage Committee* 43 (2019), 19. http://whc.unesco.org.
46. Montanari and Ceccarelli, *La viticoltura*, 78–79.
47. Rossi, "Per una geografia storica," *Là dove*, 51.
48. Patriarca, *Numbers and Nationhood*, 34.
49. Danilo Gasparini, "I custodi del paesaggio di Conegliano Valdobbiadene," *Là dove*, 101.
50. From an email from Alessandro Casellato: "Le 'case rosse' erano le case di proprietà del Conte Brandolini che le affittava ai mezzadri insieme al podere che dovevano lavorare: erano tutte tinteggiate di una particolare tonalità di rosso detto 'rosso Brandolini'. In località vicine, i grandi proprietari avevano adottato diversi modi per rendere riconoscibili le loro case dei loro mezzadri: i Conti Collalto a Susegana e nei paesi limitrofi usavano una fascia rossa orizzontale sulle pareti bianche."

51. The Metodo Bottari tied together the cultivation of grapevine and this tree; see Gasparini, "Disegnare le campagne . . . disegnare le vigne: note di storie dell'agrimensura," Gasparini and Rossi, *Disegnare*, 22.
52. See "Bachicoltura: Tecniche e prospettive per la bachicoltura italiana," online seminar July 2020 Veneto Agricoltura, Regione del Veneto, https://www.veneto agricoltura.org.
53. A century later, preparing the nursery vine cutting would be, too, along again long wooden tables aligned and concentrated pilings. See photograph from the Curto nursery in Conegliano, in *Là dove*, 58.
54. The Italian report from the Vienna exposition in 1872 dedicated a section to the silk industry. From an introduction that wonders about the "history of the silk factory and the degree of development [such production] has reached in the main nations of Europe," to a look at the numbers, 12,000 looms in Italy, of which 2,500 "in the provinces of Veneto, central and southern Italy," while Lombardia and Piemonte were counted on their own. As for the workers, thanks to the gendered language, some women appear in force: "The number of workers directly employed in this industry can be considered not less than 20,000, which number must include, in addition to the weavers, all those involved in the subsidiary operations of weaving, such as the winders, the warpers, the spoliers, those who fold, . . . who sort the silk, etc." One-third of the almost-40-million-lire-industry was in dyeing and handiwork. Original: "Prima di esporre in qual modo l'arte della seta fosse rappresentata all'Esposizione di Vienna del 1873, noi crediamo opportuno di riassumere a grandi tratti la storia del setificio e il grado di svolgimento che esso ha raggiunto presso le nazioni principali d'Europa. . . . Ci crediamo di essere assai vicini al vero assegnando in complesso all'Italia 12,000 telaj applicati alla fabbricazione delle stoffe di seta nel 1872. —Di essi circa 7,000 si troverebbero nelle provincie Lombarde, 2,500 in quelle del Piemonte, 2,500 nelle province del Veneto, dell'Italia centrale, e meridionale—*Il numero degli operai direttamente occupati in questa industria può ritenersi non inferiore ai 20,000 nel qual numero vanno compresi, oltre i tessitori, tutti gli addetti alle operazioni sussidiarie della tessitura, quali sono le operaje incannatrici, le orditrici, le spoliere, coloro che piegano e rimettono le catene, che assortiscono la seta, ecc* [itals mine and correspond to the section translated above]. Il valore medio del prodotto di 12,000 telaj, ritenuto che lavorino senza interruzione, può aggirarsi fra i 35 e i 40 milioni di lire, di cui un terzo è rappresentato dalla tintura e dalla mano d'opera." Prof. Pietro Pinchetti, Francesco Mattiuzzi, Giovanni Battista Nessi, *Relazioni dei Giurati Italiani* Fasc. 3, 1–8.
55. See Elisa Bellato, "Viti maritate e altre storie al femminile," *Là dove*, 108–9.
56. Elisa Bellato, "Viti maritate e altre storie al femminile," *Là dove*, 109.
57. Orefice, *Altre Tempi*, 86–87.
58. "Un italiano si controassicura sempre, vive tra pesi e contrapesi, è rivoluzionario e conservatore ad un tempo, e questa è la sua legge." Piovene, *Viaggio*, 40.
59. "2019 Addendum, Evaluations of Nominations of Cultural and Mixed Properties," *ICOMOS report for the World Heritage Committee* 43 (2019), 21. http://whc.unesco .org.
60. Radiotelevisione italiana, a public entity with roots to 1923.
61. Vianello and Carpenè, *La vite*, 48.
62. Ibid., 55.

63. Ibid., 43.
64. Ibid., 61.
65. Ibid.
66. Danilo Gasparini, "La fondazione della scuola enologica di Conegliano: la formazione delle élites tecnico-enologiche nazionali," *Là dove*, 93. On *rinnovazione*: Difficult in its English equivalent, *renovation*, and instead here is a definition, translated—of: "obtaining a new condition, especially in a cultural and spiritual sense, with the implicit idea of an improvement and updating or, possibly, of partial or total replacement."
67. Vianello and Carpenè, *La vite*, 62.
68. Ibid., 55.
69. Danilo Gasparini, "Nascita del Distretto enologico," *Là dove*, 58.
70. Gasparini, "La viticoltura nel Nord Italia," 12.
71. See: Danilo Gasparini, "Nascita del Distretto enologico," *Là dove*.
72. Gasparini, "La viticoltura nel Nord Italia," 12.
73. "La Storia," Adami website, https://www.adamispumanti.it/azienda-valori/. Adopted from French usage, *cru* is used widely and globally today to indicate a vineyard of particularly high quality and the subsequent noteworthy and discernible aromatic and flavor characteristics of specific grape varieties grown there.
74. Marta from NaturalMenteGuide.
75. Hints toward nineteenth-century recastings in which the contadini are not seen as the reason for Veneto's agricultural problems, but the rural structure is, appear in Berengo, *L'agricoltura veneta*, 20.
76. Dino Coltro, "Appendice: Appunti per una storia sociale e linguistica del mondo contadino," in *Paese Perduto*, vol. 3, 415.
77. Taramelli, "Cenni sul suolo e sul clima della provincia di Treviso," *La vite*, 12.
78. "The new mathematics is a sort of supplement to language, affording a means of thought about form and quantity and a means of expression, more exact, compact, and ready than ordinary language. The great body of physical science, a great deal of the essential fact of financial science, and endless social and political problems are only accessible and only thinkable to those who have had a sound training in mathematical analysis, and the time may not be very remote when it will be understood that for complete initiation as an efficient citizen of one of the new great complex worldwide States that are now developing, it is as necessary to be able to compute, to think in averages and maxima and minima, as it is now to be able to read and write." H.G. Wells, *Mankind in the Making,* 162. eBook.
79. Patriarca, *Numbers and Nationhood,* 50.
80. Ibid., 42.
81. Ibid.
82. Ibid., 43.
83. Ibid.
84. From "Cesare Correnti a Luigi Bodio," *Scritti scelti di Cesare Correnti in parte inediti e rari*, in Patriarca, *Numbers and Nationhood*, 43.
85. An echo toward chemistry in Patriarca's writing that "statistics was a study of society based on naturalistic premises . . . method of observation," *Numbers and Nationhood*, 48.
86. Patriarca, *Numbers and Nationhood*, 72.

87. Giovanni Carraro, "A piedi nel 'Pentagono d'oro,'" *Là dove*, 194.
88. See Vianello and Carpenè, *La vite*, 67 and 99 for topography and number of vineyards, respectively.
89. Noticeable, too, among these already loose bunches were, uniquely among Prosecco/Glera biotypes, differently sized berries.
90. See Vianello and Carpenè, *La vite*, 82–99.
91. *Viticultural vocation.*
92. Diego Tomasi, "Il punto di vista agronomico," *Là dove*, 72.
93. See, for example, "Grapes 101," *Appellation Cornell* 26 (August 2016), College of Agriculture and Life Sciences, Cornell University, https://grapesandwine.cals.cornell.edu/.
94. Diego Tomasi, "Il punto di vista agronomico," *Là dove*, 72.
95. Or else, too high alcohol and an unbalanced wine.
96. For setting color, longevity, refreshingness, balance among other things. Acidity is often a wine's defining feature, even when not its most noticable one.
97. As opposed to the pertica di campagna (countryside pertica) which varied from place to place and in northern Italy was used to convey the area of an agraian surface of land (about 650 square meters), one pertica censuaria was equivalent to 1,000 square meters, or 10 ara. So about 232,532.2 hectares. See: "pertica da campagna," in "Glossario italiano," ed. Filippo Camerota, *Epistemic History of Architecture*, Bibliotheca Hertziana, Max Planck Institute for Art History, http://wissensgeschichte.biblhertz.it:8080/Glossario/Glossario_Italiano/pertica_da_campagna/edit_main.
98. Taramelli, "Cenni sul suolo e sul clima della provincia di Treviso," Vianello and Carpenè, *La vite*, 6-7. "Piano, colli e montagne abbracciano un'area di pertiche censuarie 2,325,322, delle quali 1/23 giace incolto. Vi scorrono utilmente impiegati 240 corsi d'acqua. Vi pascono circa 173,000 capi di bestiame. Una popolazione di 348,473 abitanti gode dei prodotti del suolo e delle proprie industrie."
99. "Scazzone," in Italian.
100. Taramelli, "Cenni sul suolo e sul clima della provincia di Treviso," *La vite*, 5.
101. "2019 Addendum, Evaluations of Nominations of Cultural and Mixed Properties," *ICOMOS report for the World Heritage Committee* 43 (2019), 20. http://whc.unesco.org.
102. Ibid., 21.
103. Faraon, *Il sasso nello stagno*, 12.
104. In equivalent engineering literature in English, this equation—relating internal pressure *P*, internal diameter *Di*, and thickness *t*—is generally called the Barlow formula. Outside of Italy, one engineer claims, the only "Mariotte formula" recognized is the one on compressed gases.
105. See Edme Mariotte, *Traité du mouvement des eaux et autres corps fluides* (Paris: Estienne Michallet, 1686). Google Books.
106. Cutting the grass between vines.
107. Stacking hay.
108. From a presentation by Cristina Munno, AISO, Ecoscuola di storia orale nel paesaggio del prosecco superiore, July 9, 2021.
109. Faraon, *Il sasso nello stagno*, 18.
110. *Là dove*, 274.

111. Ninni, *Materiali per un vocabolario*, 11.
112. Taramelli, "Cenni sul suolo e sul clima della provincia di Treviso," *La vite*, 10.
113. Where, reported Sannino, "a third Prosecco is found, and is called Prosecco lungo, from the shape of the berry sharply elongated, delicately aromatic, of a just-pinkish color. It is not very productive, because of a deformation of its flowers, which open in a star shape and with imperfect stamens, of a green color like the petals." Sannino, "Note ampelografiche," *Rivista* 1913, 126.
114. Including of course the latter's Prosecco Balbi biotype and whose work had been a priority for Carpenè as well. In a May 29, 1871 report on the Società's work presented to its members, write De Leo and Favero, Carpenè described the technical reasons he had decided to not include some of their wines in the Esposizione Fiera Vini di Firenze, at the same time noting that he had decided to invest on the large-scale growing of Prosecco, following the count's success with the grape. "Nella relazione sul proprio operato presentata ai soci il 29 maggio 1871, Carpenè illustrava le ragioni tecniche che lo avevano condotto a escludere alcuni dei vini prodotti dalla Società dalla presentazione alla fiera, segnalando in aggiunta la decisione di investire sulla coltivazione su larga scala del Prosecco, sull'esempio di quanto fatto con successo dal consigliere e re-visore dei conti della Società, il nobile Balbi Valier, nelle sue proprietà. Quest'ultimo, infatti, ebbe il merito di individuare negli anni Sessanta una varietà pregiata di Prosecco, il cosiddetto «Prosecco tondo», in seguito rinominato «Prosecco Balbi»." De Leo and Favero, *Ca' Foscari e Carpenè Malvolti*, 56–57.
115. *Là dove*, 202–3.
116. Carpenè, *Rivista* 1877, 213. In occasion of Paris's 1878 world exposition.
117. "I vigneti," Adami website, https://www.adamispumanti.it.
118. "2019 Addendum, Evaluations of Nominations of Cultural and Mixed Properties," *ICOMOS report for the World Heritage Committee* 43 (2019), 14. http://whc.unesco.org.
119. Born in Pieve di Soligo, Zanzotto is a voice of this zone, His "L'Aria di Dolle," in his final collections of poems, famously pays homage to Rolle as a "una cartolina inviata dagli dèi," *a postcard sent from the gods*.
120. Dalmasso and Dell'Olio, "Vini bianchi tipici," 25.
121. Tomasi and Gaiotti, *The Terroirs*, 153.
122. Taramelli, "Cenni sul suolo e sul clima della provincia di Treviso," *La vite*, 9.
123. See chapter 9.
124. "In fact, only the vineyards of Bardolino, Valpolicella and Soave, along with small and partial stretches of the Trevisian hills, gave a 'base product,' which resisted bottling and could flow into exportations toward distant markets." Berengo, *L'agricoltura veneta*, 302. "Infatti, solo le vigne del Bardolino, del Valpolicella e del Soave, assieme a piccoli e parziali tratti della collina trevisana, davano un 'prodotto di fondo', che resistesse all'imbottigliamento e potesse sfociare nell'esportazione verso mercati lontani."
125. Shared with apologies for the source, as while authors Ada and Remo Dolce's book serves as a rare recording of the contadini of the Treviso province it is not a respectful one. This is also included as a thank-you to AISO, who gave each of us participants a reprinted copy as both info source and occasion to protect. I hope I've done so here. And, it serves as the occasion to include oral historian Daniela

Perco's words on giving back. The book was reprinted in 1992, with a postscript from poet Luciano Cecchinel.

126. Corn, wheat, silkworm cocoons. Vianello and Carpenè, *La vite*, 76.
127. See "Come allevare la vite a Sylvoz," *Coltivazione biologica*, https://www.coltivazionebiologica.it/.
128. Comizio agrario di Conegliano, August 31,1885. The notice signed by Cerletti notes that their experiments with this also yielded good results with the application of Latte di Calce (2 to 3 percent calce spenta in solution). The document is displayed in a school hallway.
129. "Le Colline del Prosecco di Conegliano e Valdobbiadene," *Commissione nazionale italiana dell'Unesco*, September 2, 2019, http://www.unesco.it.
130. Ibid.
131. In this zone, "Global climate change has increased the incidence of heavy rain events, surface erosion, landslides and heavy runoff. The State Party acknowledges the need for improved drainage infrastructure to respond to these pressures. It has provided a summary of mitigation interventions undertaken over the past two decades. These include inter-row grassing, avoiding rittochino systems in the steepest areas (drainage ditches dug between the rows of vines), maximum row lengths and terrace widths, conservation of hedges, and various works to establish ditches and drainage." "2019 Addendum, Evaluations of Nominations of Cultural and Mixed Properties," *ICOMOS report for the World Heritage Committee* 43 (2019), 15. http://whc.unesco.org.
132. Anonymous. From unreleased original document interview of AISO Scuola di storia orale nel paesaggio del prosecco superiore, July 2020.
133. Briefly: "I've been here for more than twenty years, and I never get tired of it. The hill, 360 degrees, I look here, I look there, it's like the sea, if you go to the sea you sit, you look at the waves, you look at the fire in the evening, you stay there for hours, That is what this panorama is like."
134. A common phrase in wine tasting and controversial for some wine scientists. E.g., "Earlier this century, rather like a new comet appearing in the night sky, the phenomenon of minerality arrived. Suddenly wines were described as tasting 'mineral,' and apparently it has now become the most widely used taste descriptor. . . . We have just seen that vines don't absorb geological minerals, let alone rocks." Maltman, *Vineyards*, 176–77. See also Mark A. Matthews, *Terroir and Other Myths of Winegrowing* (Oakland: University of California Press, 2017).
135. "Io è vent'anni e più che son qua: ho ancora da stancarmi, ho ancora da stancarmi. Perché la collina, qui, a 360 gradi . . . guardo di qua, guardo di là . . . è come il mare, io sono appassionato di mare, se andate al mare: ti siedi su uno scoglio, guardi le onde . . . guardi il caminetto, alla sera, seduto: non ti stanchi mai, stai lì ore. Così è questo panorama." Allora, "guardando questa uva . . . nei vigneti del nonno viti di tutti i tipi, uva dolce, malvasia, varietà scomparse (ma ora quasi tutte recuperate), vari cloni del prosecco. [Rolle] è vocato, A causa del sottosuolo argilloso, che provoca le frane ma allo stesso tempo conferisce al vino una mineralità molto apprezzata."
136. Montanari and Ceccarelli, *La viticoltura*, 78–79.
137. See Vianello and Carpenè, *La vite*, 37, 52–53.
138. Encompassing twelve comuni, including Monfumo, Castelcucco, Maser, and Asolo.

139. On March 19, 1915, a notice from Director M. Giunti announcing the year's forty-five-day course in silkworm and mulberry cultivation; each of eight students to be paid 2.5 lire a day plus travel expenses. Spriveri and Coradazzi, *Anniversario 130.*
140. See, e.g., Danilo Gasparini, "Bianche greggi tra verdi filari," *Là dove*, 218–21 for also a look at alongside wine and silk, the cashmere and wool production here.
141. Vianello and Carpenè, *La vite*, 69.
142. *Arms.*
143. Vianello and Carpenè, *La vite*, 40.
144. Oleaceae tree (Fraxinus ornus).
145. Maple.
146. See Vianello and Carpenè, *La vite*, 104–5.
147. In conversation with Gasparini, July 12, 2021.
148. "Periti," "agrimensori," in Rossi, "Per una geografia storica," *Là dove.*
149. Rossi, "Per una geografia storica," *Là dove*, 46.
150. See Gasparini, "Disegnare le campagne . . . disegnare le vigne: note di storie dell'agrimensura," Gasparini and Rossi, *Disegnare*, 5–17.
151. Roughly and in design and scope that matched the mentioned instruments, *mapped.*
152. Rossi, "Per una geografia storica," *Là dove*, 47.
153. "2019 Addendum, Evaluations of Nominations of Cultural and Mixed Properties," *ICOMOS report for the World Heritage Committee* 43 (2019), 14. http://whc.unesco.org.
154. Vianello and Carpenè, *La vite*, 38.
155. Ibid., 100.
156. By quantities instead they tell of a smoothing and of other categories: "A large landowner has a greater number of both or each; a medium-sized one a smaller number, the small landowner generally only one. Large and small owners differ only by how many of these farms make up their property and by nothing else." "Il grande proprietario possiede un numero maggiore di masserie e chiusure; il medio ne possede un numero minore, ed il piccolo si restringe spesso ad una sola masseria ed anche ad una sola chiusura. Il grande ed il piccolo proprietario sono diversi per il numero di masserie e di chiusure che costituiscono la loro proprietà e non per altro," Vianello and Carpenè, *La vite*, 39.
157. Ibid., 101.
158. See Miro Graziotin, "Parlano. Queste pietre: Casèi, casère, sass e piere e *umanità*," Gasparini and Rossi, *Disegnare*, 52.
159. See Vianello and Carpenè, *La vite*, 125.
160. The producer is Fato 1784.
161. Mulch.
162. *Di là dei carpini e le roveri spauriti / del sasso sdrucciolato sulla lastra levigata / in un'ombra umida di fiati sfiniti / come qualcuno che dorme e che ansima // di là dell'ultimo chiarore della strada da strascino / del sasso sdrucciolato sulla lastra levigata / come singhiozzo di strame fradicio fin nel midollo / c'è un verso arrochito che va e viene e ansima // un'anima che vuol restar viva / dietro il sasso sdrucciolato sulla lastra levigata / dove nessuno più ormai arriva / se non è da tanto che dorme e ansima // per singhiozzo di strame fradicio fin nel midollo / di là dei carpini e le roveri spauriti / di là dell'ultimo chiarore della strada da strascino / in un'ombra umida di fiati sfiniti.* Original

and this translation into Italian from Luciano Cecchinel, "de là del sanjut sofegà del stran" ("di là del singhiozzo soffocato dello strame"), *Sanjut de stran*, 40.

163. From an email with Cristina Munno, at the time of my visit in July 2021 the Assessora alla cultura of Cison del Valmarino.
164. From a woods walk with Cison del Valmarino, then-Assessora alla cultura Cristina Munno and NaturalMenteGuide guide Francesca.
165. Taramelli, "Cenni sul suolo e sul clima della provincia di Treviso," *La vite*, 11.
166. "Occurring in crystalline, amorphous, and impure forms (as in quartz, opal, and sand respectively)," "Silicone dioxide," *OED*, https://www.merriam-webster.com.
167. *Al tràgol jért: 1972–1984* (Pederobba: I.S.C.O., 1988).
168. During my visit to Tóvena with AISO, Ecoscuola di storia orale nel paesaggio del prosecco superiore, July 2021.
169. "Queste . . . malattie c'erano già ai tempi di mio nonno, poi malattie come la flavescenza che è proprio questo insetto, ai tempi di mio nonno non c'era. È uscita dopo e sì, questa malattia è dovuta anche all'abbandono dei boschi che ci sono tanti boschi che sono abbandonati. Una volta entravano i vecchi . . . , con la questione che ognuno aveva il suo bestiame e la sua stalletta, si pulivano anche i boschi, si facevano il fieno, adesso . . . è stato un po' abbandonato questo discorso qua e . . . sui boschi vecchi il legno ha tanti animali che sono infetti per le viti. . . . Una volta [i boschi] venivano più ripuliti, sistemati, sempre curati, adesso nessuno entra più, anche perché non c'è guadagno, sì, non vale la pena proprio, sì, ognuno magari si fa le sue legne per l'inverno però non è proprio come una volta che andavano, mio nonno andava anche a instradare le foglie che cadevano degli alberi per metterle dentro la stalla, per tenere pulito." AISO, Scuola di storia orale nel paesaggio del prosecco superiore, July 2020.
170. This second of his three books (the third still in production) is based on a museum he keeps of enological Italy, a collection he began with special attention to the hills: Gregoletto started the collection in 1989, covering 210 years that stop at 2010. He is not interested in what happens after that for the museum, he told me during a visit; it is not what he recognizes as Prosecco.
171. "Strumenti grandinifughi," Gregoletto, *Vite ambulante*, 44–45.
172. Pietro Laverda, "Quando in campagna si sparava alle nuvole," *Album di storia Laverda: Macchine, storie e personaggi di una grande industria italiana*, edited by Piergiorgio Laverda, *LaverdaWorld, il Giornale dello specialista del raccolto* (Breganza, Laverda Spa. 2002–2009. http://www.laverdastoria.com. Pietro Laverda, maker of Italian anti-hail cannons began production in 1898, patented one in 1901; see Centro di cultura e civiltà contadina, Biblioteca internazionale "La Vigna," https://www.lavigna.it.
173. From a conversation in Tóvena, during the Ecoscuola di storia orale nel paesaggio del prosecco superiore, July 2021.
174. "Da vari anni Livio Vanzetto aveva identificato nel Checchini l'inesauribile scrittore dialettale di «Vita del Popolo» e dell' «Ancora», degli Almanacchi e dei Lunari, autore di dialoghi, apologhi, raccontini e romanzi, tra religiosi e sociali" ("Years ago, Livio Vanzetto identified Checchini as the indefatigable dialect writer of *Vita del Popolo* and of *Ancora*, of Almanacs and Lunari ['booklet containing all the days of the year, the lunar phases, the civic and religious holidays, and the weather fore-

casts,' translated from *Grande dizionario italiano Hoepli*], author of [works] between religious and social." Isnenghi, "Introduction," *Paron Stefano Massarioto*, 9.

175. "No vedì che i contadini serve da fondamente par fabricar poateri e vualtri andè a far i muri e el querto? Vogia no vogia; se sti siori no se metarà a megiorare le condisioni del contadin col giutarlo, coragirlo, rispetarlo, e sovergliarlo e farghe cognassar che l'arte del contadin xè la pi nobile parchè tuto vien for dale man del contadin, in pochi ani i sarà anca lori materiali dela fabrica deo poareti!" From Checchini's writing in *Vita del Popolo,* as covered in Gaspari, *Grande guerra e ribellione contadina*.
176. Isnenghi, "Introduction," *Paron Stefano Massarioto*, 7.
177. Ibid., 9.
178. Ibid., 8-11.
179. "L'amicizia con Massarioto, contadino alfiere della morale," *Corriere degli Alpi*, Belluno (August 20, 2014), https://corrierealpi.gelocal.it.
180. "Qui l'eccezione è che la vita e la recita fanno tutt'uno, perché l'autore e il personaggio sono quello che dicono di essere. Si creano, per il pubblico, possibilità di adesione e immedesimazione assolutamente inconsuete. Per una volta, la parola scritta, la cultura, non rimangono esterne, non si manifestano come cosa d'altri." Isnenghi, *Paron Stefano Massarioto*, 10.
181. "Per una volta, la cultura orale si fa cultura scritta." Ibid., 9.
182. Isnenghi, "Introduction," *Paron Stefano Massarioto*, 9.
183. Of the four libraries with a copy, only the national library in Rome would lend it to me, through the New York Public Library.
184. A plot, he means, of only vines, trained on pali secchi, stakes of wood and later metal or concrete.
185. Vianello and Carpenè, *La vite,* 76-77.
186. See Tomasi and Gaiotti, *The Terroirs.*
187. *Là dove*, 274-75.
188. Agricultural field delimited by forest or a perimeter of trees.
189. See Introduction and Glossary.
190. This is a reworking of Taramelli's words, but the materials have not been moved. See Taramelli, "Cenni sul suolo e sul clima della provincia di Treviso," *La vite.*
191. Taramelli, "Cenni sul suolo e sul clima della provincia di Treviso," *La vite,* 13.
192. *Agricultural outfits.*
193. Fondazione Nord-Est, "Storie e uomini di impressa," *Là dove*, 236.
194. See Fondazione Nord-Est, "Storie e uomini di impressa," *Là dove*, 236-37.
195. Miro Graziotin, "I custodi del paesaggio," *Là dove*, 100.
196. The word he uses is *metodicamente*; methodically.
197. "Morigerata," he writes of the far-reaching moral fix: respectful of the limits imposed by decency or honesty, free from any excess and recklessness; moderate, measured, sober.
198. *Study so that you might know yourself.*
199. Carpenè's son Etile, who took over his father's winery and work. "My father Adriano's stories," Adami wrote in an email, "refered to a 'Signor Carpenè,' not identified by name, but as a close friend of Abele Adami." By timing and other identifiers, this must have been Etile.

CHAPTER 8

1. "Un'ora con Bepi Mazzotti," *Incontri*, Rai.
2. Over film footage of the well-known historical mill in Refrontolo, the narrator calls out, "Il mulino della Croda. Autentico. Ma anche questa è una invenzione di Mazzotti, ci siamo capiti." ("The mill of the Croda. Authentic. But this, too, is an invention of Mazzoti, let's be clear.") "Un'ora con Bepi Mazzotti," *Incontri*, Rai.
3. Tens of thousands, which are the foundation of the 120,000-plus images of the Fondo Fotografico Giuseppe Mazzotti. Foto archivio storico trevigiano (FAST), http://fast.provincia.treviso.it.
4. "Dry."
5. Montanari and Ceccarelli, *La viticoltura*, 371.
6. Noted on the cover of De Rosa, *Tecnica*.
7. He refers to the Latin phrase, Semel in anno licet insanire: *once a year it is acceptable to go crazy.* De Rosa, *Guida*, 138.
8. "Parlando per un momento del come vada tolto un tappo di sughero da una bottiglia di spumante (unico tipo di stappatura che viene eseguito alla presenza degli ospiti), non valido quindi per tutti gli altri tipi di tappi, sottolineiamo come sia da considerarsi gesto volgare ed insultante per lo spumante e per chi lo ha amorevolmente e sagacemente prodotto, lo sparo del tappo. Lo spumante non è un oggetto da Luna Park; è invece quanto di meglio la sapienza enologica sa preparare. Sparare il tappo può essere concesso 'semel in anno' solo nel brevissimo spazio che segna la nascita del nuovo anno. È una consuetudine, e lasciamola così. Poi, mai più. Il tappo va sbloccato a mano o con l'apposita pinza. La bottiglia va tenuta inclinata contro il petto con la destra. La sinistra afferra saldamente il tappo sbloccato e lo trattiene mentre la destra ruota lentamente e in continuazione la bottiglia stessa. Quando vi è la netta sensazione che il tappo salterebbe, allora invece lo si trattiene e lo si lascia fuoriuscire molto lentamente sino a che un soffio (ho detto 'soffio,' non 'sparo') ci segnala la buona conclusione dell'operazione." De Rosa, *Degustazione*, 138.
9. The *shot* is however helpful for illustrating the play of CO_2 in spumante, De Rosa had written earlier in *Tecnica*. "In this moment, the rite of the 'shooting' of the cork consummated, the stream of wine that immediately bursts forth raises a lively wave of foam, of a volume often several times higher than the quantity of liquid actually poured, foam that after a few seconds disperses, shrinking at last into a ring touching the walls of the glass and, in well-made products, giving rise to a 'perlage' that persists for a variable time according to the conditions that we will examine shortly." ("In questo momento, consumato il rito dello "sparo" del tappo, il fiotto di vino che subito prorompe solleva una vivace ondata di spuma, di un volume spesso superiore di parecchie volte al quantitativo di liquido effettivamente versato, spuma che dopo alcuni secondi si disperde, riducendosi infine a forma di anello a contatto con le pareti del bicchiere e, nei prodotti ben preparati, dando luogo ad un 'perlage' persistente per un tempo variabile a seconda delle condizioni che più avanti vedremo.")
10. "Adhesion in an extremely thin layer of molecules (as of gases, solutes, or liquids) to the surfaces of solid bodies or liquids with which they are in contact," Merriam-Webster, https://wwwmerriam-webster.com.

11. De Rosa, *Tecnica*, 9.
12. "Più gentile," Montanari and Ceccarelli, *La viticoltura*, 459 (from *Tiere furlane* 30, 63).
13. Montanari and Ceccarelli, *La viticoltura*, 459 (from *Tiere furlane* 30, 63).
14. The same year, remember, that the mezzadria was ended by law.
15. "The Carpenè family in these past decades applied themselves to perfecting the tank method [setting not only the Prosecco method, but in the context of the nation and its wines one of the prime variations of the Italian method] inserting a series of ingenous and radical refinements, and so rendering definitively modern and rational the system of rapid spumantizzazione in autoclave. Their influence in the modern system is substantial also due to their introducing into the cycle (by Antonio Carpenè in 1936 and then as sensational innovation) of pasteurization and controlled temperature of the spumante once it was bottled," De Rosa, *Tecnica*, 449.
16. De Rosa, *Tecnica*, 478.
17. Floor plan, De Rosa, *Tecnica*, 214.
18. As seen at producers Rugè in Valdobbiadene, July 2021.
19. De Rosa, *Tecnica*, 9.
20. De Rosa, *Tecnica*, 450.
21. "Quadro" so also four-sided frame, spumante framework being acidity, sugar, primary aromas, bubbles.
22. De Rosa, *Tecnica*, 215–16.
23. See Glossary.
24. De Rosa, *Tecnica*, 450.
25. Ibid.
26. "That pressure, in the field of spumante wines, is obviously always a positive number and is based on some atmospheric unit," De Rosa, *Tecnica*, 220.
27. Two types of stainless steel were then used in "this specific field," De Rosa notes. "The 18/8 (A.I.S.I. 304), that is an iron alloy with 18% chromium and 8% nickel, commonly used, and 18/8/2 (A.I.S.I. 316), with the addition of 2% molybdenum [a trace mineral found in milk, cereals, legumes, nuts, leafy vegetables, and organ meat], less commonly used, a little more expensive than the former but also more resistant to wine acids (it is in fact preferred for production of juices and citrus derivatives)." *Tecnica*, 221.
28. The 300 series; for example, Aliberti makes theirs of AISI 304L and 316L, https://www.sirioaliberti.it.
29. While Prosecco Superiore is meant to preserve primary, or grape, flavors and aromas, and thought of (and so encouraged) for drinking right away, these wines' high acidity levels and complexities mean that some of them will age for years, and even decades, changing into what is called tertiary flavors, those of a well-aged wine, which when white means nutty, honeyed, beeswax, spices depending on the variety. Two producers known for their ageable, and aged, examples are Ruggeri and Nino Franco (who in 2019 celebrated a century anniversary by opening bottles made across that time span).
30. See Glossary for official ranges of sugar and therefore typology type.
31. De Rosa, *Tecnica*, 8.

32. *Esposizione Universale del 1878,* 291, 296, "Le uve bianche sono coltivato nelle colline del trivigiano. Le viti sono maritate cogli alberi ed in coltura mista." First read in *Tiere furlane* 30, 63.
33. *Riduzione*—he writes in "Più vino e meno pellagra," *Rivista* 1880—or a modification that is appropriate or convenient, a simplification (*Scuola*, 28-30). There are differences in particulars and details, though not in substance, from the article pages I have from a bound edition of the 1880 *Rivista* editions. "Producendo assai più vino e assai meno granoturco, si darebbe un gran colpo alla miseria," Carpenè writes on pages 678-79 in the latter: "Producing *much more* wine and *much less* [emphasis mine] granoturco, would deal a great blow to the hills' misery." Details on the problems to be fought against abound: "Nel Veneto il contadino non beve vino metodicamente come ben pochi operai [sic] anche delle città, ma sentendo il bisogno di corroborare le sue forze, l'operaio si dà all'acquavite e ad altri liquori e l'alcoolismo [sic] si fa strada a grave danno della salute pubblica. . . . La pellagra, la stupidità, l'abbrutimento morale, sono favoriti d'assai dall'abuso dei liquori che è conseguenza della mancanza dell'uso metodico del buon vino," 678. "Anche l'enologo per migliorare deve lottare con le abitudini ed i pregiudizi dei consumatori," 680. ("In the Veneto, the contadino doesn't drink wine regularly, as do also only very few workers in the cities, but feeling the need to gather his strength, the worker turns to liquors and alcoholism advances with great damage to public health. Pellagra, stupidity, loss of morality, are very much favored by the abuse of liquors which is the consequence of the lack of regular use of good wine. . . . The enologist, too, in order to improve must battle with the habits and prejudices of consumers.") Note those operaio and contadino lines crossed, also early in the article in an overview of how Carpenè sees at that time the roles and problems in the campo, without scientists: "Pel contadino infine le abitudini sono una seconda natura e lo rendono cieco anche ai progressi della viticoltura. Ha sempre veduto le viti maritate agli alberi e non sa persuadersi della maggior convenienza d'assegnare per esse esclusivamente una porzione di terreno di educarle in modo più razionale. Se il proprietario impone al contadino l'impianto di un vigneto, a grande stento si rassegna ad obbedirlo ma trascura poscia talmente le viti che in seguito intisichiscono ed offrono al padrone lo spettacolo dell'insuccesso. D'ordinario il proprietario non gode prestigio verso il contadino perché quasi come questo e talvolta anche più, ignora l'arte dei campi, cioè non possiede la scienza agronomica e la pratica agraria necessarie per imporsi all'operajo dei campi, il quale quantunque pieno di pregiudizi non ignora l'ignoranza del padrone e spesso se ne ride di lui. Certamente vi sono dei proprietari ohe conoscono il fatto loro, stando in giornata coi progressi dell' agricoltura, ma questi sono in troppo scarso numero. Si va predicando l'istruzione fra i contadini e questa sempre un tesoro, ma urge molto più che venga impartita al proprietario, che per ben dirigere l'operaio nei campi, fa d'uopo sappia ispirargli coi fatti quella fiducia, cui non può pretendere senza essere bene istruito. Così oggidì l'agricoltura è in mano dei contadini," 674-675. ("Finally, for the contadino, habits are second nature and they blind him to the progress of viticulture. He has always seen vite maritate and cannot persuade himself of the greater convenience of assigning a section of land exclusively to vines and to train them in a way more razionale. If the owner directs the contadino to plant a vineyard, the latter with difficulty resigns himself to obeying him but then neglects the

vines to such an extent that they later dry out and offer the owner the spectacle of failure. Ordinarily the owner does not enjoy prestige in the eyes of the contadino because often like him and sometime even more so, the former ignores the art of the campo, that is he does not have the agronomic science and the agrarian practical training necessary to dictate methods to the worker of the campo, who already filled with prejudice does not ignore the proprietor's ignorance and often laughs at him. Certainly there are some owners who know their business, staying up to date with agricultural progress, but these are too few in number. Education among the peasants is being preached and this is always a treasure, but it is much more urgent that it be imparted to the owner, who, in order to well-manage the worker in the campi, must know how to inspire through trust, which he cannot expect, without being well educated. So today agriculture is in the hands of the contadini.") Abroad and drinkers elsewhere outline another organizing and productions story: "All'esportazione gli agricoltori non possono pensare, a questa provvedano le *Cantine sociali* ed appositi *Stabilimenti enotecnici*, che acquistando uve ed anche vini, li renderanno atti ai trasporti e forniti dei caratteri speciali che i forestieri esigono," 680. ("The agricoltori cannot worry about exporting, this can be handled by the Cantine sociali and Stabilimenti enotecnici, who buying grapes and also wines, will make them suitable for transport and fitted with the special characteristics that foreigners request.")

34. See Campbell, *The Botanist*.
35. Campbell, *The Botanist*, 172, 184.
36. 1851, Giulio Sandri, as covered by Agosteo, *La rivoluzione di Bacco*, 64.
37. This is colloquially what varieties long grown in an area are called; it is not an indication of era or a superlative.
38. Again, *amarognolo*. See Introduction.
39. In his poem, "A Venezia."
40. See, for example, Nicoletta Speltra, "Carnia, dove I folletti amano il prosciutto," *La Stampa*, March 4, 2010, https://www.lastampa.it.
41. "L'andar per."
42. Veronelli, *I vini d'Italia*, 173.
43. Schiratti, *Invito*, 11.
44. Schiratti, *Invito*, 56. There is in addition a quote which appears in a national Italian business publication, but for which I was not able to find a source. As it both reflects Schiratti's living through the 1917 seige (see, Michael Travan, "L'anno della fame a Pieve di Soligo e il diario del farmacista Giuseppe Schiratti nella tesi di Nicolas Donadel," *Quotidiano del Piave*, October 16, 2018, https://www.qdpnews.it) and uses his characteristic uppercase lettering for key words, I am including it in this footnote: "Turista, nel mentre ti accingi a visitare questa zona meravigliosa di bellezze e solerti iniziative, di fiorenti industrie, di ricche realizzazioni in tutti i campi del lavoro umano, ti prego, per un solo momento, di dedicare un pensiero al lontano 1918, Allora, la Grande Guerra, lungo la STRADA, aveva seminato morte e distruzione! Ma la laboriosità innata della usa gente, ha fatto che sia tutto risorto, che il male sia cancellato, che sia ritornata la vita, là dove non vi era che rovina." Gianni Soleni, "Prosecco: da 'Strada del vino bianco' a patrimonio UNESCO!" *Dirigenti nordest* 20, no. 6 (June 6, 2019), Federmanager Venezia.

45. Every March, to Col San Martino came the Mostra Mercato dei vini dei colli, in botti, botticelli, fiascetti. See, Schiratti, *Invito*, 34.
46. Founder of Venice's Harry's Bar and the Bellini cocktail, which he created from Prosecco and the local white peaches.
47. From an interview with Wilma and Walter on 13 July 2021.
48. Importantly, he uses *assimilare*: *assimilate, absorb, digest, imbibe.* "Un'ora con Bepi Mazzotti," *Incontri*, Rai.
49. A requirement for those osterie wishing to be considered for the coveted *Bottega del Vino* designation and sign for indicating to those along the Strada. And furthermore "must, by statute, be able to offer at least two of the following white wines and all at the suitable tasting temperature: Bianco dei Colli secco or amabile, Prosecco secco or amabile, Cartizze secco or amabile, Prosecco frizzante or spumante, Cartizze frizzante or spumante." Schiratti, *Invito alla Strada del Vino Bianco Conegliano Valdobbiadene* (1968), as included in Alessio Berna, Stefano Scaggiante, Luigino Barisan, "La Strada del Prosecco e Vini dei Colli Conegliano Valdobbiadene: la più antica Strada del Vino italiana," *RNN Magazine: La rivista della Rete Rurale Nazionale* 2 (December 2011), 63. http://www.pianetapsr.it/magazine.
50. "Si beve—Cartizze, che è come uno sciampagne rustico e cordiale." "Un'ora con Bepi Mazzotti," *Incontri*, Rai.
51. In this ideal location filled with sun and cooling winds, consistently ripening grapes mean fuller-bodied wines from both the resulting glycerol and more grams of sugar per liter of both juice (roughly, thought of as *potential alcohol* by winemakers) and wine, all in balance with the higher acidity from those cool nights. Vines are replaced here only when they die, and then from a cutting from an old, decades and sometime more than a century, vine that has been adapting to these specific conditions. See Glossary for Spumante typologies.
52. "Norme per la tutela delle denominazioni di origine dei mosti e dei vini," Decreto del presidente della Repubblica, July 12, 1963, no. 930. Supplement to *Gazzetta Ufficiale della Repubblica Italiana*, no. 188, July 15, 1963, https://www.gazzettaufficiale.it.
53. "Per denominazioni di origine dei vini s'intendono i nomi geografici e le qualificazioni geografiche delle corrispondenti zone di produzione—accompagnati o non con nomi di vitigni o altre indicazioni—usati per designare i vini che ne sono originari e le cui caratteristiche dipendono essenzialmente dai vitigni e dalle condizioni naturali di ambiente." "Norme per la tutela delle denominazioni di origine dei mosti e dei vini," Decreto del presidente della Repubblica, July 12, 1963, no. 930. Supplement to *Gazzetta Ufficiale della Repubblica Italiana*, no. 188, July 15, 1963, https://www.gazzettaufficiale.it.
54. See Glossary.
55. This designation reserved for the most historical and quality-defined wines carries greater production and site restrictions, too. The two DOC and DOCG designations were both created in that 1963 legislature, but the latter was used for the first time in 1980, for two tuscan wine places.
56. Veronelli, *I vini d'Italia*, 185.
57. *Anfore de Falerno o de vin greco—no le val un bicer de bon Proseco.* Veronelli, *I vini d'Italia*, 185. Falerno is mythical, the greatest of the Romans' wines; Greece's wines were legendary in quality, too, evidenced by among other things the many grapes

by the name of *malvasia* throughout Italy—a Greek name signaling, like Prosecco, *quality*.

58. Outside a meal, or fuori tavola, table set with food and people.
59. Veronelli, *I vini d'Italia*, 186.
60. Same I'd say, for the wide non-consideration of what they do, for dictionaries.
61. "Si prende come punto di partenza per la descrizione dei confini la località Fornace (q. 175) a tre chilometri circa da Valdobbiadene verso ovest, dove il confine amministrativo tra i comuni di Valdobbiadene e Segusino incontra la strada Valdobbiadene-Segusino. Segue quindi il confine amministrativo tra questi comuni fino a Col Antich, dove incontra la curva di livello di quota 500, che segue fino a Ca' Pardolin, nei pressi di Combai, da qui lascia la quota 500 e prosegue sul sentiero, che porta fino alla piazza del paese attraverso prima via Cimavilla e quindi per via Trieste. Quivi, seguendo la strada che porta alla chiesa, raggiunge la casera Duel, poi, percorrendo il crinale della collina, attraversa la strada Miane-Campea, risale per monte Tenade e, sempre seguendo il crinale del colle, raggiunge località Tre Ponti sulla strada Follina- Pieve di Soligo. Attraversata la strada, il confine risale sulla collina Croda di Zuel e percorrendo il crinale passa a monte della chiesetta di S. Lucia a q. 356 a monte di «Zuel di là», ed a monte di Resera; il confine segue quindi la strada Resera-Tarzo fino all'inserimento con la Revine-Tarzo. Dal suddetto bivio il confine, sempre seguendo tale strada, raggiunge Tarzo e quindi Corbanese fino, all'incrocio con la strada Refrontolo-Cozzuolo, in località Ponte Maset, segue quindi il confine tra il comune di Tarzo e di Vittorio Veneto fino a raggiungere la strada vicinale detta «dei Piai»." See Disciplinare di produzione della denominazione di origine controllata e garantita dei vini "Conegliano Valdobbiadene—Prosecco."
62. "In particolare tale zona è così delimitata," *Gazzetta Ufficiale della Repubblica Italiana* 160, no. 185 (Roma, August, 8 2019). ("In particular, such zone is delimated like this").
63. This opens the official delimitation in Article 3 of the Conegliano Valdobbiadene DOCG disciplinare.
64. Official località.
65. Residual sugar.
66. "Percorrendo la strada maestra Valdobbiadene-Segusino, tocca di nuovo la località Fornace chiudendo così il perimetro della zona delimitata," ends Article 3 of the Conegliano Valdobbiadene DOCG disciplinare.
67. *Soft sweetness*, and descriptor of amabile wines.
68. See De Rosa, *Andar per vino*.
69. An allusion to a string or succession of pearls, to describe the quality of streams of bubbles in a glass of sparkling wine.
70. "Flauto," De Rosa, *Degustazione*, as presented at https://www.darapri.it/come stappare/degustarederosa/degustarederosa.htm.
71. "Coppa," De Rosa, *Degustazione*, as presented at https://www.darapri.it/come stappare/degustarederosa/degustarederosa.htm. You'd actually do well to ignore this advice, too. If the coppa is at fault for its large surface area and rapid loss of both bubbles and chill, the flute impedes full aroma detection. Use a white, or universal, wine glass, for most enjoyment, or the tulip-shaped Prosecco glasses, hard to find, but that offer a sense of uncompromised and specific elegance.

72. Literally: *fell by falling*; meaning that it flowed from the filtrating sacks pulled by gravity, and not due to a centrifuge or pump in that mechanical filtration system. Many wineries have been returning to gravity-led systems to move wine around wineries.
73. Led by Luigi Manzoni, at the school in the 1920s and 1930s. Manzoni conducted two series of experiments with cross-breeding vinifera varieties. From those studies, his best known grape is Manzoni Bianco (or, numbers that indicate vineyard and experiment, Incrocio Manzoni 6.0.13—sixth row, second experiment series, plant 13, for the vine chosen for cloning as the variety representative) which is a crossing between Riesling Renano and Pinot Bianco. Today it is planted throughout Italy, and included in the Bianco dei Colli di Conegliano DOC. See http://www.manzonibianco.com. Manzoni and his work are the subject of a new permanent exhibition at the school, designed by architect Enrico de Mori, brother of Da Condo's Beatrice.
74. Brokers and buyers of wine grapes.
75. Interview with Primo and Annalisa Franco, March 2020.
76. Drusian's word; from a conversation we had when I visited in October 2019.
77. Enthuses one Puglian producer: https://www.darapri.it/approfondimenti/cronologia-personaggi-illustri-in-campo-viticolo-enologico/.
78. Planted from cuttings of existing vines, and not bought from a nursery.
79. Official term for a spumante bottle with the vintage printed on the label. Not all wines are allowed to do so.
80. Carpene, *Sunto*, vol. 1, 3rd ed., 97.

CHAPTER 9

1. In interviews with Paola Ferraro and her family, between 2018 and 2020.
2. Stark, "In Defence," 893.
3. Conceived of as, roughly, *happiness during walks in which you see the beauty of getting lost in your own thoughts as you enjoy a landscape*. In tribute to Bembo: Browning "possessed his copy of *Gli Asolani*, and followed him in punning on the name, by forming a verb 'asolare'—'to disport in the open air, amuse oneself at random,'" from "Asolando: extracts," in Bernard Richards, ed., *English Verse 1830–1890* (New York: Routledge, 1980), 279, abstract at https://www.taylorfrancis.com.
4. See The National Gallery of Art, Washington, DC.
5. Stark, "In Defence," 893.
6. "Biosphere reserves are 'learning places for sustainable development'. They are sites for testing interdisciplinary approaches to understanding and managing changes and interactions between social and ecological systems, including conflict prevention and management of biodiversity," UNESCO, https://en.unesco.org.
7. "The MAB programme is an intergovernmental scientific programme that aims to establish a scientific basis for enhancing the relationship between people and their environments. It combines the natural and social sciences with a view to improving human livelihoods and safeguarding natural and managed ecosystems, thus promoting innovative approaches to economic development that are socially

and culturally appropriate and environmentally sustainable," UNESCO, https://en.unesco.org.

8. Angelo Peretti, consultant to the Consorzio Asolo Montello; from a work conversation.
9. Primo tenente, or first lieutenant.
10. Stark, "In Defence," 892.
11. *Kriesgkarte*, vol. 1, 225.
12. "I vitigni più coltivati sono: nell'Asolano e colli di Maser-Cornuda il Verdiso e la Rabiosa dominànti; poi, in minor misura, il Prosecco e la Bianchetta." Dalmasso and Dell'Olio, "Vini bianchi tipici," 27.
13. See Glossary.
14. Vianello and Carpenè, *La vite*, 125. They did not then see the mix as a positive: "come nelle colline di Asolo e Montebelluna, dove le varietà mescolate insieme abbondano maggiormente, si vendemmia senza operar scelte, e questa è la primaria ragione per la quale ogni anno variano d'alquanto i caratteri organolettici dei vini." ("as in the hills of Asolo and Montebelluna, where varieties planted together are prevalent, and harvested without distinguishing between them, and this is the primary reason for which every year the organoleptic characters of the wine vary widely.")
15. Manuel De Landa, *A Thousand Years*, 274.
16. See Roger Sesto, "Bianchetta trevigiana: non teme la peronospora (ma la muffa sì)," *Civilita del bere* (January 17, 2019), https://www.civiltadelbere.com.
17. "Nell'educarli," "in the raising" or "educating" of them.
18. See *Durella* entry in D'Agata, *Native Wine Grapes*, 275–77.
19. Second, CO_2-capturing, fermentation done in bottle, practically synonym of the method in Champagne.
20. "Condotti d'acqua."
21. "Città."
22. Guido Cipriani, et al., "The SSR-based molecular profile of 1005 grapevine (*Vitis vinifera* L.) accessions uncovers new synonymy and parentages, and reveals a large admixture amongst varieties of different geographic origin," *Theoretical and Applied Genetics* 121, 1569–85 (2010). https://doi.org/10.1007/s00122-010-1411-9.
23. See *Durella* entry in D'Agata, *Native Wine Grapes*, 275–77.
24. *Ampelografia Generale della provincia di Treviso.*
25. *Annuario d'Italia*, 1217.
26. *Kriegskarte*, vol. 1, 235.
27. Taramelli, "Cenni sul suolo e sul clima della provincia di Treviso," *La vite*, 14–15.
28. Città Slow: "Slow Food philosophy applied to communities; ecogastronomy as daily life practice, seasonality, healthfulness, craft, landscape, religious rites," https://www.cittaslow.org.
29. Vittorio Veneto Museo di Seta, https://www.museivittorioveneto.it/museo_del_baco_da_seta/museo.html.
30. Stark, "In Defence," 893.
31. Irene Graziotto, "La Lingua delle Città: Progetto LinCi Treviso," thesis in progress.
32. Giuseppe Boerio, author of the 1867 *Dizionario del dialetto veneziano*, of the Venetians' many remakes of the Latin the Romans brought considered the best one that of the city of Venice. It was, he wrote, a language "easiest in its structure, sweet

in its inflection, in metric numbers, homogeneous to every kind of writing"; of the "metropolis" of a once "prosperous state and fertile mother of illustrious men of science, politics, poetry, fine arts and every manner of culture"; and which Latinized several "national terms" unknown in other parts of Italy. Of the works Boerio turned to to make his Italian-Venetian dictionary, among Venetian, Latin, other dialects' dictionaries, are Italian from a farming basis—the *Vocabolario agronomico italiano* of Gio. Batista Gagliardo—and from a military one—*Dizionario militare italiano* of Giuseppe Grassi di Torino. "Sorto poi e diffuso l'idioma latino e fattosi comune a tutta Italia, siccome gl'Insubri, che abitavano l'attuale Lombardia, lasciarono la lingua Celtica., ritenendone però i dittonghi, le aspre e fischianti desinenze e i suoni nasali e gutturali che rimasero sino a' dì nostri; siccome i Toscani perdettero l'antica loro lingua Etrusca; così i Veneti adottarono anch'essi il nuovo linguaggio e lo parlarono comunemente, meschiandolo e modificandolo cogli accenti e colla dolcezza dell' antica loro lingua grecanica, e latinizzando non meno una quantità di termini nazionali, che non erano conosciuti in altri luoghi d'Italia. . . . Uno solo è radicalmente il dialetto oggi comune a Lutte le provincie degli antichi Veneti, poche essendo le varietà notabili e le differenze di qualche voce da un luogo all'altro. Ma non può negarsi che il migliore, come il più facile nella struttura, dolce nell'inflessione, metrico nei numeri, omogeneo ad ogni Penere di scrittura, sia quello che parlasi propriamente dal popolo della città di Venezia, la quale per tanti secoli fu metropoli d' un floridissimo Stato e madre feconda d'uomini illustri nelle scienze, nella politica, nella nautica, nella poesia, nell'eloquenza estemporanea, nelle belle arti e in ogni maniera di coltura." Boerio, "Discorso Preliminare," *Dizionario*, 7-8.

33. "È un bellissimo colle, ricoperto da denso strato di terriccio vegetale elaborato da secolare vegetazione boschiva. Ridotto a coltura, nel suo versante meridionale sarebbe un vero giardino," Taramelli, "Cenni sul suolo e sul clima della provincia di Treviso," *La vite*, 12. Which he unfortunately follows up with, "Pared down to cultivation, on its south face it would be a true garden."
34. Intimation of *married vines.*
35. See Durella entry in D'Agata, *Native Wine Grapes*, 275-77.
36. The Venetian propensity to shorten Italian words. Colmello, says Treccani, is a gruppo familiare, suddivisione di un paese o di un comune.
37. He writes "prassi" or *praxis*, but *practice* seems a closer translation to this work about the land, even though at this point the scientists no longer speak of it as such, as Carpenè's writings still give an impression of showing. Perhaps I'm doing so in rebellion to Scienza's times, but I also think he over-terms sometimes and doesn't mean to.
38. "Breve riflessione finale. La zonazione 'perfetta' è la zonazione che 'non si conclude mai.' Essa è tanto più di successo quanto più continua a svilupparsi nel tempo, a perfezionarsi nelle metodologie e il suo 'Manuale d'uso' a radicarsi nel territorio e, soprattutto, a essere sempre più parte della cultura di impresa e della prassi quotidiana dei produttori, degli operatori, degli altri soggetti che hanno compreso l'efficacia operativa e la 'plasticità' di utilizzo di tale strumento." Scienza, *Zonazione*, 30.
39. *Sui lieviti* understood.

40. To continue on *this* path, see Angelo Costacurta and Enzo Michelet, *Boschèra: Dai Vini Selvatici Al Nobile Torchiato Del Cansiglio*, Vittorio Veneto: Kellermann Editore, 2021.

EPILOGUE

1. From *Six Memos for the Next Millennium* by Italo Calvino. Copyright © 1988 by the Estate of Italo Calvino. English translation copyright (c) 2016 by the Estate of Italo Calvino. Used by permission of HarperCollins Publishers.
2. *Giornale Vinicolo Italiano* 8 (1882), 203; very favorable review of "spumante superiore" from the Società Enologica di Conegliano, made by the metodo Carpenè which is favorably elaborated on, too; no grape variety is mentioned.
3. Mario Isnenghi, "I luoghi della cultura," *Il Veneto*, 396.
4. "Water facts," U.S. Bureau of Reclamation, https://www.usbr.gov.
5. "Ocean," National Geographic Encyclopedia, https://www.nationalgeographic.org.
6. Montanari and Ceccarelli, *La viticoltura*, 439.
7. "Acquistare del buon vino sfuso mi consentiva una frequentazione altrimenti impossibile fatta di assaggi, chiacchiere, racconti, confidenze in cantina, e qualche volta una vera merenda o una cena. Insomma un pezzetto di civiltà contadina sopravvissuto alla violenza del consumismo arrivato sin qui." Marcello Baraghini, *Le parole della terra*, Introduction.
8. Carpenè, "La verità sui vini spumanti," *L'Enotecnico*, 37.
9. For more possibilities in moving forward in winemaking, the writings of Donna J. Haraway in *Staying With the Trouble* offer new strands from here. Particularly: "Cyborgs are not machines in just any sense, nor are they machine-organism hybrids. In fact, they are not hybrids at all. They are, rather, imploded entities, dense material semiotic 'things'—articulated string figures of ontologically heterogeneous, historically situated, materially rich, virally proliferating relatings of particular sorts, not all the time everywhere, but here, there, and in between, with consequences. Particular sorts of historically situated machines signaled by the words *information* and *system* play their part in cyborg living and dying." *Staying*, 104.
10. Visentin and Vallerani, "A Countryside to Sip," 2.
11. A very few producers make frizzante in autoclave, from a base wine. See Zanichelli, *Effervescence* for this being the opposite of the ancestral method, and more.
12. "Dopo aver raccolto i documenti di tanta storia e di tante dolci testimonianze di buon vivere, ho tuttavia la sensazione di non aver ancora detto tutto, e m'accorgo che nella Marca Gioiosa esiste un'altra dimensione, che pure è alla base delle sue fortune gastronomiche. I focolari e le tavole domestiche e pubbliche sono degli altari dove si celebra l'antico rituale del gusto di vivere, del gusto dell'amicizia, dell'arte del conversare, e del ricordare rievocando, facendo sì che il passato diventi presente, che si risolva in racconti vivificanti, nell'esercizio di dolci memorie, che la lieve ebrezza del vino trasforma in racconti favolosi. 'Carpe diem, dum petis era fugit' ma certo nel trevigiano esiste il meraviglioso potere di fissare le memorie, e quel che vale per i giorni della vita vale anche per il cibo e per il bere. Ci augu-

riamo questo nostro libro serva a chi lo legge, non solo come volume di ricette, ma anche di memorie di una città e del suo territorio, a cui attingere per ritrovare ore di semplice, discreta, ma vera e attingibile felicità. Le abbiamo scritte 'con amore' per tutti e per tutte le cose nostre, e mi ritrovo come avessi raccontata la storia di una mia grande famiglia, per molti aspetti esemplare, che in un sano spirito di emulazione, e tuttavia in accordo fra tutti, ha saputo conquistare un suo primato, che tutta Italia oramai le riconosce." Maffioli, "Conclusione," *Cucina Trevigiana*.

13. "L'ultima sagra"
14. "Passeggiata sulle colline," *Là dove*, 175 (from *Veneto felice*).
15. See Gobbato, *Bortolomiol*. He was a graduate of the school in Conegliano, an enthusiastic producer and supporter of the Italian method and of the possibilities of making a drier, or Brut, Prosecco, and who "saw Carpenè as an example to follow when it came to setting up the company Fratelli Bortolomiol," 44. "I still believe [he] was the guiding hand of the Prosecco process," says friend Francesco Scudeller of Bortolomiol (16).

GLOSSARY

1. All adapted from Italy's national registry's ampelographic descriptions, with notes on their sources with each variety. Descriptions of grape and bunch are upon ripeness.
2. National Registry entry from Italo Cosmo, Mario Polsinelli, and F. Sardi, "Bianchetta trevigiana" in *Principali vitigni da vino coltivati in Italia*, vol. 2, Ministero dell'Agricoltura e delle Foreste, 1962.
3. National Registry entry from Italo Cosmo and Mario Polsinelli, "Prosecco," in *Principali vitigni da vino coltivati in Italia*, vol. 1, Ministero dell'Agricoltura e delle Foreste, 1960.
4. National Registry entry from Antonio Calò, Attilio Scienza, and Angelo Costacurta in *Vitigni d'Italia*, Edagricole, 2006.
5. Ibid.
6. National Registry entry from Italo Cosmo and Mario Polsinelli, "Prosecco," in *Principali vitigni da vino coltivati in Italia*, vol. 1, Ministero dell'Agricoltura e delle Foreste, 1960.
7. Ibid.
8. di Meo, "La storia della Chimica," 185.
9. D'Agata, *Native Wine Grapes*, 22. Entire definition adapted from this source.
10. See Treccani, http://www.treccani.it.
11. Translated from Massimo Pivetti e Grazia Gabrielli, eds, *Grande Dizionario Italiano* di Aldo Gabrielli, 4th ed. (Milano: Ulrico Hoepli Editore, 2020).
12. R. Arroyo-García, et al, "Multiple origins of cultivated grapevine (Vitis vinifera L. ssp. sativa) based on chloroplast DNA polymorphisms," *Molecular Ecology* 15, no. 12 (October 2006), 3707–14. 10.1111/j.1365-294X.2006.03049.x.
13. See D'Agata, *Native Wine Grapes*.

Bibliography

Accorti, Marco. "L'ateismo a Venezia: libero pensiero e le doti del cuore." *L'Ateo* no. 4 (2005). https://docplayer.it/968294-L-ateo-di-fine-ottocento.html.

Acerbi, Giuseppe. *Delle viti italiane o sia materiali per servire alla classificazione*. Milano: Giovanni Silvestri, 1825. Google Books.

Agosteo, Giovanni Enrico. *La rivoluzione di Bacco a cavallo dell'Unità d'Italia: storia della vitivinicoltura italiana all'incontro con i parassiti d'oltreoceano*. Canterano: Aracne Editrice, 2017.

Annuario del Comizio Agrario di Conegliano per l'Anno 1886. Treviso: L. Zoppelli, 1887. Google Books.

Annuario d'Italia: Calendario generale del regno. Vol. 7. Ministro dell'Interno. Roma: Stabilimento Bontempelli, 1892.

Baedeker, Karl. *Italy: First Part: Northern Italy. Baedeker's Handbook for Travellers*. Leipsic: Karl Baedeker, 1889. Google Books.

———. *Southern Germany and Austria, Including Hungary and Transylvania. Baedeker's Handbook for Travellers*. Leipsic: Karl Baedeker, 1887. Google Books.

———. *Austria, Including Hungary, Transylvania, Dalmatia, and Bosnia. Baedeker's Handbook for Travellers*. Leipsic: Karl Baedeker, 1896.

Bassani, Angelo, ed. *La chimica e le tecnologie chimiche nel Veneto dell'800: atti del settimo Seminario di storia delle scienze e delle tecniche nell'Ottocento Veneto*. Venezia: Istituto Veneto di Scienze, Lettere ed Arti, 2001.

Bassi, Silvia. "Un giardiniere e botanico delle lingue: Andrea Zanzotto traduttore e autotraduttore." PhD diss., Università Ca' Foscari Venezia, 2011. http://hdl.handle.net/10579/1068.

Belloni, Silvano. *Grammatica Veneta*. 2nd ed. Padova: Essedra Editrice, 2009. Google Books.

Ben-Ghiat, Ruth, and Mia Fuller, eds. *Italian Colonialism*. New York: Palgrave Macmillan, 2002.

Bensaude-Vincent, Bernadette, and Isabelle Stengers. *A History of Chemistry*. Translated by Deborah van Dam. Cambridge: Harvard University Press, 1996.

Berengo, Marino. *L'Agricoltura veneta dalla caduta della repubblica all'unità*. Milano: Banca Commerciale Italiana, 1963.

Bioversità del Veneto. http://biodiversita.provincia.vicenza.it.

Boerio, Giuseppe. *Dizionario del dialetto veneziano*. 3rd ed. Venezia: Reale Tipografia di Giovanni Checchino, 1867. Google Books.

Bolle d'annata, 2019. Technical report of the district of Conegliano-Valdobbiadene DOCG in collaboration with UvaSapiens, Farra di Soligo (TV).

Boriani, Luigi. *Introduzione alla storia della farmacia in Italia*. Bologna: Mareggiani, 1897. Google Books.

Boschiero, Giovanni. *Relazione sulla industria dei vini.* Alessandria: Gazzotti, 1873. Google Books.

Botteon, Luisa. "Nobili, osti e furfanti: aspetti economici e sociali nelle osterie del Coneglianese tra Seicento e Settecento." In *Convegno Economia e commerci nelle Prealpi Venete (sec. XIII–sec. XX): Convegno nazionale 24 maggio 2014, Vittorio Veneto.* Edited by Circolo Vittoriese di Ricerche Storiche, 223–87. Vittorio Veneto: Dario De Bastiani, 2014.

Brock, William H. *The History of Chemistry: A Very Short Introduction.* Oxford: Oxford University Press, 2016.

Brugnatelli, Tullio. *Trattato di Chimica Generale.* Vol. 1. Milano: Vallardi, 1861. Google Books.

Brunello, Piero. *Ribelli, questanti e banditi: proteste contadine in Veneto e in Friuli 1814–1866.* Sommacampagna: Cierre edizioni, 2011.

Bruno, Giuseppe. *Di là del Piave: Sulla Strada del Vino Bianco.* Mestre-Venezia: Edizioni Fidesarte, 1981.

Calò, Antonio, and Francesco Maria De Franco. *Scripta manent: Raccolta del pensiero di Antonio Carpenè*. Godega Sant'Urbano: Dario De Bastiani Editore, 2006.

Campbell, Christy. *The Botanist and the Vintner: How Wine Was Saved for the World*. Chapel Hill: Algonquin Books, 2006.

Cancellier, Severina, Aldo Coletti, Matteo Coletti, Stefano Soligo, and Enzo Michelet. "Quattro 'nuovi' vitigni per la viticoltura Veneta." *Quaderni di scienze viticole ed enologiche*. No. 29 (2007): 5–25. Dipartimento di Scienze Agrarie, Forestali e Alimentari. https://www.disafa.unito.it/do/documenti.pl.

Caniato, Luciano, and Giuseppe Palugan. *Conegliano tra Ottocento e Novecento: gente, palazzi e strade*. Treviso: Canova, 1998.

Cannizzaro, Stanislao. *Sunto di un corso di filosofia chimica.* Palermo: Sellerio, 1991.

Cappelleri, Giuseppe. "L'evoluzione delle tecnologie chimiche in ambito vitivinicolo." In *La chimica e le tecnologie chimiche nel Veneto dell'800: atti del settimo Seminario di storia delle scienze e delle tecniche nell'Ottocento Veneto.* Edited by Angelo Bassani. Venezia: Istituto Veneto di Scienze, Lettere ed Arti, 2001.

Carbonelli, Giovanni. *Sulle fonti storiche della chimica e dell'alchimia in Italia.* Lavis: La Finestra Editrice, 2003. Google Books.

Carpenè, Antonio. Various letters and articles in the *Annali di Viticoltura ed Enologia Italiana. Scuola di Viticoltura ed Enologia di Conegliano* 1872–1877; *Rivista di Viticoltura ed Enologia Italiana, Scuola di Viticoltura ed Enologia di Conegliano* 1877–1886; *Rassegna di Viticoltura ed Enologia Italiana, Scuola di Viticoltura ed Enologia di Conegliano* 1887–1900. Conegliano: Grava-Cagnani.

——. *Cenni sull'Alluminio.* Thesis, R. Università di Pavia, 1861. Pavia: Tipografia dei Fratelli Fusi (dedicated: "allo zio Federico in segno di stima e riconoscenza").

——. *Sull'acido lattico.* Doc. thesis., R. Università di Pavia, 1862. Pavia: Ditta Tipografica Eredi Bizzoni (dedicated, surrounded by flourishes, "alla mia famiglia").

——. *Sunto teorico-pratico di enologia*. 3rd ed. 2 vols. Torino: Ermanno Loescher, 1878. Google Books.

——. *Sunto teorico-pratico di enologia,* vol. 1. 5th ed. Torino: Ermanno Loescher, 1888. Google Books.

——. *Sunto Teorico-Pratico di Enologia* vol. 2. 2nd ed. Torino: Ermano Loescher, 1882. Google Books.

Carter, Nick, and Simon Martin, eds. *Special Issue: The Difficult Heritage of Italian Fascism. Modern Italy* 24, no. 2 (May 2019). https://www.cambridge.org.

Casellato, Alessandro. "Sestieri popolari." In *Storia di Venezia: l'Ottocento e il Novecento.* Edited by Mario Isnenghi and Stuart Woolf. Roma: Istituto dell'Enciclopedia Italiana, 2002. https://www.treccani.it.

Cecchinel, Luciano. *Sanjut de stran.* Venezia: Marsilio Editore: 2012. eBook.

Cerruti, Luigi, ed. *Bella e Potente: La chimica del Novecento fra scienza e società.* Roma: Editori Riuniti, 2003.

Cobb, Harold M. *The History of Stainless Steel.* Materials Park: ASM International, 2010. Proquest.

Coltro, Dino. *La nostra polenta quotidiana: una storia contadina al femminile.* Sommacampagna: Cierre Edizioni, 2002.

——. *Paese perduto: La cultura dei contadini veneti.* 3 vols. Verona: Bertani Editore, 1982.

Comboni, Enrico. *Trattato di enochimica.* Vol. 2. Milano: Tipografia Fratelli Rechiedei, 1882. Google Books.

——. "Vagoni serbatoi perfezionati pel più sicuro ed economico trasporto dei vini." *Rassegna* 1888, 401–8. Google Books.

Costantini, Enos. "Che Prossec cu plâs a Tite e a Cec . . . : Dai pàstini triestini ai colli della Marca passando per la Patria," *Tiere Furlane/Terra Friulana* 1, no. 3, 2009, 7–20. https://www.regione.fvg.it.

D'Agata, Ian. *Native Wine Grapes of Italy.* Berkeley: University of California Press, 2014.

——. Italy's *Native Wine Grape Terroirs*. Berkeley: University of California Press, 2019.

Dalmasso, Giovanni, and Giuseppe Dell'Olio. "Vini bianchi tipici dei Colli Trevigiani: sottozone di Conegliano e Valdobbiadene." *Annali della sperimentazione agraria*. Vol. 25 (1937): 1–93.

De' Crescenzi, Pietro. *Trattato della agricoltura di Piero De' Crescenzi traslatato nella favella fiorentina*. Vol. 2. Milano: Dalla società tipografica de' classici italiani, 1805. Google Books.

De Landa, Manuel. *A Thousand Years Of Nonlinear History*. Brooklyn: Zone Books, 1997.

De Leo, Carolina, and Giovanni Favero. *Ca' Foscari e Carpenè Malvolti. Il Risorgimento dell'economia nel Veneto dell'Ottocento*. Venezia: Edizioni Ca' Foscari, 2018.

De Rosa, Tullio. *Andar per vini.* Fagagna: Arti Grafiche M. Tonutti, 1970.

———. *Guida alla degustazione del vino: la valutazione edonistica*. Faenza: Faenza Editore, 2011.

———. *Tecnica dei vini spumanti.* Conegliano: edited by the *Rivista di Viticoltura e di Enologia*, 1964.

———. *Tecnologia dei vini spumanti*. 2nd ed. Brescia: Aeb, 1987.

di Meo, Antonio. "La storia della chimica in Italia al momento della fondazione del GNFSC." In *Memorie di Scienze Fisiche e Naturali, no.* 134. Vol. 40, part 2, 179–87. Verona: Accademia Nazionale delle Scienze detta dei XL, 2016. https://www.accademiaxl.it/pubblicazioni-2/rendiconti/.

di Meo, Antonio, ed. *Storia della chimica in Italia.* Roma: Vignola, 1996.

Dickie, John. *Delizia! The Epic History of the Italians and Their Food.* London: Sceptre, 2010.

Disciplinare Asolo DOCG 2014. Ministero delle politiche agricole alimentari e forestali. https://www.politicheagricole.it/.

Disciplinare Conegliano-Valdobbiadene DOCG 2019. Ministero delle politiche agricole alimentari e forestali. https://www.politicheagricole.it/.

Dolce, Ada, and Remo Dolce. *Tradizioni popolari della Marca Trevisana* (1938). Vittorio Veneto: H. Kellermann, 1992.

Duggan, Christopher. *Force of Destiny: History of Italy Since 1796*. Boston: Houghton Mifflin, 2008.

———. *A Concise History of Italy.* Cambridge: Cambridge University Press, 1994.

Esercizi di Paesaggio. Quaderno 4, Regione del Veneto, Direzione Urbanistica e Paesaggio. Venezia, 2011. https://issuu.com/urbanisticaepaesaggio/docs/quaderno_4.

Esposizione Universale del 1878 in Parigi: Sezione italiana. Catalogo generale. Roma: Tipografia Barbèra, 1878. Google Books.

Faraon, Alessio. *Il Sasso nello Stagno: Segni d'Acqua nelle "Terre Alte" di Cison di Valmarino.* Crocetta del Montello: Grafiche Antiga, 2010.

Fiabe e leggende della Marca. Various authors, grades 1–5. *Ciclamini* 14. Treviso: Editrice Santi Quaranta, 2015.

Finelli, Michele. "Il dialogo invisibile: Mazzini e Ricasoli ai tempi del governo provvisorio." In *Pisa dal Granducato al Regno D'Italia.* Edited by Alessandro Breccia. Firenze: Edizioni Polistampa, 2011.

Forgacs, David, and Robert Lumley, eds. *Italian Cultural Studies: An Introduction*. Oxford: Oxford University Press, 1996.

Fornasari, Massimo, and Vera Zamagni. *Il movimento cooperativo in Italia: un profilo storico-economico (1854–1992)*. Firenze: Vallecchi, 1997.

"France, Austria and Italy." *New York Times.* March 15, 1860.
Gaspari, Paolo. *Grande guerra e ribellione contadina: Chiesa e Stato, possidenti e contadini in Veneto e Friuli, 1866-1921.* Udine: Istituto editoriale veneto friulano, 1995.
Gasparini, Danilo. "La viticoltura nel Nord Italia: il caso Veneto." In *Terra e storia,* no. 5. 2014.
Gasparini, Danilo, and Massimo Rossi, eds. *Disegnare le vigne: La descrizione del paesaggio nelle mappe e nei testi antichi.* Crocetta del Montello: Grafiche Antiga, 2019.
Gera, Francesco. *Nuovo Dizionario Universale e Ragionato di Agricoltura.* Vol. 1. Venezia: Giuseppe Antonelli, 1834. Google Books.
Giavedoni, Fabio, and Maurizio Gily, eds. *Guida ai vitigni d'Italia.* Bra: Slow Food Editore, 2005.
Giua, Michele. "Per un interpretazione dialettica della storia delle scienze sperimentali in Italia dopo Galilei." *Belfagor* 9, no. 3 (31 May 1954): 334–44. https://www.jstor.org/stable/26068354.
Gobbato, Ettore. *Bortolomiol: Dreaming of Prosecco.* Veronelli Editore, 2009.
Gordon, Susan H. "What a Little Hilltop in Abruzzo Can Tell Us About Words for Place." *Gastronomica* 20. no. 2 (1 May 2020): 37–44.
Gregoletto, Giovanni. *Vite Ambulante: Nuove cattedre di enologia e viticoltura.* Follina: Edizioni SUV, 2014.
Jefford, Andrew. "Wine and Astonishment." https://www.andrewjefford.com.
Johnson, Hugh. *Wine.* New York: Simon & Schuster, 1974.
Kernot, Federigo. *Storia della Farmacia e dei Farmacisti.* Napoli: Tip. del Giornale di Napoli, 1871.
King, Bolton. *The Life of Mazzini.* London: Everyman's Library, 1938.
Là dove si coltiva la vite: 50 storie al Conegliano Valdobbiadene Prosecco Superiore DOCG. Crocetta del Montello: Antiga Edizioni, 2019.
Lanaro, Silvio, ed. *Il Veneto. Storia d'Italia: Le Regioni dall'unità a oggi.* Torino: Einaudi, 1984.
Lapham's Quarterly: Technology 16, no. 1 (Winter 2021). New York: American Agora Foundation, 2021.
L' Enotecnico. Vol. 1. Carpenè, Antonio, dir., and Arturo Marescalchi, ed. Conegliano: Cagnani, 1893–1894. Google Books.
Levi, Primo. *The Periodic Table.* Translated by Raymond Rosenthal. New York: Schocken Books, 1995.
Lomolino, Giovanna, Federico Zocca, Paolo Spettoli, Giampaolo Zanin, and Anna Lante. "A preliminary study on changes in phenolic content during Bianchetta Trevigiana winemaking." *Journal of Food Composition and Analysis* 23, no. 6 (September 2010). https://doi.org/10.1016/j.jfca.2010.04.001.
Lugato, Graziella. *I nobili Malvolti: Da Conegliano a Carpenedo.* Mestre: Litostampa Veneta, 2021.

Luzzatto, Federico. "La comunità ebraica di Conegliano Veneto ed i suoi monumenti." *La Rassegna Mensile di Israel* 22, no. 5 (1956): 227–38. https://www.jstor.org/stable/41278153.

Maffioli, Giuseppe. *La cucina trevigiana: Storia e ricette.* Padova: F. Muzzio. 1988. eBook.

Maltman, Alex. *Vineyards, Rocks, & Soils: The Wine Lover's Guide to Geology.* New York: Oxford University Press, 2018.

Martinotti, Federico. Annuario della Regia Stazione Enologica, Asti, 1921.

Mazzini, Giuseppe. *Doveri dell'uomo.* London: 1860. Google Books.

Mazzotti, Giuseppe. *Immagini della Marca Trevigiana.* 3rd ed. Milano: Silvana, 1957.

McGee, Harold. *Nose Dive: A Field Guide to the World's Smells.* New York: Penguin Press, 2020.

McNeely, Ian F., and Lisa Wolverton. *Reinventing Knowledge: From Alexandria to the Internet.* New York: W.W. Norton, 2008.

Ministero d'Agricoltura, Industria e Commercio. *Ampelografia Italiana: compilata per cura del comitato centrale ampelografico con la cooperazione delle commissioni provinciali.* Torino: Lit. Fratelli Doyen, 1879.

Molmenti, P. G. *La storia di Venezia nella vita privata dalle origini alla caduta della repubblica.* Torino: Roux e Favale, 1880. Google Books.

Montanari, Massimo. *Il mito delle origini: breve storia degli spaghetti al pomodoro.* Bari: Edizioni Laterza, 2019.

Montanari, Massimo, ed. *La cultura del vino in Italia.* Roma: Skira Editore, 2013.

Montanari, Viscardo, and Giuseppe Ceccarelli. *La viticoltura e la enologia nelle Tre Venezie.* Treviso: Longo & Zoppelli, 1950.

Morris, Jonathan, and Robert Lumley, eds. *The New History of the Italian South: The Mezzogiorno Revisited.* Exeter: University of Exeter Press, 1997.

Negri, Giovanni, and Elisabetta Petrini. *Roma Caput Vini.* Milano: Arnoldo Mondadori, 2011. eBook.

Ninni, Alessandro Pericle. *Materiali per un vocabolario della lingua rusticana del contado di Treviso.* Bologna: Forni, 1984.

Ong, Walter J. *Interfaces of the Word: Studies in the Evolution of Consciousness and Culture.* Ithaca: Cornell University Press, 2012. Project Muse.

Orefice, Mario Anton, ed. *Altre tempi: Caffè, osterie e cartoline di una volta.* Noventa Padovana: Villaggio Grafico, 2013.

Orosi, Giuseppe. *Farmacologia teorica e pratica.* 2nd ed. Livorno: Vincenzo Mansi Editore, 1851. Google Books.

Oxford English Dictionary. https://www.oed.com.

Pasteur, Louis. *Memoire sur la fermentation alcoolique.* Paris: Imprimerie de Mallet-Bachelier, 1860. Google Books.

Patriarca, Silvana. *Italian Vices: Nation and Character from the Risorgimento to the Republic.* Cambridge: Cambridge University Press, 2013.

———. *Numbers and Nationhood: Writing Statistics in Nineteenth-Century Italy.* Cambridge: Cambridge University Press, 2002.
Pellegrini, Giovan Battista. *Toponomastica italiana: 10000 nomi di città, paesi, frazioni, regioni, contrade, fiumi, monti spegati nella lora origine e storia.* Milano: Ulrico Hoepli Editore, 1990.
Perullo, Nicola. *Epistenology: Wine as Experience.* New York: Columbia University Press, 2020.
Petteno, Alice. "Il Prosecco colfondo: Una riflessione antropologica sul territorio, vini e persone." Thesis, Università degli Studi di Torino, 2014. https://www.academia.edu/8751332/Il_prosecco_colfondo_una_riflessione_antropologica_su_territorio_vini_e_persone.
Piovene, Guido. *Viaggio in Italia.* Milano: Mondadori, 1961.
Pirovano, Gaetano. *Trattato di vinificazione: con una rivista dei principali enologisti dal 1500 fino al 1862.* Milano: Gaetano Brigola, 1863. Google Books.
Pliny the Elder. *The Natural History of Pliny.* Translated by John Bostock and Henry T. Riley. London: H. G. Bohn, 1855. https://www.biodiversitylibrary.org/bibliography/8126.
"Population." Istituto Nazionale di Statistica (ISTAT). Serie Storiche. http://seriestoriche.istat.it/fileadmin/documenti/2_Storia%20delle%20fonti_Popolazione.pdf.
Porciani, Ilaria, ed. *Food Heritage and Nationalism in Europe.* Abingdon: Routledge, 2020.
Potocnik, Michele. *Conegliano: città murata.* Treviso: Vianello, 1993.
Potocnik, Michele, and Giovanni Carlo Federico Villa, eds. *Il paesaggio di Cima.* Crocetta del Montello: Antiga Edizioni, 2013.
Protocollo viticolo del Conegliano Valdobbiadene Prosecco Superiore DOCG. 2019, 2020.
Rapporto economico, Consorzio di Tutela Conegliano-Valdobbiadene DOCG. Annual reports 2018, 2019, 2020, 2021.
Rassegna di agricoltura, industria e commercio. Padova: Fratelli Salmin Editori, 1873. Google Books.
Relazioni dei giurati italiani sulla Esposizione Universale di Vienna del 1873. Fasc. 3 and 5. Milano: Regia Stamperia, 1873. Google Books.
Rigoni Stern, Mario. *Storia di Tönle* and *L'anno della vittoria.* Torino: Einaudi, 1993.
Rockmore, Dan. "Prove It!" *New York Review of Books.* January 13, 2022, 39–41.
Romeo, Rosario. *Risorgimento e Capitalismo.* Roma: Laterza, 1998.
Rossi, Massimo. *L'officina della Kriegskarte: Anton von Zach e le cartografie degli stati veneti, 1796-1805.* Treviso: Ed. Fondazione Benetton, 2007.
Rossi, Massimo, ed. *Kriegskarte 1798–1805: Il Ducato di Venezia nella carta di Anton von Zach.* 2 vols. Treviso: Ed. Fondazione Benetton Studi Ricerche, 2005.
———. *Kriegskarte 1798–1805: Il Ducato di Venezia nella carta di Anton von Zach, Introduzione e guida alla consultazione.* Treviso: Ed. Fondazione Benetton Studi Ricerche: 2005.

Rovasenda, Giuseppe. *Saggio di una ampelografia universale*. Torino: Ermanno Loescher, 1877. Google Books.

Sannino, F. A. "Note ampelografiche: le tribù del Prosecco e del Verdiso." *Rivista di Viticoltura e di Enologia di Conegliano*, 1913.

———. "Sui caratteri differenziali tra spumanti naturali e spumanti artificiali," *Rivista* 1907, 251–54.

———. *Trattato Completo di Enologia.* Torino: Vincenzo Bona, 1920. https://catalog.hathitrust.org/Record/012435328.

Scarpellini, Emanuela. *L'Italia dei consumi: dalla Belle Epoque al nuovo millennio.* Bari: Editori Laterza, 2008.

Schiratti, Giuseppe. *Invito alla Strada del Vino Bianco: Conegliano Valdobbiadene*, 2nd ed. Treviso: Editrice Trevigiana, 1972.

Schmitt, Patrick. "It's Official: Champagne Shipments Down 18% in 2020." *The Drinks Business*, January 26, 2021. https://www.thedrinksbusiness.com.

Scienza, Attilio. *Atlante geologico dei vini d'Italia*. Firenze: Giunti Editore, 2019.

Scienza, Attilio, et al. *La zonazione viticola del Montello-Colli Asolani.* Biadene di Montebelluna: Consorzio Tutela Vini Montello-Colli Asolani, 2012.

Selmi, Francesco, ed. *Enciclopedia di chimica scientifica e industriale ossia Dizionario generale di chimica colle applicazioni.* Vol. 2. Torino: Società l'unione tipografico-editrice torinese, 1869. Google Books.

Soldati, Mario. *Vino al vino: Alla ricerca dei vini genuini.* Firenze: Bompiani, 2017.

Sormanni, Giacomo. *Catalogo ragionato delle opere di viticoltura ed enologia: pubblicato in Italia o italiano dal principio della stampa sino a tutto l'anno 1881.* Milano: Rasebiana, 1883. Reprint by Arnaldo Forni. Google Books.

Spriveri, S., and M. Coradazzi with Domenica Scocchia, eds. *Anniversario 130 Scuola Enologica Conegliano: le immagini raccontano.* CD-ROM. ISISS G.B. Cerletti, 2006.

Stark, Freya. "In Defence of Asolo." *Country Life* 150 (October 7, 1971), 892–95.

Tomasi, Diego, and Federica Gaiotti, eds. *The Terroirs of Conegliano Valdobbiadene Prosecco Wines.* Crocetta del Montello: Antiga Edizioni, 2020.

Tomasi, Diego, Gianni Moriani, and Attilio Scienza. *La bellussera: storia di un'invenzione che innovò la viticoltura.* Crocetta del Montello: Antiga Edizioni, 2017.

Tomasi, Giuseppe. "Arriva il Progresso!" *I Distretti di Ceneda e Serravalle in Epoca Napoleonica e Austriaca 1797–1866: Convegno Nazionale 22 Maggio 2010, Vittorio Veneto*. Edited by Circolo vittoriese di ricerche storiche. Vittorio Veneto: Dario De Bastiani, 2010. Academia.com.

Tripaldi, Laura. *Parallel Minds.* Falmouth: Urbanomic Media, 2022.

Tsing, Anna Lowenhaupt. *The Mushroom at the End of the World: On the Possibility of Life in Capitalist Ruins*. Princeton: Princeton University Press, 2021.

Ulliana, Mario. *La scuola enologica di Conegliano.* Treviso: Canova, 1992.

"Un'ora con Bepi Mazzotti: testimonianza di Bepi Mazzotti e Andrea Zanzotto, in compagnia di Toti Dal Monte e Valter Balliana," *Incontri*, Rai, 1972.

Vallery-Radot, René. *Louis Pasteur: His Life and Labours.* Translated by Lady Claud Hamilton. New York: D. Appleton and Co., 1886. Project Gutenberg.
Vanzetto, Livio. "L'antagonismo popolare." *Terra d'Este: Rivista di storia e cultura* 7, no. 14, 7–24.
Vanzetto, Livio, ed. *Paron Stefano Massarioto: la crisi della società contadina nel Veneto di fine Ottocento.* Vicenza: Odeonlibri Editrice, 1982.
Venuti, Lawrence. *The Scandals of Translation: Towards an Ethics of Difference.* London: Routledge, 1998.
———. *The Translator's Invisibility: A History of Translation.* London: Routledge, 2018.
Veronelli, Luigi, Piero Accolti, Giovanni Arpino, and Giorgio Caproni. *I vini d'Italia.* Roma: Canesi, 1964.
Veronelli, Luigi, and Pablo Echaurren. *Le parole della terra: manuale per enodissidenti e gastroribelli.* Viterbo: Stampa Alternativa/Nuovi Equilibri, 2013.
Vianello, Angelo. *Studi sperimentali di economia rurale.* Treviso: Andreola-Medesin, 1862. Google Books.
Vianello, Angelo, and Antonio Carpenè. *La vite ed il vino nella provincia di Treviso.* Roma: Ermanno Loescher, 1874. Reprint: De Bastiani, 2002.
Visentin, Francesco, and Francesco Vallerani. "A Countryside to Sip: Venice Inland and the Prosecco's Uneasy Relationship with Wine Tourism and Rural Exploitation." *Sustainability* 10, no. 7 (July 2018): 2195. https://doi.org/10.3390/su10072195.
Zanichelli, Massimo. *Effervescenze: storie e interpreti di vini vivi.* Milano: Edizioni Bietti, 2017.
Zitara, Nicola. *L'unità d'Italia: nascita di una colonia.* Milano: Jaca Book, 1971.

INSPIRATION AND GUIDANCE

Banti, Anna. *Artemisia.* Translated by Shirley D'Ardia Caracciolo. Lincoln: University of Nebraska Press, 1995.
Bardt, Christopher. *Material and Mind.* Cambridge: MIT Press, 2019.
Calvino, Italo. *Six memos for the next millennium.* Translated by Geoffrey Brock. Boston: Mariner Books, 2016.
Cowan, James. *A Mapmaker's Dream: The Meditations of Fra Mauro, Cartographer to the Court of Venice.* Boston: Shambhala, 2007.
D'Agata, John, *The Lifespan of a Fact.* New York: W. W. Norton, 2012.
Damasio, Antonio. *Descartes' Error.* New York: Penguin Books, 2005.
Fukuoka, Masanobu. *The One-Straw Revolution.* Translated by Larry Korn. New York: NYRB, 2009.
Gordon, Susan H. Personal notes from UArts PhD in Creativity immersion courses in June 2019 and January 2020.
Haraway, Donna J. *Staying With the Trouble: Making Kin in the Chthulucene.* Durham: Duke University Press, 2016.

Hellman, Lillian. *An Unfinished Woman: A Memoir*. New York: Back Bay Books, 1999.
Howe, Susan. *My Emily Dickinson.* New York: New Directions, 2007.
The Joy of X, a podcast hosted by Steven Strogatz.
Loftus, Simon. *Puligny-Montrachet: Journal of a Village in Burgundy.* New York: Alfred A. Knopf, 1993.
Luiselli, Valeria. *Lost Children Archive: A Novel*. New York: Vintage Books, 2020.
McDonald, Helen. *H Is for Hawk*. New York: Grove Press, 2015.
Newton, Isaac. *The Principia: Mathematical Principles of Natural Philosophy.* Translated by I. Bernard Cohen and Anne Whitman. Berkeley: University of California Press, 2016.
Odili, Donald Odita. Studio visit on January 11, 2020.
Ording, Philip. *99 Variations on a Proof.* Princeton: Princeton University Press, 2019.
Questlove and Ben Greenman. *Creative Quest*. Audiobook. New York: Harper Audio, 2019.
Twitty, Michael W. *The Cooking Gene: A Journey through African American Culinary History in the Old South*. New York: Amistad, 2018.
The Unreliable Bestiary: Tiger. Written and performed by Deke Weaver. University of the Arts Connelly Auditorium, Philadelphia, PA, January 12, 2020.
Wood, Denis. *Everything Sings: Maps for a Narrative Atlas.* Los Angeles: Siglio, 2010.
Yergeau, Melanie. *Authoring Autism: On Rhetoric and Neurological Queerness.* Durham: Duke University Press, 2018.
Yourcenar, Marguerite. *Memoirs of Hadrian.* Translated by Grace Frick. New York: Farrar, Straus and Giroux, 2005.

Acknowledgments

In no order, here is my chance to say some small thanks to those to whom I actually owe much larger ones. Anyone who should be listed here but is not is no less deeply appreciated and crucial to this work.

To Innocente Nardi, who first planted the seed, and, later, for understanding my amateur's enthusiasm and my ambitiously heartfelt project. To Giovanni Gregoletto for sharing his extraordinary museum and writings with me. To Antonella Gregoletto for kindness, showing me the mulberry tree, and offering me a sudden opportunity to speak with their father, Luigi Gregoletto. To the Carpenè Malvolti winery, as graciously represented by Domenico Scimone and Roberta Agnoli, whose consistent reliability and generosity brightened my way further. To Franco Adami for in-depth family history and production explanations—enactment of the ways in which these two remain bound together in the hills. To the Ferraro family for sharing their lives' stories and those of their parents' and grandparents' worlds. Extra gratitude to Paola for answering my many, many follow-up questions. To the producers I was able to meet with or talk to before and during the difficult times of the pandemic, and who shared knowledge of hills and techniques and poured their Prosecco wines, golden muses, and reason for this book. To the Marca's many restaurants that welcomed me a tavola, and that keep the stories of the hills alive. To the numerosissimi producers of wine, spumante, cheese, wool, silk, botti, and olle, who have for centuries been making these hills into a golden spell.

To Oriana Balliana, who I met first as a driver through the hills, and was lucky enough to share some journeys with her again in the last months of this project. She is a proud, generous, deeply connected denizen of the Alta Marca, and among the very best of what the hills have produced. To her parents Wilma and Walter Balliana for an equally enthusiastic and insightful sharing of their lives as much-loved osti. To Enrico Canel and Beatrice De Mori for sharing stories and photographs of the Locanda and nearby hills through the twentieth century and for a nourishing meal that I can't wait to return to. To Danilo Gasparini for sharing his thoughts and words about the Marca Trevigiana and the hills' Prosecco with me; his knowledge is a precise balance of visceral, personal, and academic, and it has been an honor to learn from him. To Irene Graziotto, for support from this project's very first days and for sharing flowers and forests with me along the way; that I can understand any parts of Treviso's dialects at all is a privilege that is thanks to her.

To the Consorzio of Conegliano-Valdobbiadene for always supporting my curiosity and questions about these hills, before and then during both this book and the pandemic. They are true professionals equipped with layered knowledge, inquisitiveness, and a belief in the area which must be made of the same stuff Antonio Carpenè's was. While I know much more now than I did on that first evening, they are the representatives today of something old and strong and bold and devoted which I remain in awe of. To Giulia Pussini for her indispensable help and support for three years, consistently matched by her own large sense of the world, deep-felt curiosity, and love of spumanti. To Cristina Michielini at CREA Conegliano, for warm welcome and coffee as I raced through library materials on two cold days while the pandemic began, and later for so generously and warmly supplying me across an ocean with the pages I needed to fill in gaps in what I'd recorded back then. To Giorgio Milani at C. B. Cerletti for an unforgettable first and at-long-last walk through the School's buildings, rooms, hallways, and bountiful ephemera and who with a true librarian's heart supplied me with readings and images that deepened not only my knowledge but also the questions I was asking him.

To Alessandro Casellato, whose kind and serious passion for the practice of oral history was the beginning of my understanding speaking as a text. And to him and AISO for being so welcoming to an outsider who really wanted to be a part of what they do. To Sandro Portelli for talking to me about the practice of oral history and referring me to Casellato. And to those I subsequently met among Italy's oral historians for being the definitions of listening, caring, and generosity. To the research librarians at the Schwarzman branch of the New York Public Library, for crucial support every time I had a question or could not find a book and especially for being a lifeline during the pandemic. They were among other things conduits for many of the books that let me imagine my way back to the hills and the late nineteenth century; I am grateful as well for shelf 28 in the Wertheim study room, through the Center for Research in the Humanities scholar program. To the librarians of my branch library of the NYPL for the never-ending, and sometimes very heavy, torrent of books and maps from (and back to) libraries in the United States and Italy; a lifeline, too. To Sara MacDonald and Josh Roberts at the University of the Arts Library for a combined three years of profoundly helpful responses to my endless questions.

To Marilyn Krieger, Robert Lumley, and William McIlhenny, whose unwaning support, recommendations, suggestions, enthusiasms, insights, creativity, and guidance during my attempts to write into some kind of understanding of these wines can only be described as co-conspiratorial. Marilyn, for steady reminders about the contadini and the places where I could say more; Bob, for helping me find that bent book right here in NYC and for making me see both framework and wild imagination at the exact same time; Bill, for understanding from the messiest beginnings and for somehow also landing in my childhood and ancestral lands. Each were consistently some step ahead of me on what I'd started out thinking was my own journey. Together they made me work in ways I would never have dared to otherwise. To my crew at UArts: sparkling wine, law, Shakespeare, art education, creativity, nursing and printmaking, climate change, space, and self: how is it possible we all had so many paths cross in our very different disciplines and projects? I could not have thought in all the ways I needed to without reading and hearing and thinking about your projects, too. To our leaders, who placed such faith in us and trained us perfectly for an intensely transformative journey. To Odili Donald Odita for showing me that a composition and creation is something you come back to every day and work out. To Deke Weaver for work that reminded me and brightly illustrated that truths are told through the creation of many kinds of stories and telling.

To Ben, who manages to be a good reader, a good listener, and a good drinker while remaining the steadiest, most patient, and warmest person I know. To my brother and father for deeply caring about this work. To my mother and sister, who care about it from further away. To the people making the nineteenth-century materials available online, especially important during a pandemic.

The light that kept on (at times brighter than I realized) as I worked on this story throughout the pandemic was made up of every single one of you.

To the kind and smooth-operating team at Rowman & Littlefield: editor Charles Harmon whose surefooteded sense of wonder guided me from manuscript to solid book proposal and beyond; assistant acquisitions editor Lauren Moynihan for teaching me how to navigate permissions and warmly answering my many questions; to production editor Taylor Breeding and team whose rigor (and interest in wine) directed a mass of final minutia into a real, live book.

Index

About the Author

Susan H. Gordon was born in Rome, Italy, with family roots in Abruzzo and Campania on her father's side. She has lived on the US East Coast for many years and holds an MFA in creative writing from the New School in New York City, a PhD in Creativity from the University of the Arts in Philadelphia, and a Level-3 certification from the Wine and Spirits Trust Foundation. She has written about wine as a *Forbes* contributor since 2017 and has covered East Coast wines for *Hugh Johnson's Pocket Wine Book* since 2018. Her deep interests in writing about Italy and its wines can be read also in *Gastronomica* and *Trink* magazine. She divides her time between Rome and the Bronx.

www.ingramcontent.com/pod-product-compliance
Ingram Content Group UK Ltd.
Pitfield, Milton Keynes, MK11 3LW, UK
UKHW020219290926
15470UKWH00001B/284

9 781538 191262